This

INJECTION MOULD DESIGN
A DESIGN MANUAL FOR THE THERMOPLASTICS INDUSTRY

INJECTION MOULD DESIGN

A design manual for the thermoplastics industry

R.G.W. PYE

THIRD EDITION

in association with
The Plastics and Rubber Institute

Copublished in the United States with
John Wiley & Sons, Inc., New York

Longman Scientific & Technical
Longman Group UK Limited,
Longman House, Burnt Mill, Harlow,
Essex CM20 2JE, England
and Associated Companies throughout the world

*Copublished in the United States with
John Wiley & Sons, Inc., 605 Third Avenue, New York, NY 10158*

© The Plastics and Rubber Institute 1968; 1978
© R. G. W. Pye and The Plastics and Rubber Institute 1983

All rights reserved; no part of this publication may be reproduced, stored in a retrieval system, or transmitted in any form or by any means, electronic, mechanical, photocopying, recording, or otherwise without either the prior written permission of the Publishers or a licence permitting restricted copying issued by the Copyright Licensing Agency Ltd, 33–34 Alfred Place, London, WC1E 7DP.

First published 1968
Second edition 1978
Third edition by George Godwin 1983
Reprinted by Longman Scientific & Technical 1986, 1988

British Library Cataloguing in Publication Data
Pye, R. G. W.
 Injection mould design.—3rd ed.
 1. Injection moulding of plastics
 I. Title
 668.4'12   TP1150

ISBN 0-582-01886-2

Library of Congress Cataloging in Publication Data
Pye, R. G. W. (Ronald George William)
 Injection mould design.
 "Published in association with the Plastics and Rubber Institute."
 Includes index.
 1. Injection molding of plastics. I. Plastics and Rubber Institute. II. Title.
 TP1150.P9   1982   668.4'12   82-3132
 ISBN 0-470-20569-5 (USA only) AACR2

Printed in Great Britain at The Bath Press, Avon

Contents

PREFACE ix

ACKNOWLEDGEMENTS x

PART ONE   ELEMENTARY MOULD DESIGN

1 Mould Making 3

  1.1 General   1.2 Machine tools   1.3 Castings   1.4 Electro-deposition
  1.5 Cold hobbing   1.6 Pressure casting   1.7 Spark machining
  1.8 Bench fitting

2 General Mould Construction 31

  2.1 Basic terminology   2.2 Mould cavities and cores   2.3 Bolsters
  2.4 Ancillary items   2.5 Attachment of mould to platen

3 Ejection 63

  3.1 General   3.2 Ejector grid   3.3 Ejector plate assembly
  3.4 Ejection techniques   3.5 Ejection from fixed half   3.6 Sprue pullers

4 Feed System 112

  4.1 General   4.2 Runner   4.3 Gates

5 Parting Surface 141

  5.1 General   5.2 Flat parting surface   5.3 Non-flat parting surface
  5.4 Relief of parting surfaces   5.5 Venting

6 Mould Cooling   **153**

    6.1 General   6.2 Cooling integer-type mould plates   6.3 Cooling insert-bolster assembly   6.4 Cooling other mould parts   6.5 Water connections

7 Procedure for Designing an Injection Mould   **185**

    7.1 General   7.2 Stage A (primary positioning of inserts)   7.3 Stage B (the ejector system)   7.4 Stage C (the ejector grid)   7.5 Stage D (complete the top half of the drawing)   7.6 Stage E (complete the plan view)   7.7 Stage F (complete the cross-section)   7.8 Stage G (complete the drawing)

PART TWO   INTERMEDIATE MOULD DESIGN

8 Splits   **207**

    8.1 General   8.2 Sliding splits   8.3 Angled-lift splits   8.4 Summary

9 Side Cores and Side Cavities   **244**

    9.1 General   9.2 Design features   9.3 Types of side core and side cavity

10 Moulding Internal Undercuts   **277**

    10.1 General   10.2 Form pin   10.3 Split cores   10.4 Side cores   10.5 Stripping internal undercuts

11 Moulds for Threaded Components   **290**

    11.1 General   11.2 Moulds for internally threaded components   11.3 Moulds for externally threaded components   11.4 Mould construction

12 Multi-daylight Moulds   **330**

    12.1 General   12.2 Underfeed moulds   12.3 Triple-daylight moulds

13 Runnerless Moulds   **346**

    13.1 General   13.2 Nozzle types   13.3 Hot-runner unit moulds
    13.4 Insulated runner moulds   13.5 Hot-runner plate moulds

14 Standard Mould Units   **408**

    14.1 General considerations   14.2 Standard two-part mould units
    14.3 Deviations from the standard mould unit   14.4 Comparative terminology

15 Checking Mould Drawings   **444**

    15.1 General   15.2 Pin ejection mould   15.3 Sleeve ejection mould   15.4 Stripper plate mould   15.5 Splits type mould   15.6 Side core type mould   15.7 Underfeed type mould   15.8 Hot-runner type mould   15.9 Stepped parting surface mould design

16 Worked Examples of Simple Injection Moulds   **464**

    16.1 General   16.2 Example 1 (pin ejection type mould)   16.3 Example 2 (sleeve ejection)   16.4 Example 3 (stripper plate ejection)

TERMS USED IN INJECTION MOULD DESIGN   **477**

INDEX   **490**

Preface

The primary object of the first edition of this book was to provide a handbook on design for mould draughtsmen and designers in industry. In addition to fulfilling this function, the book has been used increasingly by the novice as an introductory guide, because it progresses in simple stages from the consideration of basic principles and components to more detailed explanation of the more complex types of special purpose mould.

When George Godwin (the publishers) asked me to consider an extended third edition I had no hesitation in recommending that the extension be used primarily to help the beginner. In this respect two new chapters have been added and a third has been completely rewritten.

The major difficulty which beginners normally experience is the decision of where to start and the procedure to adopt when designing a mould. A new chapter has been included as an introductory guide, to permit the reader to follow a definite approach in order to establish a sequential technique of his own.

Another new chapter is on mould checking. To assist the novice in this important aspect, a number of drawings have been prepared in which design errors have been incorporated. Some of these errors are quite obvious, but the more subtle errors require a very careful scrutiny of the design.

The book is divided into two sections, elementary and intermediate mould design respectively. Part One covers mould-making methods, general mould construction and ejection methods. Other primary considerations such as feed systems, parting surfaces and mould cooling are also covered. Part Two progresses to specific designs, and includes chapters on standard systems and worked examples.

London School of Polymer Technology   R.G.W.P.
The Polytechnic of North London
1983

Acknowledgements

The author wishes to place on record his sincere thanks to a number of people:

to Mr R.M. Ogorkiewicz for his advice and his constructive criticism of the original manuscript;

to Mr R. Baker, Mr P. Bullivant, Mr J. Collins, Miss B. Humphries, Dr R. Phillips and Mr J. Robinson for their valuable assistance in the preparation of the first edition; to Mr A. Byford, Mr L. Davenport, Mr J. Harris and Mr J. Haylar, who between them taught the author the fundamentals of mould design;

to Mr J. Robinson for his advice and comments on all three editions;
to his wife for her help and forbearance during the writing of this work.

Thanks are also due to the following companies for permission to reproduce photographs:

Bakelite Xylonite Limited
DME Europe
E. Elliot Limited
Fox and Offord Limited
Alfred Herbert Limited
H.B. Sale Limited
J.E. Snow (Plastics) Limited
Tooling Products (Langrish) Limited

Wherever possible, the individual designer and/or company has been credited with a specific design or device mentioned in the book. This leaves the author to acknowledge the contribution of the many thousands of anonymous designers who between them, over several decades, have formed the basis of modern injection mould design.

It should be pointed out that many of the designs and devices described and illustrated in this book are the subjects of valid patents.

PART ONE

Elementary Mould Design

1

Mould Making

1.1 GENERAL

A competent mould designer must have a thorough knowledge of the principles of mould making as the design of the various parts of the mould depends on the technique adopted for its manufacture.

This chapter is included primarily for the beginner who does not have a background knowledge of the various machining and other mould making techniques. To cover the topic of mould making thoroughly would require a companion work equal in size to this monograph and therefore this introduction to the subject must, of necessity, be superficial. However, we hope that the very fact that it is included in a monograph on design will emphasise the importance of mould making as a subject and will also encourage the beginner to a further and more complete study in this field.

The majority of moulds are manufactured by the use of conventional machine tools found in most modern toolrooms. From the manufacturing viewpoint we classify the mould into two parts (i) the *cavity* and *core*, and (ii) the remainder of the mould. The latter part is commonly referred to as *bolster work*.

The work on the cavity and core is by far the most important as it is from these members that the plastics moulding takes its form (see Chapter 2 for definitions).

The work on the cavity and core can further be classified depending upon whether the form is of a simple or a complex nature. For example, the cavity and core for a circular or rectangular box-type moulding is far simpler to make than a cavity and core to produce, say, a telephone handset moulding. The mould parts for the simple form are produced on such machine tools as the lathe and the milling machine, whereas the more complex form requires the use of some kind of copying machine.

The bolster work is not as critical as the manufacture of the cavity and core forms but, nevertheless, accuracy in the manufacture of the various parts is necessary to ensure that the mould can be assembled by the fitter without an undue amount of bench-work.

Now, while the bolster work is always produced on conventional machine tools, the cavity and core, particularly the former, can be produced by one of a number of other techniques. These include investment casting, electro-deposition, cold hobbing, pressure casting and spark machining.

MOULD MAKING

1.2 MACHINE TOOLS

The purpose of any machine tool is to remove metal. Each machine tool removes metal in a different way. For example, in one type (the lathe) metal is removed by a single point tool as the work is rotated, whereas in another type (the milling machine) a cutter is rotated and metal is removed as the work is progressed beneath it.

Which machine tool is to be used for a particular job depends to a large extent upon the type of machining required. There is, however, a certain amount of overlapping and some machine tools can be utilised for several different operations. In the illustrations which follow typical machining operations are illustrated but it must not be assumed that the particular machine tool is restricted to the operation shown.

The machine tools which will be found in the modern toolroom are as follows:
   (i) *Lathes* for turning, boring and screwcutting, etc.
   (ii) *Cylindrical grinding machines* for the production of precision cylindrical surfaces.
   (iii) *Shaping and planing machines* for the reduction of steel blocks and plates to the required thickness and for 'squaring up' these plates.
   (iv) *Surface grinding machines* for the production of precision flat surfaces.
   (v) *Milling machines* for the rapid removal of metal, for machining slots, recesses, boring holes, machining splines, etc.
   (vi) Tracer-controlled milling machines for the accurate reproduction of complex cavity and core forms.

In addition to the above list of major machine tools there is, of course, ancillary equipment without which no toolroom would be complete. This includes power saws, drilling machines, toolpost grinders, hardening and polishing facilities, etc.

1.2.1 Lathe. The primary purpose of the lathe is to machine cylindrical forms. The contour is generated by rotating the work with respect to a single-point cutting tool. For machining the outside surface, the cutter is moved parallel to the axis of rotation. This operation is called *turning*. Alternatively, metal may be removed from the inside of the work in which case the operation is called *boring*. When the tool is moved across the face of the work it is called *facing*.

The principal parts of the lathe are illustrated in Figure 1.1. The workpiece is secured at one end in a chuck and supported at the other end by a *centre*, fitted in the tailstock. The chuck is mounted on the headstock spindle and driven by an electric motor via a gearbox and transmission system (the last two items are not shown). The headstock and tailstock are both attached to the machine bed, and the position of the tailstock is adjustable to accommodate various lengths of workpiece.

The cutting action is by means of a single-point cutting tool mounted in a toolholder. The cutting tool is positioned, prior to the commencement of the

MACHINE TOOLS

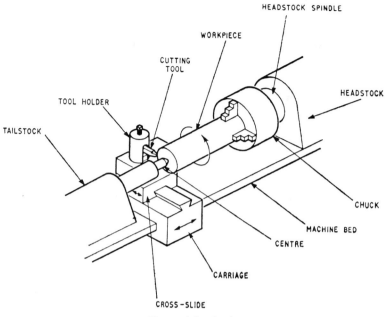

Figure 1.1—Lathe

cutting operation, so that the cutting point is in line with the axis of the work. The tool can be moved, primarily, in two directions. For normal turning and boring, a longitudinal movement is required and this is achieved by moving the carriage along the slideways of the bed. For facing the end of the workpiece, a transverse movement is required and this is achieved by moving a cross-slide along the slideways of the carriage. Note that these slideways are at right angles to the bed slideways. Both the longitudinal and transverse movements can be power operated.

The speed of the carriage (or cross-slide) relative to the rotational speed of the work is adjustable and this, together with the depth of cut chosen, determines the finish obtained on the work. For rough machining a relatively deep cut with a fast feed is used, but for finishing a shallow cut with a fine feed is required. When a large amount of metal has to be removed, a number of successive roughing cuts are made until the required diameter is approached. The part is then finished to size with one or two finishing cuts.

The lathe is extremely versatile and is used for making a large variety of mould parts. For example, guide pillars, guide bushes, circular support blocks, ejector rods, ejector rod bushes, push-back pins, etc., are all manufactured on the lathe. In addition to this bolster work the cavity and core are also produced on a lathe if the moulding form is cylindrical. Turning is a relatively fast machining operation and for this reason moulds for circular components are cheaper to produce than corresponding moulds for components of any other form.

MOULD MAKING

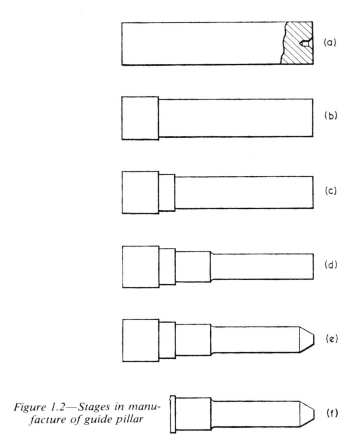

Figure 1.2—Stages in manufacture of guide pillar

Internal and external thread forms are also easily generated, when required, as for example on the end of an ejector rod (Figure 3.7). A slight complication arises if the thread is required on the core (see Figure 11.18) or in the cavity (Figure 11.38) to produce a complementary moulded thread as shown. In these cases it is necessary to make some allowance for the plastics material shrinkage on the mould thread pitch (i.e. the mould thread pitch must be machined slightly larger than required to allow for the material shrinkage on cooling).

To describe the manufacture of a typical mould part, we take a guide pillar (Figure 1.2) to illustrate the sequence of operations.

The first step is to cut off a suitable length of steel bar and mount this in the lathe chuck. The end of the bar is faced and subsequently centre-drilled (Figure 1.2a). The tailstock centre is then positioned in this centre-drilling as illustrated in Figure 1.1.

The bar is now progressively reduced in diameter by a turning operation in a number of stages (Figure 1.2b-e). Finally, the guide pillar is parted off (Figure 1.2f) and the scrap piece of steel removed from the chuck. To permit

MACHINE TOOLS

the head of the guide pillar to be faced and reduced to the required thickness, the workpiece is again mounted in the chuck but this time it is reversed and held on the fitting diameter.

This completes the lathe work. The guide pillar is then hardened and the important fitting diameters are subsequently ground to size on a cylindrical grinding machine.

1.2.2 Cylindrical Grinding Machine. This machine tool is used for precision grinding cylindrical mould parts. Metal is removed by the action of a rotating abrasive grinding wheel which is brought into contact with a contra-rotating workpiece. The axes of both the grinding wheel and the workpiece are parallel for normal operation.

An important feature of the grinding machine is that it can cut hardened metal. This characteristic, together with the close tolerances and the high surface finish obtainable, makes this machine tool an essential piece of toolroom equipment.

A simplified drawing of a cylindrical grinding machine is shown in Figure 1.3. The workpiece is mounted at one end in a chuck and is supported at the other end by a centre fitted in the tailstock. The chuck is mounted on the headstock spindle and driven by an electric motor, attached to the headstock. Both the headstock and the tailstock are mounted on a table; the tailstock is adjustable to accommodate various lengths of work. The table, fitted on slideways of the machine bed, is a driven reciprocating member. The length of

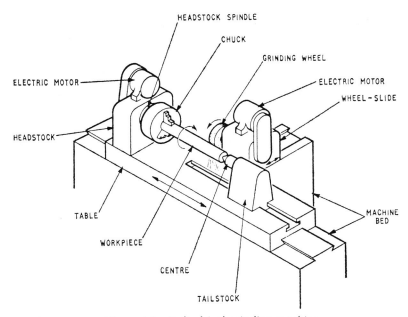

Figure 1.3—Cylindrical grinding machine

MOULD MAKING

stroke is adjustable. The workpiece, therefore, has a rotary motion and a longitudinal motion with respect to the grinding wheel. The grinding wheel spindle (not visible) is driven via a belt transmission by an electric motor mounted on top of the wheel-slide. The latter member is fitted on the slideways of the machine bed. These slideways are at right angles to the axis of the workpiece, and therefore movement of the wheel-slide moves the grinding wheel towards or away from the workpiece.

In normal grinding of cylindrical parts, the wheel-slide is adjusted forward until the rotating grinding wheel just contacts the contra-rotating workpiece. Further forward movement of the wheel-slide sets the depth of cut. The table is then caused to reciprocate, thereby grinding the outside surface of the workpiece over a preset distance. This operation is repeated and the depth of cut progressively increased until the required diameter on the workpiece is obtained.

1.2.3 Shaping and Planing Machines. A mould normally includes a number of steel plates suitably secured together. Each of these plates must have parallel faces and, ideally, the four sides should be square. Now, as the primary purpose of a shaping machine is to produce flat surfaces, this machine tool is used in the initial preparation of mould plates and mould blocks.

The principle of the *shaping machine* is illustrated in Figure 1.4. The workpiece is mounted on a table and a reciprocating single-point tool removes metal in a series of straight cuts. After each forward stroke, the table is traversed a preset increment in preparation for the next cut.

The ram is a driven reciprocating member which is guided in slideways at the top of a vertical column. The length of stroke of the ram is adjustable. The

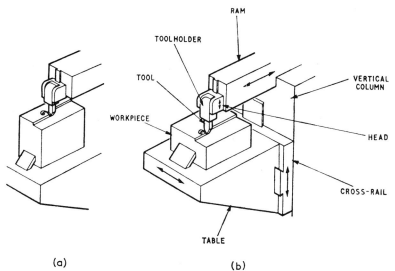

Figure 1.4—Shaping machine: (a) squaring up mould plate; (b) reducing thickness of mould plate

tool is attached to the ram via a tool-holder and head. The depth of cut by the tool is preset by vertical adjustment of the head.

The table, to which the work is securely attached, is mounted on slideways on the cross-rail. Movement of the table across the face of the cross-rail is effected by the rotation of a lead screw (not shown), actuated by the ram via a simple mechanism.

The cross-rail can be adjusted vertically on slideways at the front of the column. Adjustment of this member is carried out only during the setting-up operation. Once the workpiece is in close proximity to the tool the cross-rail is clamped in the desired position before the cutting operation commences.

Because of the single-stroke cutting action of this machine, the surface finish obtained on the work is always in the form of a series of fine grooves. The depth and width of the groove depend on the depth of cut and the traverse increment chosen. While this surface finish is suitable for certain mould parts (for example, the sides of mould plates) it is normal practice to follow the shaping operation with a surface grinding operation to produce a finer surface finish.

In Figure 1.4 the sides of a mould block are shown being squared up (a), and the block is shown in the process of being reduced in thickness (b).

For the squaring up and surface facing of large blocks of steel an alternative machine tool, called a *planing machine* is often used. This machine is similar to the shaping machine in that it employs a reciprocating action and produces a flat surface by a series of straight cuts using a single-point tool. However, with the planing machine it is the work which is reciprocated, the tool being fixed in a head above the workpiece.

Figure 1.5 illustrates a simplified version of a planing machine. The work is rigidly attached to the table which is reciprocated on the slideways of the bed. The single-point cutting tool is mounted in the cutter head and this member is adjustable in the vertical direction on the slideways of the cross-slide. This latter member, in turn, is mounted on the slideways of the cross-rail, permitting the tool to be traversed across the face of the workpiece. The cross-rail is supported above the table upon the side columns.

In operation the depth of cut is preset by suitable adjustment of the cutter-head. The table is caused to reciprocate and the single-point tool slices a shaving of steel from the workpiece. At the end of the cutting stroke the head is automatically traversed by a preset increment in preparation for the next cutting stroke. This continues until the complete surface has been planed. If necessary the depth of cut is increased and the above procedure repeated, until the required thickness of plate is obtained.

1.2.4 Surface Grinding Machine. We have previously discussed the cylindrical grinding machine for the grinding of cylindrical surfaces (Section 1.2.2). Now the surface grinding machine performs a similar function for flat surfaces, and grinding normally follows the shaping or planing operation. An excellent surface finish combined with accuracy can be achieved on hard or soft steel with this machine tool.

MOULD MAKING

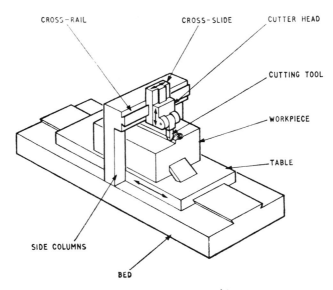

Figure 1.5—Planing machine

There are several basic designs of surface grinder and the principle of one type is shown in Figure 1.6. The workpiece is mounted on a table which is reciprocated beneath a rotating abrasive grinding wheel. Metal is removed in a series of straight cuts, the table being traversed a preset increment after each cutting stroke until finally the entire surface has been ground.

The grinding wheel is mounted on a shaft which is parallel to the surface of the workpiece. The shaft is mounted in bearings within the wheel head and

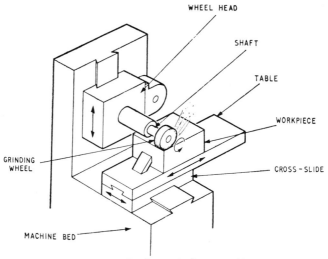

Figure 1.6—Surface grinding machine

MACHINE TOOLS

the depth of cut is preset by vertical movement of this member.

The length of stroke of the table is adjusted to suit the length of work before the cutting operation commences. A simple mechanism (not shown), operated by the table movement, actuates the cross-slide in the transverse direction. The cross-slide is mounted in slideways on the machine bed.

1.2.5 Milling Machine. Milling is an operation in which metal is removed from a workpiece by a rotating milling cutter. The workpiece can be moved in three directions at right angles to each other, with respect to the cutter. The three directions are longitudinal, transverse and vertical, respectively. There are two basic types of milling machine. In one the axis of the cutter is perpendicular to the surface of the workpiece (Figure 1.7a) and this is called a *vertical machine*. In the other, the axis of the milling cutter is parallel to the surface of the workpiece and this is called a *horizontal milling machine* (Figure 1.7b).

Both types are used extensively in the manufacture of various parts of the mould.

The table assembly is identical for both machines and we will therefore discuss this feature first. The table is mounted on the slideways of the saddle, which allows the table to be moved longitudinally. The saddle is mounted on slideways of the knee and these slideways are at right angles to the saddle slide-

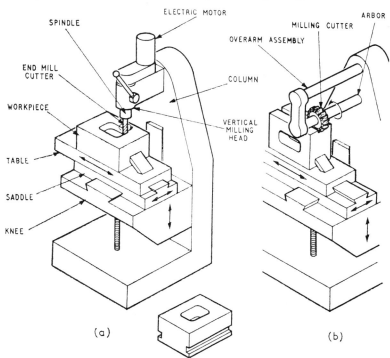

Figure 1.7—Milling machine: (a) vertical milling machine; (b) horizontal milling machine

MOULD MAKING

ways, which permits the workpiece to be traversed. Finally, the knee, which is the main supporting member, is mounted on the vertical slideways of the column. This allows for the workpiece to be adjusted vertically. In the horizontal machine the depth of cut is preset by this vertical movement. However, with the vertical machine the vertical movement of the knee is used primarily to bring the workpiece close to the milling cutter, the actual depth of cut being preset by vertical movement of the spindle on which the cutter is mounted.

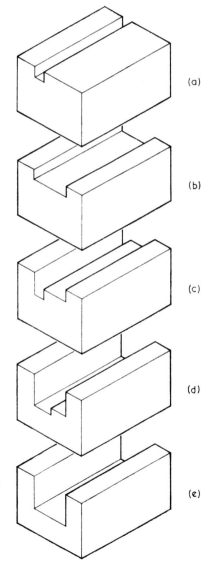

Figure 1.8—Sequence of operations in machining deep slot on horizontal milling machine

MACHINE TOOLS

The vertical milling machine (Figure 1.7a) incorporates a vertical milling head mounted on a horizontal extension of the column. The head comprises a spindle suitably driven by an electric motor via a belt transmission system. The spindle, which is rotated at high speed, can be fed in the vertical direction either manually, or automatically via the drive system. In Figure 1.7a an *endmill* cutter is fitted into the spindle. This type of cutter incorporates cutting edges both on its periphery and on its underside. Thus the cutter can be sunk into the workpiece which can then be moved either longitudinally or transversely depending upon the form required. Figure 2.19 illustrates a pocket being machined in a mould plate by an endmill cutter.

The horizontal milling machine (Figure 1.7b) incorporates a driven horizontal arbor upon which is mounted the milling cutter. Deflection of the arbor during the cutting stroke is prevented by the overarm assembly. The most common type of milling cutter used on this machine is the side and face cutter which is essentially a disk of tool steel which has cutting teeth on the periphery (i.e. the face) and on the sides. This type of cutter is used both for cutting slots and for surface machining. By using milling cutters with a contoured face various profiles can be machined on the workpiece.

The operation of both types of milling machine is similar. The movement of the work in all three directions can be manually or automatically operated. It is normal practice to operate the machine in only one direction at a time. For example, consider the machining of a large slot in a bolster.

The sequence of the operations is shown in Figure 1.8. The operator presets the depth of cut and uses a longitudinal cutting stroke to machine the first slot (a). At the end of this cutting stroke, the work is traversed a preset amount and another longitudinal cut made, and so on, until the required width of slot is obtained (b). The depth of the cut is then increased and again a longitudinal cut made (c). At the end of this and of each subsequent stroke the work is traversed in preparation for the next cut. The procedure is repeated — (d) shows an intermediate position — until the required depth of slot is obtained. The depth of cut, the traverse increment adopted and the cutting rate depend upon several factors, which include the type of steel being machined, the surface finish required and the type of cutter used.

While, as we have stated, these machines are used extensively in the manufacture of various parts of a mould, they cannot easily be used for the manufacture of three-dimensional form, which is often required for the cavity and core. For these complex shapes some form of copy milling machine is needed.

1.2.6 Manufacture of Simple Mould Plate. Now we have considered the basic machine tools, we will give an example of the procedure for the manufacture of a simple mould part. Let us assume that we want to make a cavity plate for a box-type component. We would proceed as follows:
  (i) Cut off a suitable length of steel from a bar on the power saw.
  (ii) Square up the block and machine both sides on the shaping machine (Figure 1.4a and b).

MOULD MAKING

(iii) Grind both surfaces to the specified size on the surface grinding machine (Figure 1.6).
(iv) End-mill the cavity form on the vertical milling machine (Figure 1.7a).
(v) Mill a slot in the two sides of the mould plate (for clamping purposes) on the horizontal milling machine (Figure 1.7b).

This completes the first stage of the machining operation. The mould plate now passes to the bench fitter who does a certain amount of hand finishing (Section 1.8.1) and marking out for subsequent ancillary machining operations which in this example would include the following:

(vi) Bore and counterbore a hole in the centre of the plate for the sprue bush on the vertical milling machine or on the lathe.
(vii) Bore and counterbore the holes to accommodate the guide bushes on the vertical milling machine or on a jig boring machine (a special type of vertical milling machine).
(viii) Bore the water cooling channels on a radial drilling machine.

From this point the mould plate passes back to the bench fitter whose responsibility it is to hand finish the mould plate and subsequently to assemble the mould.

1.2.7 Tracer-Controlled Milling. The principle of this type of machine tool is similar to that of the vertical milling machine in that an end mill cutter is used to remove metal in a series of cuts. With tracer-controlled milling, however, the required form is generated by causing a tracer, directly coupled to a cutting head, to follow a template or a model. The master form is accurately reproduced.

One machine of this type is the Pratt, Whitney and Herbert 'Keller', and a simplified drawing of the machine is shown in Figure 1.9. The machine consists essentially of an angle plate, mounted on a stationary work table. The template, or model, and the work are mounted one above the other on the angle plate as shown. An end mill cutter is mounted in a spindle which is individually driven by a motor housed in the spindle head.

The traverse (in-and-out) motion of the cutter is obtained by having the spindle mounted on the horizontal slideways of the slide. Similarly, the vertical (up-and-down) motion of the cutter is obtained by having the slide mounted on the vertical slideways of the column. Finally, the longitudinal (forward-and-back) motion of the cutter is obtained by having the column mounted on slideways of the bed. Each of these three motions is produced by an independent drive and the cutter can therefore be moved in three directions simultaneously.

The tracer is mounted in a tracer head vertically above and in line with the cutter. Before the cutting operation commences the tracer head slide is adjusted on the vertical slideways of the spindle head so that the respective positions of the tracer and master, and the cutter and work, are identical. The transverse position of the tracer with respect to the cutter is also preset by adjustment of the tracer head which is mounted in slideways on the tracer head slide.

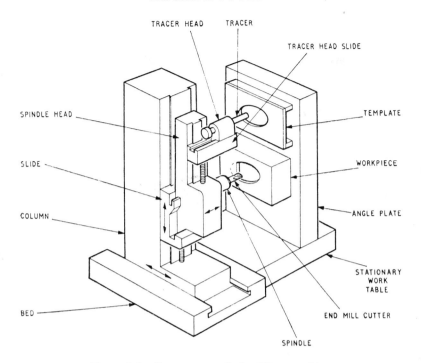

Figure 1.9—Tracer-controlled milling machine

There are two types of control on the basic machine, namely *profile tracing* and *three-dimensional tracing*. In either case a small deflection of the tracer point will make and break electrical contacts to alter automatically the cutting direction via magnetic clutches and lead screws.

PROFILE TRACING. This is a technique by which the tracer controls only the horizontal and vertical movements of the spindle head. The required cutting depth is preset before the operation starts. The tracer is caused to follow a master, normally a thin metal template, which guides the cutter in two dimensions to reproduce the master shape exactly. The set-up of the machine for the profiling operation is shown in Figure 1.9.

In operation the tracer and cutter assembly is caused to move, say, vertically upwards. Immediately the tracer strikes the template form, the direction of movement is automatically reversed and the assembly moves vertically downwards. At the end of each stroke the assembly is moved longitudinally a preset increment in preparation for the next cutting stroke. This procedure continues until the entire form of the template has been swept. The cutter can only remove a certain amount of steel during each cutting stroke so when the template form has been swept once at a particular cutting depth, the depth of cut is increased and the procedure repeated. By adopting this technique a metal template can be used to produce a relatively complex two-dimensional

MOULD MAKING

form. Figure 1.10 shows the sequence of operations in the machining of a deep elliptical cavity. Note that while the template is made from relatively thin sheet metal (Figure 1.9) any depth of cavity can be milled out that is within the capacity of the machine.

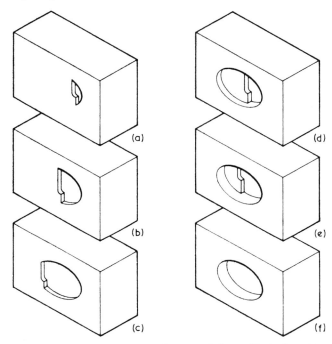

Figure 1.10—Stages in manufacture of deep elliptical cavity

THREE-DIMENSIONAL TRACING. For complex cavity and core forms which cannot be machined using the profiling technique, the profile tracer head is replaced by a three-dimensional tracing head. In operation the tracer is caused to travel over the surface of the model in a series of parallel horizontal sweeps. At the end of each sweep an automatic vertical step feed occurs. The operator can alternatively select vertical sweeps combined with a horizontal step feed.

The tracer head is designed so that the cutter, the movement of which it controls, is moving at all times. When the tracer is not in contact with the model, the tracer and cutter automatically move inwards. Immediately contact with the model is made the tracer is deflected slightly which, via electrical contacts, energises the horizontal (or vertical) travel circuit or the traverse (in-and-out) circuit, or both circuits simultaneously. The precise direction of the motion depends on the direction in which the tracer is deflected.

It is not practicable for the cutter to reproduce the contours of a deep master form in one sweep of the model. Instead this must be done in stages and

CASTINGS

the maximum depth of cut is preset before the operation commences. Figure 1.11 illustrates the sequence of machining operations of a core for a toy boat hull. On the first sweep (a,b) only the top portion of the form is reproduced in the steel. Successive sweeps at progressively greater cutter depths (c,d,e) see the emergence of the required form, until finally an accurate reproduction of the model is produced in steel (f).

Plate 1 shows a cavity form in the process of being copied.

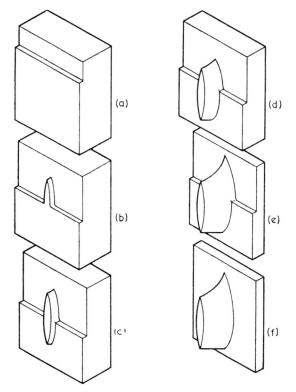

Figure 1.11—Sequence of machining operations of three-dimensional core

1.3 CASTINGS

The manufacture of cavities and cores in steel by the conventional casting method using sand moulds is not satisfactory owing to the poor finish obtained and to the porosity which occurs on, or just below, the surface of the casting. The expenditure involved in plugging, machining and finishing these conventional castings makes this method of mould making uneconomic.

MOULD MAKING

The Shaw investment casting process does not, however, share the disadvantages associated with sand casting and is therefore applicable to the manufacture of cavities and cores. The process is carried out by specialists and the mouldmaker supplies the company with a pattern of the required mould part. As the final casting will be an accurate reproduction of the pattern supplied, this must be manufactured to close tolerances and have a good surface finish. To allow for the contraction of the steel on cooling the pattern is made approximately 0.020 mm/mm (in/in) oversize.

The procedure adopted in the manufacture of a mould part by this process is illustrated diagrammatically in Figure 1.12.

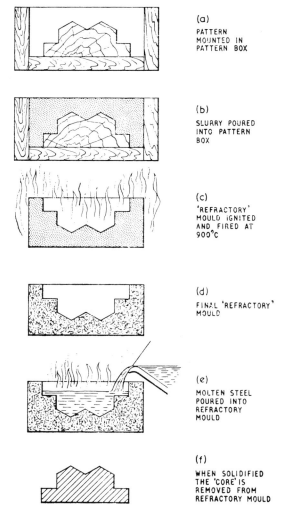

(a) PATTERN MOUNTED IN PATTERN BOX

(b) SLURRY POURED INTO PATTERN BOX

(c) 'REFRACTORY' MOULD IGNITED AND FIRED AT 900°C

(d) FINAL 'REFRACTORY' MOULD

(e) MOLTEN STEEL POURED INTO REFRACTORY MOULD

(f) WHEN SOLIDIFIED THE 'CORE' IS REMOVED FROM REFRACTORY MOULD

Figure 1.12—Stages in manufacture of core by Shaw investment casting process

ELECTRO-DEPOSITION

(a) The prepared pattern (in this case a core) is mounted in a pattern box. The internal dimensions of the pattern box correspond to the required outside dimensions of the mould plate plus the shrinkage allowance.

(b) A slurry consisting of highly refractory particles suspended in a bonding material (ethyl silicate) is poured into the pattern box. The slurry progressively hardens and, at a suitable point, the pattern and pattern box are removed.

(c) The refractory mould is immediately ignited and this is followed by a firing in a furnace at approximately 900°C, which eliminates alcohol and water evolved during the process.

(d) When the refractory mould is removed from the furnace its structure consists of refractory particles bound together with residual silica. The elimination of the alcohol leaves behind a network of very fine cracks.

(e) Molten steel at 1600°C is now poured into the refractory mould.

(f) When the steel solidifies the cast mould plate is removed from the refractory mould and the mould plate is returned to the mouldmaker for the final finishing and fitting operations.

The reasons why the refractory mould is suitable for the manufacture of mould parts, whereas the sand mould is not, can be summarised as follows:

(i) The structure of the refractory mould allows for expansion as the steel is poured without cracking.

(ii) Entrapped gas within the molten steel is allowed to escape through the fine hairline cracks; this results in a non-porous surface.

(iii) As the refractory mould is an identical replica of the pattern, the surface of the pattern is faithfully reproduced on the casting.

(iv) The low value of thermal conductivity of the refractory mould allows the steel to cool at a very slow rate.

The major limitation associated with the process is that it is not possible to guarantee an overall tolerance better than 0.005 mm/mm (in/in) which means that for large mould castings there is the possibility of considerable error. For example, the discrepancy on a component 500 mm (20 in) long may be as much as 2.5 mm (0.1 in) which would certainly necessitate a subsequent machining operation. However, in many cases, as for example in the toy industry, extreme accuracy in the dimension of the final product is not as important as the form, in which case the process often shows considerable saving over conventional machining methods.

1.4 ELECTRO-DEPOSITION

Electro-deposition is an electrochemical process used to reproduce accurately a cavity or core form from a given pattern. The pattern can be made in an easily worked material and is the reverse form to that required. That is, a male pattern is required for a cavity and a female pattern for a core. Normally it is much easier to machine a male pattern than the reverse cavity form and it is for this reason that most applications for this technique are for intricate cavity work.

MOULD MAKING

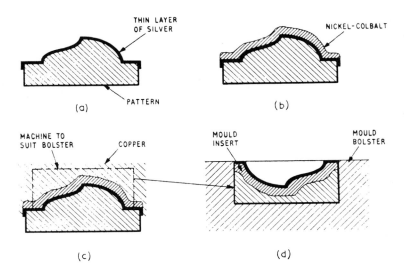

Figure 1.13—Stages in manufacture of cavity by electro-deposition technique; (a) pattern is made conductive with thin coating of silver; (b) nickel-cobalt deposited on to pattern; (c) hard copper deposited on to nickel shell; (d) 'electroformed' insert machined, and fitted to bolster

The principle of electro-deposition is illustrated in Figure 1.13.

(a) A pattern of the required form is made, normally, from either brass, an acrylic or an epoxy resin. The process requires that the pattern conducts electricity and the surfaces of the non-conducting materials must therefore be made conductive. One method, is to coat the pattern with a thin film of silver (1.5×10^{-4} mm thickness).

(b) Nickel-cobalt is deposited on to the electrically conductive pattern in a plating vat until a shell of about 4 mm (5/32 in) thickness is obtained.

(c) Hard copper is deposited on to the nickel shell in a second plating vat. As copper can be deposited at a much faster rate than nickel, a considerable thickness of this backing material is built up relatively quickly.

(d) After the pattern is removed, the outside of the electroformed insert is machined to a shape and size that can be accommodated in a bolster. The insert is held in position either by drilling and tapping it and holding back with screws through the bolster, or by machining a flange on the bottom of it and using a frame-type bolster (Section 2.3.3).

This process has advantages over some other mould making techniques; these advantages may be summarised as follows:

    (i) A male pattern in a soft material is simpler to manufacture than a die-sunk cavity in nickel-chrome steel.

    (ii) Complex parting surfaces may be achieved relatively easily with this technique.

COLD HOBBING

(iii) Nickel-cobalt is non-corrosive and it is therefore suitable for all plastics materials.
(iv) Provided a highly polished master is used the finishing costs on the electroformed insert are nominal.
(v) The electroformed cavity or core insert is in one piece which offers considerable advantages in production with free-flowing plastics materials (e.g. nylon). These materials tend to flow very readily into minute cracks and crevices of a built-up cavity or core assembly.

The main limitations of the process are in size and strength. The maximum size of electroformed insert which can be produced is limited to the size of vat available. Moreover, while nickel-cobalt is relatively hard (44-55 Rockwell C), the copper backing is less hard and this might allow a depression to form on the impression face due to a localised high force.

1.5 COLD HOBBING

Cold hobbing is a process in which a hardened steel master hob is forced into a soft steel blank under considerable pressure.

Hobbing is used for the production of cavities which by virtue of their shape would be difficult to die-sink on conventional machine tools. The basic principle of hobbing is illustrated in Figure 1.14.

(a) The master hob is mounted above a hobbing blank in a hobbing press (the platens only of which are illustrated). The soft steel hobbing blank is fitted in a substantial chase.

(b) and (c) The top platen of the hobbing press is caused to move downwards and the hardened steel hob is progressively forced into the soft steel blank. The pressure exerted by the hob causes the soft steel to flow in the cold state.

(d) When the required depth is reached, the top platen is raised, leaving the hob behind in the hobbing blank.

(e) The hobbing blank must now be removed from the chase and one method of doing this is illustrated. The chase is placed upside-down on two parallels and a knock-out bar is positioned above the hobbing blank. The top platen of the press is then lowered to release the tightly fitting blank from the bolster.

(f) The next operation is to remove the hob from the hobbing blank. Here again there are several methods, one of which is illustrated. A screw extractor is used to apply the necessary force to withdraw the hob.

(g) All that remains now is to machine the hobbing blank to a suitable shape, to case-harden the hobbed insert and finally to give the impression a light polish before mounting the insert in its bolster.

The hobbing process is particularly applicable for the production of cavities on multi-impression moulds. One master hob can often be used to produce a number of cavities, thereby saving considerable machining time. Providing the hob is highly polished, the hobbings too will have a good surface finish, which saves bench finishing and polishing time.

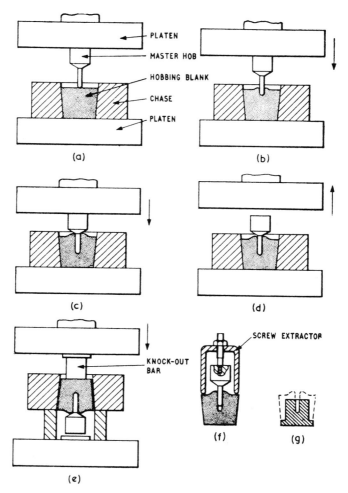

Figure 1.14—Stages in manufacture of cavity insert by cold hobbing technique

In general a male hob is easier to manufacture than a corresponding cavity of the same form; however some shapes are very difficult to hob and careful consideration must be given to a project before attempting to use this technique (Bebb (*op. cit.*)).

For ease of hobbing, a low-carbon steel is usually used for the hobbing blank. As stated above, after the hobbing operation the hobbed insert is case-hardened. This means that the outside surface (including the cavity form) is extremely hard whereas the internal core of the steel remains relatively soft. Thus if in production a localised high force is developed there is a possibility that the surface of the impression may be depressed.

1.6 PRESSURE CASTING

Beryllium-copper is a material which is increasingly being used in mould construction because it possesses several desirable characteristics. In particular it has a high thermal conductivity combined with a reasonable hardness (Brinell Hardness Number of about 250) which makes it suitable for certain types of cavity and core, and for other mould parts, such as hot-runner unit secondary nozzles.

Its high thermal conductivity means that when beryllium-copper is used for a cavity or a core the heat from the melt will be transferred away from the impression faster than if a corresponding steel cavity and core are used, and this often results in a shorter moulding cycle.

Beryllium-copper can be machined, in which case the conventional machine tools are used, and it can be cold-hobbed, hot-hobbed or pressure-cast. The last technique offers certain advantages over the hobbing methods, in that cold or hot hobbing of beryllium-copper tends to work harden the material which results in the development of stress concentrations.

Pressure casting (or liquid hobbing) is used mainly for the production of cavities but it can be used, where applicable, for the production of the cores as well. As the terms suggest, it is basically a process which combines the casting and hobbing techniques.

The basic principle of the pressure casting process is illustrated in Figure 1.15.

(a) A master hob is made from a good-quality steel. Beryllium-copper has a shrinkage of approximately 0.004 mm/mm (in/in), therefore the hob is manufactured oversize to allow for this.

(b) The master hob is attached to a plate and mounted in a chase. The assembly is preheated and then fitted on to parallels on the lower platen of the hobbing press.

(c) Molten beryllium-copper is then poured into the recess formed by the hob and the chase. A shield is often used to protect the hob during the pouring stage.

(d) A plunger attached to the moving platen of the hobbing press is brought down on top of the molten beryllium-copper and a force applied. The plunger is a good slide fit in the chase.

(e) When the beryllium-copper has solidified the plunger is withdrawn, and the hobbing and hob is removed from the chase. Subsequently the hob is extracted from the hobbing, which is then machined externally to suit the bolster.

(f) The hobbing is then annealed, hardened, lightly polished and finally fitted into the bolster.

The advantages of this technique are basically the same as those described for the cold hobbing of steel. However there is an important difference. Beryllium-copper is poured around the master hob in the pressure casting process, whereas in the cold hobbing process the master hob is forced into the steel. More intricate and delicate forms can therefore be produced with the pressure casting technique, without the risk of hob failure.

MOULD MAKING

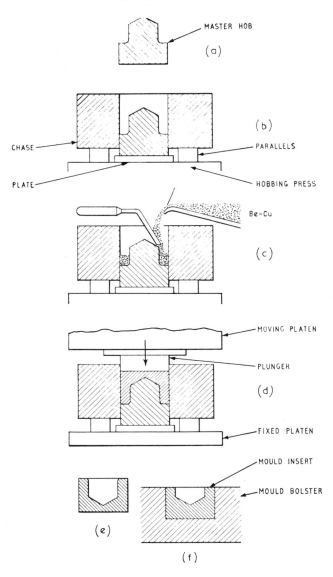

Figure 1.15—Stages in manufacture of cavity insert by pressure casting technique

It should be noted, however, that the major advantage is not so much the mould making process as the properties obtainable from beryllium-copper.

The main limitation is in the size of hobbing obtainable. This is controlled by the size of hobbing press available and the material melting capacity.

SPARK MACHINING

1.7 SPARK MACHINING

This is one of the more recent additions to mould making methods and, strictly speaking, it should come under the machine tool section. However, as the principle of operation is different from that of all other basic machine tools it is preferable to discuss this technique separately.

Spark machining is a process in which steel, or other metals, can be machined by the application of an electrical discharge spark. The spark is localised and metal is progressively removed in small quantities over a period of time.

First consider the spark erosion machine. This is illustrated diagrammatically in Figure 1.16. The workpiece, which in our case is a mould insert blank,

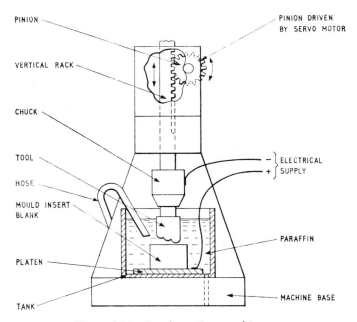

Figure 1.16—Spark erosion machine

is mounted on a platen and submerged in a tank containing a dielectric fluid (normally paraffin). The tank is mounted on the machine base. The tool, which has the required complementary form of the cavity (i.e. similar to a hob), is mounted in a chuck which is attached to a vertical rack. A servomotor (not shown) actuates the rack via the pinion. Thus the tool can be moved in the vertical direction with respect to the workpiece. Both the tool and the workpiece are connected to an electrical supply, and the tool becomes a negative electrode and the workpiece a positive electrode.

The sequence in the machining of a mould insert is shown in Figure 1.17. This diagram illustrates a cross-section through a tool and workpiece.

(a) The rack of the machine is moved downwards by the servomotor until

25

MOULD MAKING

a specific distance between the tool and the workpiece is reached at which point the dielectric separating the two electrodes breaks down and a spark occurs. This spark results in a small particle of the workpiece being eroded away. At the same time a similar but less severe erosion takes place from the tool.

(b) A jet of dielectric fluid is directed at the workpiece via the hose (Figure 1.16) and the eroded particles are washed away as the tool is momentarily lifted. The tool again descends but, this time, because of the erosion, the tool descends a minute amount more than the previous stroke. Again the spark occurs, but in a different place, and another particle of the workpiece is eroded away.

(c) So it continues, the tool lifts, the eroded particles are washed away, the tool descends and another spark occurs at the point of minimum gap.

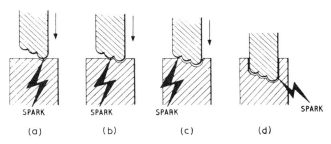

Figure 1.17—Stages in manufacture of cavity by spark erosion technique

(d) We have mentioned previously that not only is the workpiece eroded, but so is the tool. This means that for a relatively deep cavity several tools will be required. The first two or three tools will perform the roughing out operation and the final tool will perform the finishing operation. The last sketch shows the final tool at the maximum depth required.

The dielectric fluid is constantly being circulated; the fluid containing the eroded sludge is withdrawn from the base of the tank, passed through filters and pumped back to the tank via the hose.

Hardened steel can be machined by this technique so that intricately shaped hardened cavities (which might be impracticable by conventional machine-tool methods because of possible distortion during hardening) can be formed.

Spark machining finds many applications in the repairing and modifications of hardened cavities and cores. For instance, a small pocket can be machined into the mould plate and a suitable insert fitted, which saves the necessity of softening the mould plate with the possibility of distortion, etc.

The major limitation of the process is that several tools are required to produce one cavity. If the cavity form is complex, the cost of machining these tools may make the use of this technique uneconomic.

1.8 BENCH FITTING

Irrespective of the machine tool or technique used to manufacture the various parts of the mould, the final responsibility for the finishing of the individual parts and for fitting them together lies with the bench fitter. The mould finishing and assembly procedure adopted by the bench fitter varies from toolroom to toolroom and quite often between individual toolmakers working in the same toolroom; it is therefore impossible to set down a standard pattern for the work. In consequence, we intend only to indicate the general approach to this problem without going into details. We will do this by considering the various stages in the bench fitting involved in the manufacture of a simple mould. The various stages are illustrated in Figure 1.18.

1.8.1 Stage 1. Finishing the Impression. When the mould plates are received from the machine tool section, the impression form (on both plates) is in the rough machined state. Cutter marks, burrs, etc., are very apparent on the surface. The bench fitter's first job is to produce a cavity and core free of machine marks and to the shape and dimensions specified on the mould detail drawing.

Basic hand tools, such as files, scrapers and chisels are used for this operation, various sizes, grades and shapes being used as and when applicable. In addition, wherever possible, power driven flexible shaft equipment is used to speed up this operation. This equipment incorporates various heads which accommodate special needle files or scrapers. The heads can have a rotary or a reciprocating motion.

Once the cavity (or core) is free of machine marks the next stage is to remove the marks left by the file and the scraper. This is achieved by one or more of several techniques depending on the shape of the cavity (or core). These techniques are honing, lapping (including diamond lapping) and emery-cloth finishing. The last is the most common method used for simple forms. A medium emery cloth is used initially with a suitable backing tool to remove the deep scratches left by the file and scraper. A slightly finer grade is then used in a different direction of motion to remove the scratches left by the preceding emery cloth. This procedure is continued, using progressively finer grades of emery cloths and emery papers until, finally, a scratch-free surface is obtained.

At a somewhat later stage, after all other work on the cavity or core is complete, the impression form is polished. This is accomplished with polishing cloths, mops or bobs, in conjunction with a polishing compound such as polishing soap or rouge. Polishing, which is a lengthy, time-consuming operation, must generally continue until the impression form has a mirror-like finish.

Whereas a considerable amount of the bench fitter's time is spent on finishing the impression when supplied with a mould plate or insert which has been produced by the machine tool technique, far less time is required if certain of the other mould making techniques are used, such as electro-deposition or hobbing, etc.

MOULD MAKING

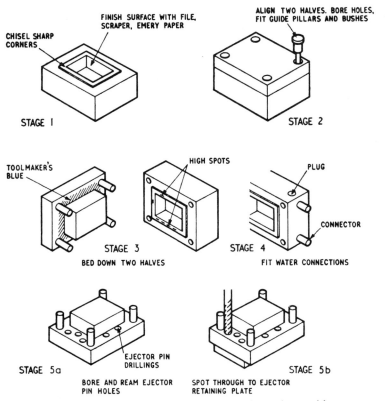

Figure 1.18—Stages in bench fitting of simple mould

If the cavity (or core) plate is in the form of a built-up assembly (Chapter 2) then the insert and bolster must be drilled, tapped and counterbored (where applicable) to permit socket-headed screws to be fitted so that the assembly of parts, in fact, functions as a single mould plate.

1.8.2 Stage 2. Aligning Cavity and Core. Once the cavity and core have been semi-finished, the next operation is to align the two parts with respect to each other so that the moulding produced will have the correct wall section. This is achieved by using packing pieces between the cavity and core. The two mould plates are clamped together and returned to the milling or jig boring machine to have guide holes bored through both plates. When this operation is complete, the clamps are removed, the mould plates separated and the guide pillars and guide bushes fitted. (For a cross-sectional view see Figure 2.22.) The two mould plates are again brought together and checked to ensure that the core is in alignment with the cavity. A dummy moulding is often made at this stage, using wax, so that the wall section of the product can be checked. Any slight inaccuracies need to be corrected, of course.

BENCH FITTING

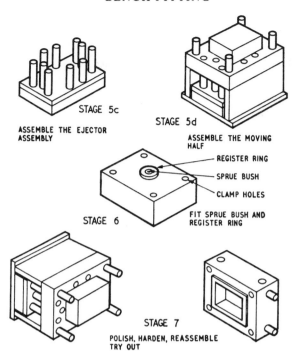

Figure 1.18—Stages in bench fitting of simple mould (contd)

1.8.3 Stage 3. Bedding Down. The next stage is to bed down the two mould halves. This is the process of 'marrying' the two opposing mould halves together to prevent the plastic material escaping between the two surfaces when the material is injected into the impression. Basically the process of bedding-down is simple. One surface (the core plate in our example) is given a very fine coating of toolmaker's blue. The two plates are then momentarily brought together and where there are high spots on the second mould plate (the cavity plate in our example), blue will be picked up. These high spots are removed by scraping and filing. This procedure is repeated until an even film of blue is transferred from one plate to the other. (See Chapter 5 for further details on parting surfaces.)

1.8.4 Stage 4. Water Cooling Circuit. The holes drilled for the water circulation in the mould plates are tapped and plugs, baffles, or connectors fitted as appropriate. The circuit is checked to ensure that the flow is unidirectional and that no leakage occurs. (For further details on mould cooling see Chapter 6.)

1.8.5 Stage 5. Fitting Ejector System.
(a) The holes to accommodate the ejector pins and push-back pins are marked out on the mould plate and subsequently bored and reamed.

MOULD MAKING

(b) The retaining plate is nominally clamped in position below the mould plate. The ejector holes, etc., are spotted through to this plate.

(c) The retaining plate is drilled and counterbored to accommodate the ejector pins and push back pins. The ejector plate and retaining plate are marked out, drilled, counterbored, and tapped where appropriate to permit the two plates to be held rigidly together by socket-headed screws. Assemble the ejector plate assembly.

(d) The ejection half of the mould consisting of the mould plate, support blocks and back plate are marked out according to the mould detail drawing, drilled, counterbored and tapped where specified. The entire moving half of the mould is assembled. (For nomenclature and further details about ejector systems see Chapter 3.)

1.8.6 Stage 6. Fitting Sprue Bush and Register Ring. Turning to the fixed mould half, the sprue bush and the register ring are located and fitted. (For a cross-sectional view see Figure 2.4.). Clamping holes are marked out (with respect to the register ring), drilled and tapped.

1.8.7 Stage 7. Polishing, Hardening and Try-Out. The mould is disassembled and the cavity and core form polished (see Stage 1). All parts which require heat treatment are sent for hardening. When this operation is complete the mould is reassembled and the cavity and core form given a final polish. The mould is then sent for try-out on an injection machine to produce a sample moulding. This is checked and, if necessary, adjustments are made.

The mould is now ready for production.

2

General Mould Construction

2.1 BASIC TERMINOLOGY

2.1.1 Impression. The injection mould is an assembly of parts containing within it an 'impression' into which plastic material is injected and cooled. It is the impression which gives the moulding its form. The impression may, therefore, be defined as that part of the mould which imparts shape to the moulding.

The impression is formed by two mould members:
 (i) The *cavity*, which is the female portion of the mould, gives the moulding its external form.
 (ii) The *core*, which is the male portion of the mould, forms the internal shape of the moulding.

2.1.2 Cavity and Core Plates. This is illustrated for a simple hexagonal container in Figure 2.1. The basic mould in this case consists of two plates. Into

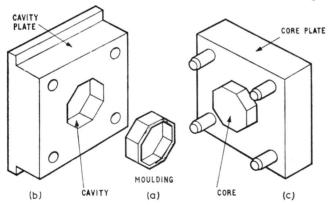

Figure 2.1—Basic mould consisting of cavity and core plate

one plate is sunk the cavity which shapes the outside form of the moulding and is therefore known as the *cavity plate*. Similarly, the core which projects from the *core plate* forms the inside shape of the moulding. When the mould is closed, the two plates come together forming a space between the cavity and core which is the *impression*.

2.1.3 Sprue Bush. During the injection process plastic material is delivered

GENERAL MOULD CONSTRUCTION

to the nozzle of the machine as a melt; it is then transferred to the impression through a passage. In the simplest case this passage is a tapered hole within a bush as shown in Figure 2.2. The material in this passage is termed the *sprue*, and the bush is called a *sprue bush*.

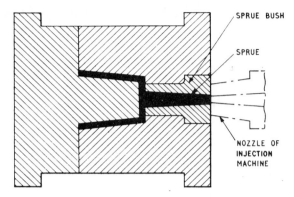

Figure 2.2—Feed system for single-impression mould

2.1.4 Runner and Gate Systems. The material may be directly injected into the impression through the sprue bush (Figure 2.2) or for moulds containing several impressions (multi-impression moulds) it may pass from the sprue bush hole through a *runner* and *gate* system (Figure 2.3) before entering the impression.

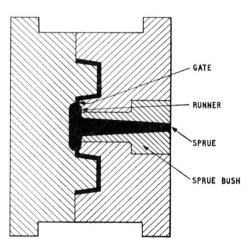

Figure 2.3—Feed system for multi-impression mould

2.1.5 Register Ring. If the material is to pass without hindrance into the mould the nozzle and sprue must be correctly aligned. To ensure that this is so the mould must be central to the machine and this can be achieved by including a *register ring* (Figure 2.4).

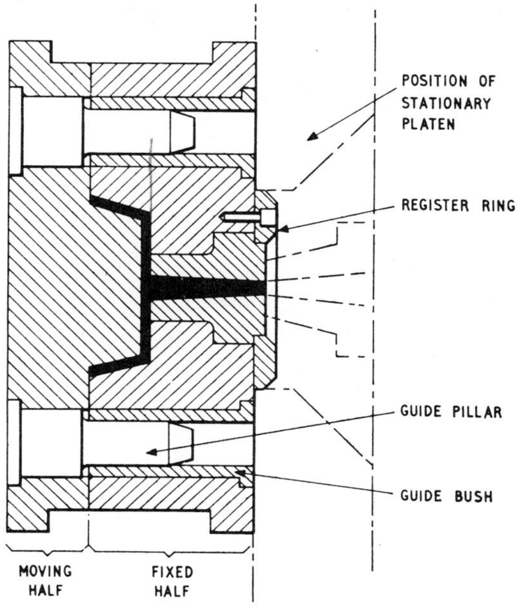

BASIC TERMINOLOGY

Figure 2.4—Basic mould incorporating sprue bush, register ring, guide pillars, guide bushes

2.1.6 Guide Pillars and Bushes. To mould an even-walled article it is necessary to ensure that the cavity and core are kept in alignment. This is done by incorporating *guide pillars* on one mould plate which then enter corresponding *guide bushes* in the other mould plate as the mould closes. An example with guide pillars mounted on the core side and corresponding guide bushes in the cavity side is shown in Figure 2.4. The size of the guide pillars should be such that they maintain alignment irrespective of the applied moulding force; this they are normally able to do (for exceptions see Section 2.4.1). All the constituent parts of the basic mould have now been described and a cross-section drawing of the assembled mould illustrated (Figure 2.4).

2.1.7 Fixed Half and Moving Half. It can be seen (Figure 2.4) that the various mould parts fall naturally into two sections or *halves*. Hence, that half attached to the stationary platen of the machine (indicated by the chain dotted line) is termed the *fixed half*. The other half of the mould attached to the moving platen of the machine is known simply as the *moving half*. Now it has to be decided in which of the two halves the cavity or core is to be situated. Generally the core is situated in the moving half and the overriding reason why this is so, is as follows:

The moulding, as it cools, will shrink on to the core and remain with it as the mould opens. This will occur irrespective of whether the core is in the *fixed half* or the *moving half*. However, this shrinkage on to the core means that some form of ejector system is almost certainly necessary. Motivation for this ejector system is easily provided if the core is in the moving half. (See

33

GENERAL MOULD CONSTRUCTION

Chapter 3.) Moreover, in the case of our single-impression basic mould, where a direct sprue feed to the underside of the moulding is desired the cavity must be in the fixed half and the core in the moving half (Figure 2.4).

2.1.8 Methods of Incorporating Cavity and Core. We have now seen that in general the core is incorporated in the moving half and the cavity in the fixed half. However, there are various methods by which the cavity and core can be incorporated in their respective halves of the mould. These represent two basic alternatives: (*a*) the *integer method* where the cavity and core can be machined from steel plates which become part of the structural build-up of the mould, or (*b*) the cavity and core can be machined from small blocks of steel, termed *inserts*, and subsequently *bolstered*. The choice between these alternatives constitutes an important decision on the part of the mould designer. The final result, nevertheless, will be the same whichever method of manufacture is chosen. In either design the plate or assembly which contains the core is termed the *core plate* and the plate or assembly which contains the cavity is termed the *cavity plate*.

2.2 MOULD CAVITIES AND CORES

So far we have discussed the formation of the mould impression from the relative positions of the cavity and core. These give the moulding its external and internal shapes respectively, the impression imparting the whole of the form to the moulding. We then proceeded to indicate alternative ways by which the cavity and core could be incorporated into the mould and found that these alternatives fell under two main headings, namely the integer method and the insert method. Another method by which the cavity can be incorporated is by means of split inserts or *splits*. This has not been mentioned previously but is a variant of the insert method and is discussed in Chapter 8. We now go on to discuss the integer and insert methods separately in detail.

2.2.1 Integer Cavity and Core Plates. When the cavity or core is machined from a large plate or block of steel, or is cast in one piece, and used without bolstering as one of the mould plates, it is termed an *integer cavity plate or integer core plate*. This design is preferred for single-impression moulds because of the strength, smaller size and lower cost characteristics. It is not used as much for multi-impression moulds as there are other factors such as alignment which must be taken into consideration. Typical mould designs which incorporate an integer cavity and core are shown in Figure 2.5.

MANUFACTURE OF INTEGER CAVITY AND CORE. Of the many manufacturing processes available for preparing moulds only two are normally used in this case. These are (*a*) a direct machining operation on a rough steel forging or blank using the conventional machine tools, or (*b*) the 'precision investment casting' technique in which a master pattern is made of the cavity and core. The pattern is then used to prepare a casting of the cavity or core by a special

MOULD CAVITIES AND CORES

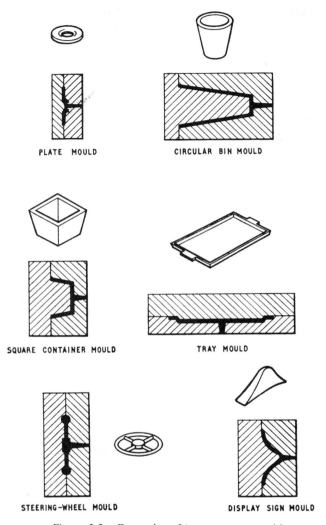

Figure 2.5—Examples of integer type moulds

process (Section 1.3). A 4¼% nickel - chrome - molybdenum steel (BS 970-835 M30) is normally specified for integer mould plates which are to be made by the direct machining method. The precision investment casting method usually utilises a high-chrome steel.

USE OF LOCAL INSERTS. These may be incorporated in the integer block in order to simplify the process of mould making. If they are used, a recess or hole is made in the cavity or core plate to accommodate the insert which is then securely fitted into position. Some examples of the judicious use of local inserts in the integer type of mould are given below.

GENERAL MOULD CONSTRUCTION

Example 1. The cavity form for a bucket which has a rim at the base to stand on. In this case the narrow groove in the base of the cavity, necessary for forming this rim, represents a reasonably difficult machining problem. This difficulty can be overcome by making the base of the cavity in the form of a local insert, as shown in Figure 2.6. This method has the additional advan-

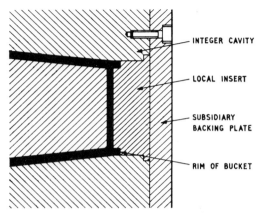

Figure 2.6—Local insert fitted to an integer cavity to facilitate machining

tage that the cavity can be formed by a straight-through machining operation which cuts down the overall machining time. The local insert is machined separately and then supported at the base by a subsidiary backing plate.

Example 2. A local variation in an otherwise constant form is required in a moulding. The example chosen here is of a bath with carrying handles, one of which is shown in Figure 2.7a. The inside form of the bath rim requires a complementary male form around the top of the cavity (b). Now, while most of it can be produced by a simple turning operation, the presence of the handle projection prevents any simple machining operation if a wholly integer mould is attempted. The general form of the rim and the impression is shown at (c) and the local section at the handle (d). Without a local insert all of the projecting male form on the cavity side will have to be made by copy milling which is time-consuming and costly. However, by incorporating the local insert shown at (e) the rim can be turned and a recess made for the insert by a simple milling operation. The local insert, incorporating the male form for the handle is then fitted into the recess in the cavity plate. So far as the core side is concerned, it is not necessary to use an insert. In this case the external mould form for the handle is made by removing metal from the general form of the core plate and not, as in the previous case, by adding metal to it.

Example 3. A mould contains slender projections which may get damaged and require replacement. Any small projection that forms a moulding recess or hole and which is liable to damage because of its proportions rela-

MOULD CAVITIES AND CORES

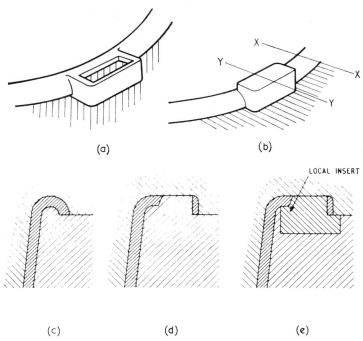

Figure 2.7—*Illustrating desirability of fitting local insert into an integer cavity: (a) part of component (a bath)—note projecting handle; (b) sketch of corresponding portion of cavity—note projection to form handle; (c) cross-section through mould at 'X-X'; (d) cross-section through mould at 'Y-Y' without local insert; (e) cross-section through mould at 'Y-Y' with local insert*

tive to the rest of the mould should be incorporated as a local insert. This will allow easy replacement of the damaged part. An example is illustrated in Figure 2.8. This shows a sketch and part section of a mould for a toothbrush stock. A hole is required in the handle of the component. The core which forms this is a slender rectangular projection, and while this can be made by

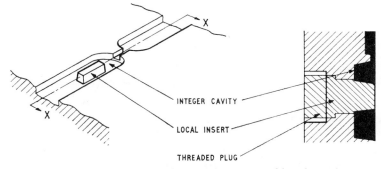

Figure 2.8—*Local insert forming hole in toothbrush stock*

37

GENERAL MOULD CONSTRUCTION

machining from the solid cavity plate it is more practical to let in a local insert as shown in the sectional drawing. The local insert is held in position by a threaded plug.

Example 4. A round hole is required in the moulding. The reasons for adopting a local insert in this case will be similar to the last example if the size of the hole necessitates a slender core. However, in general, the reason for using local inserts is that the male form required to mould the hole is round and a round, local insert, core pin is very much simpler to produce than the corresponding pin machined from the solid. When round holes are specified in the component form, the mould designer should always consider the use of local inserts. These round pins fit into holes machined into the mould member (cavity or core plate).

The various methods of securing circular local inserts to a mould plate are shown in Figure 2.9. Relatively large-diameter local inserts can be secured

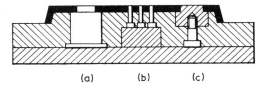

Figure 2.9—Various methods for securing local inserts

either by a flange fitting (a) or by the *screw-down* method (c). For small-diameter local inserts the shoulder method is always adopted (b). Note that the overall lengths of the local inserts in the latter design have been shortened for ease of manufacture.

Example 5. An engraving is to be included in the impression. This is usually best incorporated as a local insert *pad* to allow for change of engraving if this is required. There are other reasons however. First, if the engraving were to be made at the bottom of a deep cavity, the engraving operation would be extremely difficult. Second, as engraving machines are usually of light construction, they will not accommodate large mould plates. An engraved insert pad is shown in Figure 2.10. In this case the pad is incorporated in a pocket machined in the bottom of the cavity, and secured with socket-headed screws.

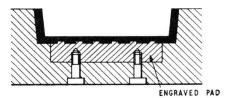

Figure 2.10—Engraved (local insert) pad fitted to integer cavity

2.2.2 Inserts: Cavity and Core. For moulds containing intricate impressions, and for multi-impression moulds, it is not satisfactory to attempt to

MOULD CAVITIES AND CORES

machine the cavity and core plates from single blocks of steel as with integer moulds. The machining sequences and operation would be altogether too complicated and costly. The insert-bolster assembly method is therefore used instead.

The method consists in machining the impression out of small blocks of steel. These small blocks of steel are known, after machining, as *inserts*, and the one which forms the male part is termed the *core insert* and, conversely, the one which forms the female part the *cavity insert*. These are then inserted and securely fitted into holes in a substantial block or plate of steel called a *bolster*. These holes are either sunk part way or are machined right through

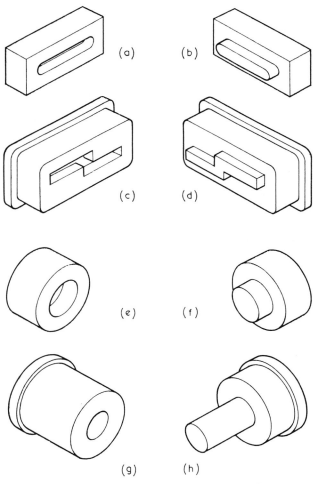

Figure 2.11—Types of cavity and core inserts: (a,b) rectangular, screwdown; (c, d) rectangular, flanged; (e, f) circular, screw-down; (g, h) circular, flanged

GENERAL MOULD CONSTRUCTION

the bolster plate. In the latter case there will be a plate fastened behind the bolster and this secures the insert in position.

SHAPE AND TYPE OF INSERT. To simplify machining the designer should make the insert either circular or rectangular in shape. Which of these two shapes is to be used depends on the shape of the moulding. It is convenient to make circular or near circular mouldings in correspondingly shaped inserts and all other shaped mouldings in rectangular inserts. Examples are illustrated in Figure 2.11. Circular inserts are fitted into holes in the bolster (Figure 2.12). A bolster for a multi-impression mould will, therefore, have a

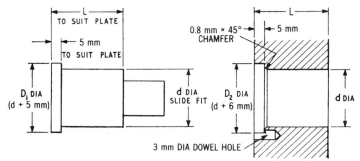

Figure 2.12—Fitting details for small core insert

number of holes either in lines or on a pitch circle diameter. The latter system is illustrated in Figure 2.13. An important economic aspect of a multi-impres-

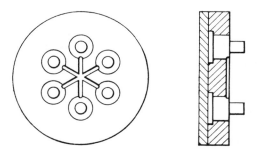

Figure 2.13—Circular core inserts (flanged) fitted to frame-type bolster

sion mould containing circular inserts is that the manufacture of the inserts, the turning and the machining of the bolster holes by boring are both cheap machining operations.

As previously stated, for components with shapes other than circular it is generally desirable to make the insert rectangular. The reason for this is best explained by an example. Consider the case of a long narrow component, say a box. If a circular insert is used, the diameter of the insert will be related to the length of the component, so that the circular insert will be relatively large.

MOULD CAVITIES AND CORES

If a multi-impression mould is required, the mould will also have to be large to accommodate these inserts. This will offset the economic advantages gained by using circular inserts in the first place. Rectangular inserts are fitted into a multi-impression mould in a different manner to the circular inserts just mentioned. In general, they are placed together to form one large insert (Figure 2.14). This illustrates the core plate of a six-impression mould in

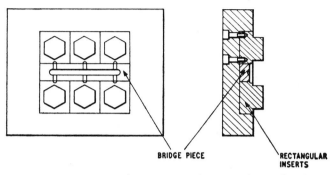

Figure 2.14—Rectangular core inserts (screw-down) fitted to solid bolster

which each core has been machined as a separate insert. However, the rectangular inserts have been placed side by side in two sets of three and between the two sets is inserted a central block of steel called a *bridge piece*. This central block is incorporated to allow one large pocket to be machined out of the bolster plate rather than two smaller pockets each to contain three inserts only. It may be felt that the bridge piece could be dispensed with by making the inserts larger and meet, in fact, on the mould's centre line. However, this creates difficulties for both the sprue bush and the feed system. Without the bridge piece the sprue bush would enter the mould on a line where the edges of two inserts meet and the main runner would be along the line also. If there is a slight gap between the inserts the material will creep down and tend to force the inserts apart.

Although in many cases the multi-impression mould will have the same impression repeated several times, there are examples when it will contain a number of differently shaped impressions. This happens quite frequently in the toy industry, where all the component parts of a toy are incorporated in the same mould. One moulding *shot* then gives all the components necessary to complete the article.

The dimensions adopted for each insert will depend primarily on the size and shape of the moulding. However, the individual insert dimensions are adjusted so that the final, overall, form of the complete insert assembly is rectilinear. This procedure facilitates the machining of the bolster and in fitting the inserts to the bolster.

Local inserts can again be used in cavity and core inserts. Previously we saw their use in integer cavities and cores, and their function here is similar.

Another variation in the make-up of a cavity or core insert is that either one

GENERAL MOULD CONSTRUCTION

or both can be made in more than one piece. An example (Figure 2.15) is of a component called a rocker arm (a). The slender wings on the component make the machining of a one-piece cavity insert difficult. By making the cavity insert in two parts (b) the machining is simplified. The outside form of the composite insert is made a convenient shape for bolstering purposes.

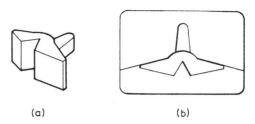

(a) (b)

Figure 2.15—Composite rectangular cavity insert:
(a) component (rocker arm); (b) composite cavity insert

METHODS OF FITTING INSERTS. There are two methods by which inserts can be securely fitted into a bolster. Which method is used will depend on the type of insert and the problem is outlined below.

Method (i) uses the screw-down technique. In this case the insert fits into a blind recess in the bolster and is secured by socket-headed screws from the underside through holes bored in the bolster. Note that the threaded holes in the insert which receive the screws should not shown drilled right through the insert, because the resultant gap left above the end of the screw may well become a material trap for any flash (Figure 2.16a). Similarly, if the insert is

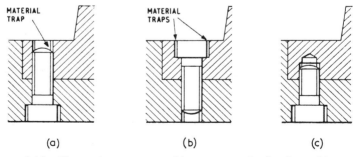

Figure 2.16—Illustrating correct and incorrect methods of attaching screw-down type insert to bolster. Methods (a) and (b) both result in crevices in which plastic material may subsequently be trapped. Method (c) is correct design

secured in the reverse manner by screws from the top face of the insert (Figure 2.16b) then an undesirable crevice is again created, around the head of the screw, which again may act as a material trap. The correct technique is illustrated in Figure 2.16c.

Rectangular inserts (Figure 2.11a, b) are normally fitted by this method

MOULD CAVITIES AND CORES

because they are usually sufficiently large to accommodate screws between the impression and the edge of the insert. Small circular inserts on the other hand generally cannot accommodate screws and for this reason, as well as others, are usually fitted by method (ii).

Method (ii) utilises a flange which is incorporated in the insert design (Figure 2.11g, h) and which fits into a mating recess in the bolster. In this case a hole is made right through the bolster, the mating recess being on the underside, the insert is located so that the flange sits in the mating recess and is secured by a backing plate screwed on to the underside of the bolster (Figure 2.13). One advantage of this method is that it is simpler to machine a circular hole right through the bolster than a blind recess. Another advantage is that it is relatively simple to machine a flange and mating recess for a circular insert, which is not the case for a rectangular insert. However, the overriding advantage of this method lies in the case of a multi-impression mould where inserts are used in both halves. The fact that the holes for the inserts go right through the bolster in each half allows both bolsters to be machined together, if corresponding cavity and core inserts can be made of the same size. Perfect alignment of cavities and cores should now result from simply fitting the inserts into the two bolsters. In practice this is found to apply particularly to circular inserts because, for reasons to be given, rectangular inserts are rarely flanged. The disadvantages in using flange-type rectangular inserts (Figure 2.11c, d) are: (i) extra machining is necessary to incorporate the flange on the insert, and (ii) the recess in the bolster constitutes more work and increases the cost. This objection would only be overruled either because of the increased ease of alignment or in the case of small rectangular inserts because there is insufficient steel for the screw-down technique to be used.

2.2.3 For and Against Integer and Insert-Bolster Methods. Both the integer and the insert-bolster methods have their advantages depending upon the size, the shape of the moulding, the complexity of the mould, whether a single impression or a multi-impression mould is desired, the cost of making the mould, etc. It can therefore be said that in general, once the characteristics of the mould required to do a particular job have been weighed up, the decision as to which design to adopt can be made. Some of these considerations have already been discussed under various broad headings, but to enable the reader to weigh them up more easily, when faced with a particular problem, the comparison of the relative advantages of each system is discussed under a number of headings.

(*i*) *Cost*. The total cost is derived from (*a*) the cost of the mould material and (*b*) the cost of machining and fitting. (*a*) The integer method requires the whole mould plate to be made of expensive mould steel, whereas the insert-bolster method needs only that part which forms the impression to be made of mould steel, the bolster being of the considerably cheaper, mild steel. In material cost, then, the integer mould would usually be more expensive. (*b*) The machining and fitting of a single-impression integer type of mould is less

GENERAL MOULD CONSTRUCTION

costly in time and in the number of operations as compared with the insert-bolster combination. However, for multi-impression moulds there are other factors to be considered.

(*ii*) *Number of impressions.* The difficulty in machining and aligning the cavities and cores in an integer type mould increases with the number of impressions the mould contains. Therefore for multi-impression moulds it is usually preferable to use the insert-bolster system.

(*iii*) *Multi-impression mould alignment.* The ease with which adjustments to the cavity and core positioning can be made is of particular significance in the making of multi-impression moulds. For example, after the completed mould is tried out and the mouldings are examined, some non-uniformity in wall section may be detected in certain mouldings. While this can usually be rectified in an insert-bolster assembly, it is far more difficult in an integer mould.

(*iv*) *Mould size.* The very fact that the integer mould is a whole unit whereas the insert-bolster mould is an assembly means that the overall size of an integer mould will be smaller than a corresponding insert - bolster mould. However, if we consider the individual impressions, by confining the moulding form to inserts, the difficulties in handling and in machining the blocks are minimised. The integer mould requires that quite heavy steel blocks are handled during the manufacturing stage. This not only increases the difficulties of manipulation of these blocks but also increases the capital cost of the machinery used.

(*v*) *Heat treatment.* It is often desirable to heat-treat that part of the mould which contains the impression to give a hard, wear-resisting surface. During this heat treatment the possibility exists that the steel may distort. The smaller the block of steel the less likely is this distortion to occur. Thus, from the hardening stand-point, the insert method is preferred to the integer mould.

(*vi*) *Replacement of damaged parts.* With the insert - bolster system it is possible to repair a damaged impression while continuing to operate the mould with the remaining impressions, the runner feeding the damaged impression being suitably blocked with the result that there is a minimum interruption to production.

(*vii*) *Cooling system.* This is usually far simpler to design for an integer cavity or core plate because the designer can place his cooling system close to the cavity walls without the sealing complications that arise when attempting to cool cavity and core inserts. (See Chapter 6.)

(*viii*) *Conclusions.* Unquestionably for single impression moulds the integer design is to be preferred irrespective of whether the component form is a simple or a complex one. The resulting mould will be stronger, smaller, less costly, and generally incorporate a less elaborate cooling system than the insert - bolster design. It should be borne in mind that local inserts can be judiciously used to simplify the general manufacture of the mould impression.

For multi-impression moulds the choice is not so clear-cut. In the majority of cases the insert-bolster method of construction is used, the ease of manufacture, mould alignment, and resulting lower mould costs being the overriding factors affecting the choice.

For components of very simple form it is often advantageous to use one design for one of the mould plates and the alternative design for the other. For example, consider a multi-impression mould for a box-type component. The cavity plate could be of the integer design to gain the advantages of strength, thereby allowing a smaller mould plate, while the core plate could be of the insert-bolster design which will simplify machining of the plate and allow for adjustments for mould alignment.

2.3 BOLSTERS

We have seen that when it is decided to incorporate the cavity and core into a mould design as inserts, they must be securely retained in the mould. This is achieved by fitting the inserts into a bolster, which, when fitted with suitable guiding arrangements, ensures that alignment of the cavity and core is maintained.

The fundamental requirements of a bolster can be summed up as follows:
 (i) It must provide a suitable pocket into which the insert can be fitted.
 (ii) It must provide some means for securing the insert after it is fitted in position.
 (iii) It must have sufficient strength to withstand the applied moulding forces.

Bolster material. The bolster is normally made from mild steel plate to the BS 970-040 A15 specification. In certain cases, however, a medium carbon steel (BS 970-080 M40) is to be preferred, as, for instance, in the case of small area inserts where the moulding has a projected area which is relatively large compared with the base of the insert. In such a case the pressure developed in the impression by the melt will tend to 'hob' the insert into the bolster. The use of a better-quality bolster steel would obviously be an advantage in such a case, as it minimises this tendency.

Type of bolster. Various types of bolster have been evolved by designers in attempts to simplify either the fitting of the insert or the machining of the bolster. There are five main types of bolster to consider, as follows:
 (i) Solid bolster (Figure 2.17). This is suitable for use with both rectangular and circular inserts.
 (ii) Strip-type bolster (Figure 2.18a). Suitable only for rectangular inserts.
 (iii) Frame-type bolster (Figure 2.18b, c). Although this can be used for

GENERAL MOULD CONSTRUCTION

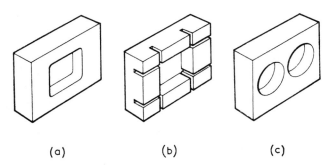

Figure 2.17—Solid bolsters: (a) basic rectangular pocket type; (b) pocket made by slotting technique; (c) basic circular pocket type

both types of inserts, it is particularly suitable for circular inserts.
 (iv) Chase-bolster (Figure 2.18d, e). This type is used in conjunction with 'splits' (split inserts).
 (v) Bolster plate (Figure 2.18f). This is used in particular circumstances with certain types of both rectangular and circular inserts.

2.3.1 Solid Bolster (Figure 2.17). This bolster is made by squaring up a block of suitable steel; then, by a direct machining operation, a pocket is sunk into the top surface to a predetermined depth. The shape of the pocket is either rectangular or circular to suit the shape of the mould inserts.

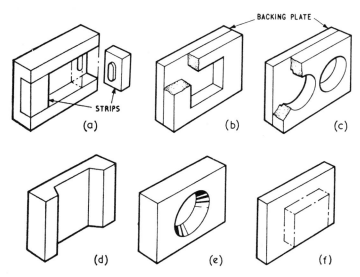

Figure 2.18—Alternative types of bolster: (a) strip type; (b) rectangular frame type; (c) circular frame type; (d) open channel type; (e) enclosed chase bolster; (f) bolster plate

BOLSTERS

The circular pocket is the simplest to manufacture; straight-forward boring and grinding operations provide a pocket into which the circular insert is easily fitted, thus providing accurate positioning in the mould. A typical solid bolster suitable for circular inserts is shown in Figure 2.17c. The inserts are retained by suitable screws from the underside of the bolster and, as can be seen, this particular example has been designed as a two-impression mould.

A pocket to suit the rectangular insert is more difficult and expensive to machine in the bolster than that described above. In this case the bolster block is mounted on to a vertical milling machine, and the rectangular form is sunk by means of an end mill type of cutter as shown in Figure 2.19. The circu-

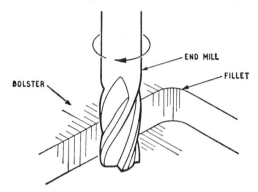

Figure 2.19—End-milling pocket in solid bolster

lar form of the end mill, however, leaves a radius in each corner of the cavity. Because of this the designer should stipulate as large a radius as is practicable in these corners to permit large-diameter end mills to be used for the cutting operation. The reason for this is that the smaller the radius, the smaller is the end mill required and hence the higher the cost of the operation because of the longer machining time necessary. The rectangular type of mould insert will have square corners resulting from its basic manufacture and it is necessary, therefore, to decide whether to modify the insert to suit the radius in each corner of the bolster pocket or vice versa. Both ways will now be considered, the first two methods being modifications of the insert and the last two modifications of the bolster pocket:

(i) Radii are incorporated on each corner of the insert which is then accurately fitted to the bolster. This method provides a smooth unbroken mould surface which is desirable but, because of the accurate fitting required, is expensive (Figure 2.20a).

(ii) The corners of the insert can be chamfered at 45° (Figure 2.20b) This obviates the necessity of accurately matching the radius as described for (i) and it is cheaper to make.

(iii) The bolster cavity can be recessed at the corners to accommodate the square-cornered insert (Figure 2.20c). This involves quite a simple operation, the end mill used to machine out the pocket being sunk in to each corner to machine away the unwanted fillet.

GENERAL MOULD CONSTRUCTION

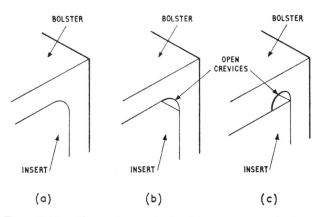

Figure 2.20—Alternative methods of fitting rectangular insert to solid bolster

Methods (ii) and (iii), while cheaper than (i), suffer from the disadvantage that they result in open crevices which may become material traps if the mould flashes. Build-up of plastic material in these traps is difficult to remove and can, if allowed to become excessive, damage the opposite face of the mould.

All three designs have a common disadvantage in that a simple grinding operation of the sides of the pocket is impossible. Due to this, more bench work is necessary in fitting the insert to the bolster.

   (iv) Figure 2.17b shows a method of machining a bolster that allows the sides of the pocket to be ground and also provides square corners for the pocket so that the rectangular insert is easily fitted. The bolster block is initially slotted, as shown, on a horizontal milling machine using a side and face cutter. The centre portion is then removed as before by an end milling operation.

An advantage of method (iv) is that the sides can be ground with a saucer-shaped grinding wheel which passes through the slots. However, this bolster suffers the same disadvantage as (ii) and (iii) in that the slots may become material traps unless they are fitted with small strips of mild steel. While the illustrations show the outside form of the bolster as rectangular they could equally well be made circular.

2.3.2 Strip-Type Bolster. This is an alternative method for making a bolster to suit rectangular inserts and it overcomes some of the disadvantages listed for the solid bolster. The pocket is made by machining a slot completely through the bolster block. Steel strips are then fitted at either end of the slot to complete a frame for the inserts as shown (Figure 2.18a). To prevent the strips from moving under possible side thrust, a projection extends from the underside of the strip and this fits into a mating recess in the bolster. The strips are also securely bolted to the bolster with socket-headed screws.

The advantage of this method is that both the sides and base of the pocket

BOLSTERS

can be ground as also can the inner edge of the strips. This means that all the important surfaces are ground and the subsequent fitting of the rectangular insert is simplified. However, this bolster is not as strong as the solid bolster type as the supporting walls are non-continuous.

The proportions of the strip should be such that it has a width to depth ratio of 1½ : 1. For example a cavity insert 50 mm (2 in) deep will have a strip 75 mm (3 in) wide. This may well be a disadvantage when designing a bolster to accommodate a deep insert in that the overall length of the mould (when viewed in plan) may become excessive.

2.3.3 Frame-Type Bolster. This bolster consists of two parts, namely a frame and a backing plate. The frame is made by machining an aperture of the required shape completely through the bolster plate as illustrated (Figure 2.18b, c) for use with both rectangular- and circular-shaped inserts, respectively. The bottom of the insert is supported by a backing plate secured to the frame with a number of socket-headed screws. The inserts themselves may be secured either in the same manner by screws through the backing plate or alternatively (and this is the more usual) by the use of flanged inserts (see Figure 2.13). In the latter case a recess is machined at the bottom of the aperture in the bolster plate, and the insert is fitted and secured in place by the backing plate. This type of bolster is particularly useful with small inserts where there is often insufficient room in which to position screws. It is also usually included in the design of multi-impression moulds containing circular inserts.

Note that the backing plate, which is ground to ensure a perfect seating, must be of adequate thickness to withstand the force transmitted to it by the insert during the injection of the melt.

2.3.4 Chase-Bolster. When splits (split inserts) are to be incorporated in the mould design it is necessary for one of the bolsters to lock the splits in their closed position. Although this topic as a whole is discussed in Chapter 8, the various designs of chase-bolster are included here for completeness.

The open channel. This is used for shallow rectangular splits and is made by machining a channel across the width of the bolster plate. The sides of the channel are sloped or angled as illustrated (Figure 2.18d). The machining of this bolster is fairly simple and grinding is possible on all faces.

The sloping sides or wedge faces of the channel are extremely susceptible to wear. For this reason they are usually faced with hardened 'wear strips', made from a low-carbon steel suitably carburised to give a hard wear-resisting surface. This permits the use of a lower-quality steel for the main bolster than would otherwise be practicable.

Enclosed chase-bolster. For deep splits the chase-bolster is normally of the enclosed type (Figure 2.18e) as against the open-channel type just discussed. It is machined from a solid block and the pocket which is to accommodate the splits may be of a tapered circular or a tapered rectangular form. From the cost standpoint the circular form is to be preferred, but its use is normally limited to single-impression moulds. (Plate 8.)

GENERAL MOULD CONSTRUCTION

2.3.5 Bolster Plate. Inserts can be mounted directly on to a plain bolster plate. This system provides no side support or location and the walls of the inserts must therefore be of sufficient thickness to withstand the applied moulding pressure without undue deflection. The insert must also be securely screwed and dowelled in position to prevent misalignment. Figure 2.18f shows the bolster plate with an insert indicated in position by a chain dotted line. Generally this design is only used for a limited pre-production run to prove a component design prior to a more extensive tooling programme being embarked upon.

2.3.6 Bolsters—Comparative Survey. The comparative table below (Table 1) indicates the relative merits of each bolster type, with regard to the ease with which it is machined, ground and the insert fitted. The relative strengths are also indicated. The rating values range from 1 to 10, a rating of 1 indicating great difficulty in machining, grinding and fitting inserts, respectively, and with regard to strength—that it is extremely poor. The opposite will be indicated, of course, for a rating of 10.

It can easily be seen that whereas a bolster type may even have the top rating from some considerations, from others it may have the lowest. In fact, no bolster type is perfect in all respects, and the designer has to decide which is the most suitable for the particular application in hand.

2.4 ANCILLARY ITEMS

In addition to the main constructional parts of the mould discussed in the

TABLE 1
BOLSTER—COMPARATIVE SURVEY

Type	Fig. No.	Ease of machining	Ease of grinding	Ease of fitting	Relative strength
1. Basic solid rectangular	2.17a	4	1	4	10
2. Basic solid circular	2.17c	8	6	8	10
3. Slotted type	2.17b	5	6	6	5
4. Strip type	2.18a	6	9	6	6
5. Rectangular frame type	2.18b	5	4	4	8
6. Circular frame type	2.18c	7	8	8	8
7. Open channel type	2.18d	7	9	7	6
8. Enclosed chase bolster (circular)	2.18e	8	6	8	10
9. Bolster plate	2.18f	10	10	10	1

ANCILLARY ITEMS

previous sections there are certain essential ancillary items that are vital if the mould is to function satisfactorily. These ancillary parts ensure that the two mould halves remain in correct alignment, that a path for the plastic material is provided from the machine nozzle to the mould face, and that the mould is accurately positioned on the machine. These ancillary items are: (i) the guide pillars and bushes, (ii) the sprue bush, and (iii) the register ring.

2.4.1 Guide Pillars and Guide Bushes. A description of a simple guide pillar and bush, and their use, has already been given (Section 2.1.6).

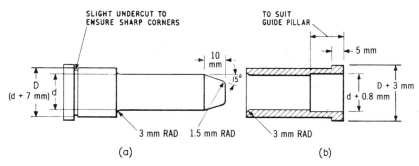

Figure 2.21—Guide pillar and guide bush design

TYPES OF GUIDE PILLARS, MATERIALS AND FITTING. The detail of the fitting of a typical guide pillar will now be given, and is illustrated in Figure 2.21. It should be noted that the guide pillar (Figure 2.21a) is designed so that the working diameter d is smaller than the fitting diameter D by a minimum of 7 mm (¼ in). This introduces a shoulder on to the pillar where it emerges from the mould plate, and while this is initially more costly to machine than a constant diameter pillar, it nevertheless has certain advantages. The fitting diameter of the guide pillar can be made the same as the guide bush; thus the same diameter hole can be bored and ground through both mould plates when clamped together. This allows perfect alignment (Figure 2.22). Also, if the

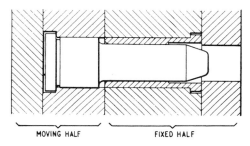

Figure 2.22—Illustrating the fitting of guide pillar and guide bush to two-plate mould

GENERAL MOULD CONSTRUCTION

working diameter d of the guide pillar is bent it may be more easily removed without damaging the fitting hole. To illustrate this point a bent constant-diameter guide pillar is shown in Figure 2.23. Attempts to remove this will damage the fitting hole which will consequently have to be reground before another pillar can be fitted. On the other hand, the removal of a bent pillar which incorporates a shoulder (Figure 2.23b) will present no difficulties.

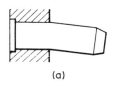
(a)
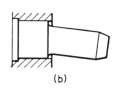
(b)

Figure 2.23—Illustrating the advantage of using stepped guide pillar: (a) bent constant-diameter guide pillar, difficult to remove; (b) bent stepped guide pillar, relatively simple to remove

It is essential that the guide pillar be securely held in the mould plate and diameter D (Figure 2.21) is a press fit. There is a danger that the pillar may be pulled from its seating as the mould opens. To overcome this a flange is normally incorporated at the bottom of the pillar and this is accommodated in a recess in the mould plate as shown in the assembly drawing (Figure 2.22). A clearance is permissible around the flange to facilitate fitting. When the mould plate is deep, say over 100 mm (4 in), it is often convenient to alter the design, omitting the flange and securing the base of the pillar in position by means of a grubscrew (Figure 2.24).

The surface of the pillar must be hard and wear-resisting. This is achieved by machining the pillar from a low-carbon steel (BS 970-080 M15) which is then case-hardened. This process gives a surface which resists pick-up and scoring as the pillar continually enters and leaves the guide bush.

If the guide pillar is likely to be subjected to bending forces the use of a carburising nickel-chrome steel (BS 970-835 M15) is to be preferred.

The normal size range of guide pillars obtainable as standard parts (to reduce mould cost these should be used wherever possible) is between 10 mm (⅜ in) and 38 mm (1½ in) working diameter. Some very large moulds may require guide pillars outside this range. The decision as to which size to use depends on the size of the mould and whether or not a side force is likely to be exerted on the guide pillar. As a rough guide, Table 2 has been compiled from a large number of working moulds. As an additional reference, Table 3 shows

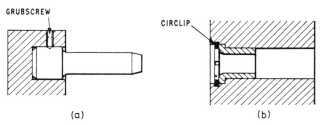

Figure 2.24—Methods for securing guide pillars and guide bushes in deep mould plates

ANCILLARY ITEMS

the range of sizes adopted in the DMS European standard mould base system.

TABLE 2

Working diameter (mm)	Size of Mould (mm)
10	125 x 125 — 125 x 156
14	156 x 156 — 196 x 396
18	246 x 246 — 246 x 496
22	296 x 296 — 346 x 594
30	396 x 396 — 544 x 594
38	594 x 594

TABLE 3

Working diameter (d) (mm)	(in)	Size of Mould (mm)	(in)
10	⅜	100 x 100	4 x 4
13	½	100 x 150	4 x 6
16	⅝	150 x 200	6 x 8
19	¾	200 x 250	8 x 10
22	⅞	250 x 300	10 x 12
25	1	300 x 400	12 x 16
32	1¼	400 x 600	16 x 24
38	1½	600 x 700	24 x 28

FUNCTION OF GUIDE PILLARS. The design of mould, as well as the size and shape of the moulding, will affect the size, number and disposition of guide pillars on the mould plate.

Although the guide pillar system is primarily concerned with alignment of the mould faces as they close during the moulding cycle, the pillars have also the subsidiary functions of protecting the core and acting as locating pins when the mould is being assembled.

Guide pillars are usually necessary to ensure that both halves of the mould are kept in alignment while the mould is closing, though the necessity for this will depend on the design of the component. For instance the impression can,

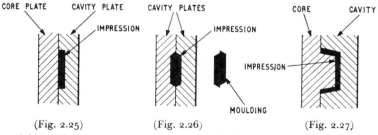

(Fig. 2.25)   (Fig. 2.26)   (Fig. 2.27)

Figure 2.25—When impression is in one mould plate, only nominal alignment between mould plates is required

Figure 2.26—When impression is formed by both mould plates, accurate alignment is essential. Note resulting component from inaccurate alignment

Figure 2.27—Misalignment between the cavity and core for box-shaped component; note resulting undesirable unequal wall section

GENERAL MOULD CONSTRUCTION

as in case (i), be wholly in one mould half with no actual core (Figure 2.25). In this case only nominal location of the two halves is required, as relative movement between the two plates will not affect the dimensions of the moulding.

On the other hand, as in case (ii), where the impression is formed by cavities in both halves of the mould (Figure 2.26), good alignment between the two halves is essential. Even slight misalignment produces a moulding that is 'out of mitre' as shown.

In the more usual case (iii), a cavity and core form the impression. Any misalignment (Figure 2.27) here is particularly unfortunate as the resulting moulding will have one wall thinner than the other. In addition to this, material entering the mould will take the easiest route and flow down the thicker section first. This will tend to move the core further out of alignment, and result in unacceptable mouldings. An extension of this case to large mouldings is considered in the next section.

On two-plate moulds the guide pillars are normally fitted on the moving half so that they provide some protection for the core when the mould is off the machine. However, care must be taken in their positioning to ensure that they do not restrict the fall of the moulding after ejection. The alternative method is to fit the guide pillars on the fixed mould plate which, irrespective of their position, allows a free fall-away path for the moulding. In either design the length of the guide pillar should be such that both mould halves are positively aligned before the core enters the cavity. Without this precaution any slight misalignment, perhaps due to wear of the platen bushes, may cause the core to strike the cavity wall with disastrous results (Figure 2.28a). To safeguard against this possibility the guide pillar should be of sufficient length to enter the guide bush before the core enters the cavity (Figure 2.28b).

Note that the taper on the guide pillar tip provides a lead-in into the bush.

GUIDE PILLARS REINFORCED BY A TAPERED LOCATION. For most small moulds the guide pillars are normally able to maintain alignment of the two halves, within reasonably close tolerances, irrespective of the force applied during injection. However, moulds for thin-walled articles require special treatment as even a slight discrepancy cannot be tolerated in this case. With some very large mouldings the size of pillar necessary to maintain alignment becomes excessively large. Therefore, in either case, the designer must consider means of maintaining alignment without relying entirely on guide pillars. A satisfactory way of achieving this is by the incorporation of a tapered location, normally in addition to the guide pillars.

Consider, for example, a mould for a thin-walled bucket. If the design (Figure 2.29) relies entirely on guide pillars for alignment, difficulties may arise due to the melt tending to flow down one side of the impression first. This will occur if the dimensions of the impression vary due to a slight discrepancy in the shape of the cavity or core. The melt will then flow first through that part of the impression which has the thickest cross-section. The result of this is that a differential force will then be applied on face A which is resisted by the guide pillars B. The forces involved in the moulding of such an article are extremely large and will result in increased misalignment which will, in

ANCILLARY ITEMS

turn, cause variation in wall thickness of the moulding.

One method of holding the core in alignment in the case of such a large moulding is to provide a recess in the cavity plate into which fits a tapered location turned from the core plate at C (Figure 2.29b). The taper ensures

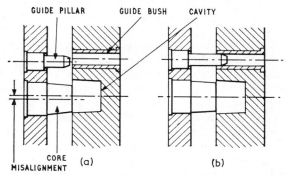

Figure 2.28—Incorrect (a) and correct (b) length for guide pillar

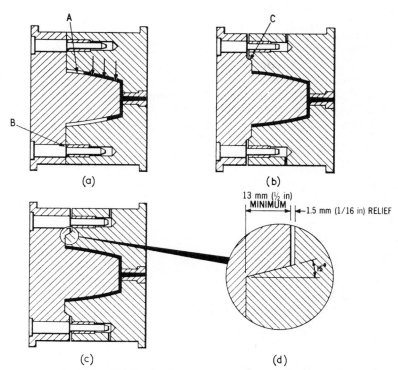

Figure 2.29—Tapered location: (a) unbalanced force on core of large mould being resisted by guide pillars only; (b) tapered location recess in cavity plate; (c) tapered location recess in core plate; (d) enlarged view of tapered location

GENERAL MOULD CONSTRUCTION

that the two surfaces do not rub and cause wear during the normal operating stroke of the mould. The guide pillars, which are still incorporated, provide approximate alignment while the tapered location assures final alignment.

This method of incorporating a tapered location, however, has two disadvantages: (i) it necessitates machining the cavity from a deeper block of steel to accommodate the depth of the recess, (ii) the applied internal forces may force the cavity to expand and hence registering between both mould halves is immediately lost. The preferred method is to provide a tapered location on the cavity side, fitting into a recess on the core side (Figure 2.29c). This achieves the same result as in the previous case but overcomes both disadvantages. It also has the additional advantage of acting as a form of 'chase' for the cavity, helping it to resist the expansion forces. An enlarged view of the tapered location is shown (Figure 2.29d). The example given was for a circular component which permitted a circular tapered location to be used. Providing that the length - breadth dimensions do not vary too much, the circular tapered location can be adopted for rectangular components also. Where there is a wide variation of width - breadth dimensions a straight tapered location must be included on all sides.

POSITIONING OF GUIDE PILLARS. The number of guide pillars incorporated in a mould varies from two (Figure 2.30a), for the simplest type, to four (Figure 2.30c), which is the preferred number for rectangular moulds. Some circular moulds have three guide pillars (Figure 2.30b).

In addition to aligning the two mould halves, the guide pillar is used to prevent the mould from being assembled in the wrong way, and in this sense acts as a locating pin. The design of the two-pillar system (Figure 2.30a) is such that one guide pillar is made larger than the other. The three- and four-pillar system however is designed so that one or more of the pillars are offset. Figure 2.30b shows a three-pillar system in which two pillar holes are offset by 7 mm (¼ in) from the basic 120° marking plane, whereas with the rectangular bolster (Figure 2.30c) the two bottom holes are set 13 mm (½ in) further in than the top two. All these methods guarantee that the mould cannot be assembled incorrectly.

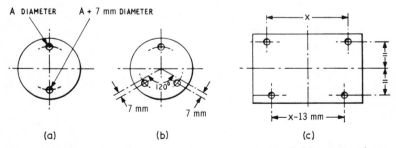

Figure 2.30—Guide pillar positioning. Positioning details for two-(a), three-(b), and four- (c) guide pillar systems

ANCILLARY ITEMS

GUIDE BUSHES. A guide bush is incorporated in the mould to provide a suitable wear-resisting working surface for the guide pillar and to permit replacement in the event of wear or damage.

A typical bush is shown (Figure 2.21b). Its internal bore is designed as a slide fit on the guide pillar, while the external diameter is a press fit into the mould plate. A radius is made at the front end of the bore (as shown) to provide a lead-in for the guide pillar. The rear end of the bush is counterbored to a greater diameter than the working diameter (Figure 2.22).

On each stroke the guide pillar should pass right through the working diameter of the bush, its end passing well into the counterbore. If the counterbore is not incorporated, the pillar will operate only over a limited part of the internal bore and cause uneven wear which in time may create a ridge inside the bush.

Bushes are usually machined from a low-carbon steel (BS 970-080 M15) and case-hardened. This ensures that the required wear-resisting surface is obtained. Figure 2.21b shows the bush with the usual flange type of fitting. For deep mould plates the bush can be fitted from the top surface of the plate and held in position with a circlip (Figure 2.24b).

2.4.2 Sprue Bush. The sprue bush is defined as that part of the mould in which the sprue is formed. In practice the sprue bush is the connecting member between the machine nozzle and the mould face, and provides a suitable aperture through which the material can travel on its way to the impressions or to the start of the runner system in 'multi-impression' moulds.

The sprue bush is fairly highly stressed in some applications and should therefore be made from a 1½% nickel chrome steel (BS 970-817 M40) and should always be hardened. Note that while initially the nozzle may have been set up correctly (according to the machine maker's instructions), an increase in the injection cylinder temperature will cause a fairly high force to be applied to the sprue bush because of the expansion of the injection cylinder.

When a direct sprue feed into the impression is contemplated it is advantageous to prevent possible rearward movement of the sprue bush by stepping the end and fitting a substantial register ring as shown in Figure 2.34. The register ring is designed so that the applied force is transmitted to the platen of the injection machine.

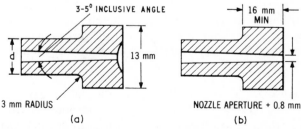

Figure 2.31—Sprue bushes: (a) spherical seating; (b) flat seating

GENERAL MOULD CONSTRUCTION

The internal aperture of the sprue bush has between 3° and 5° included taper, which facilitates removal of the sprue from the mould at the end of each moulding cycle.

There are two basic designs of sprue bush which differ only with respect to the form of seating between the sprue bush and the nozzle of the machine. Both designs are shown (Figure 2.32a, c). The first of these is a sprue bush with a spherical recess which is used in conjunction with a spherical front ended nozzle as illustrated (Figure 2.32a). The second has a perfectly flat rear face, the seating between it and its corresponding nozzle being shown (Figure 2.32c).

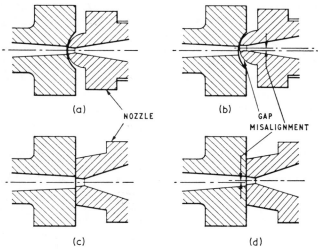

Figure 2.32—Illustrating effect of misalignment between sprue bush and nozzle

Providing the alignment between the nozzle and the bush apertures is perfect, a leak-free joint is achieved with the spherical seating. It should be noted that the radius on the nozzle is slightly less than that in the bush, which ensures both parts are in physical contact at the apertures. However, if the nozzle and sprue bush are slightly out of line (due perhaps to wear on platen bushes) a gap will result and leakage behind the mould will develop (Figure 2.32b). This drawback is not present if slight misalignment occurs at the seating of the sprue bush and flat-faced nozzle (Figure 2.32d). As can be seen, no leakage can occur if the two apertures are slightly out of line and no restriction to flow will occur either, providing that the sprue bush has an aperture slightly larger than that of the nozzle. An allowance of 0.8 mm (1/32 in) on diameter is usually made.

2.4.3 Register Ring. The *register ring*, also called locating ring, is a flat circular member fitted on to the front face (and sometimes also to the rear face) of the mould. Its purpose is to locate the mould in its correct position on the injection machine platen.

ANCILLARY ITEMS

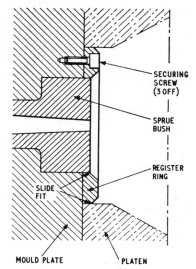

Figure 2.33—Location of mould on platen by register ring

When the mould is mounted on the machine the front mounted register ring fits into a circular hole which is accurately machined in the injection platen on the cylinder-nozzle axis. This ensures that the small aperture in the nozzle is in direct alignment with the sprue bush hole. Now, since the sprue bush is the connecting member between the machine nozzle and the mould face, this alignment of nozzle aperture and sprue bush hole permits an uninterrupted flow of material from the cylinder, through the nozzle and sprue hole into the mould runner system. The register ring, in fact, forms a direct connection between the sprue bush and the hole in the injection platen of the machine.

In Figure 2.33 the register ring is shown to be located in two members, on the outside diameter of the sprue bush, and in the platen aperture. A similar arrangement is shown in Figure 2.34. In this design the register ring is also used to secure the sprue bush in position.

As the register ring is permanently attached to the mould (by socket-headed screws as shown), correct alignment follows automatically whenever the mould is set up on the machine; no adjustment by the setter is therefore necessary.

Figure 2.35 shows a part-section of a mould where the register ring cannot be located on the sprue bush itself because the sprue bush is sunk into the injection bolster to reduce the overall length of the sprue. (Note—this necessitates an extension nozzle.) In this design the register ring is located in a recess in the rear of the injection bolster and it is essential that this recess and the bored hole for the sprue bush be concentric. For this reason the boring of both the recess and the sprue bush hole is generally carried out at one setting.

Details of a typical register ring are shown in Figure 2.36.

GENERAL MOULD CONSTRUCTION

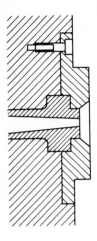

Figure 2.34—Register ring being used for dual purpose of mould location and nominally securing sprue bush

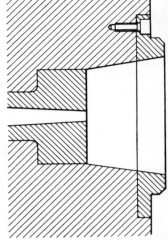

Figure 2.35—Alternative arrangement for use with long-reach nozzle

2.5 ATTACHMENT OF MOULD TO PLATEN

There are two ways by which the mould halves may be attached to the platen of the injection moulding machine and these are either directly by means of bolts, or indirectly by means of bolts via clamps.

2.5.1 Direct Bolting Method. In principle, holes are provided in each mould half to correspond with the holes tapped in the machine's platen. Bolts are then used to directly secure the mould to the platen. Many machine manufacturers adopt a standard hole layout and therefore a mould design for one machine can normally be fitted to another of similar size. However, it is good practice always to check the layout in question before commencing the mould design.

Various alternative direct bolting designs are shown in Figure 2.37. At (a) the front plate (or back plate) incorporates a projection which extends beyond the remainder of the mould half. Clearance holes are provided on suitable centres through this portion. The projection may be produced either by machining from the solid (as shown) or, when a two-plate construction is practicable, the rearmost plate is extended. A similar design is shown at (b), the only difference being that slots are provided through the projection instead of holes. Slots facilitate the setting of the mould on the platen, and this is particularly important for shallow-type moulds. With the mould closed there may be insufficient space left between the projections of both mould halves to permit a bolt to be inserted into a hole, whereas the bolt can

ATTACHMENT OF MOULD TO PLATEN

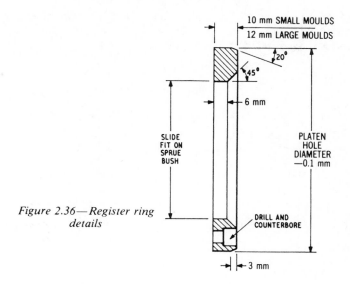

Figure 2.36—Register ring details

normally be fitted, relatively easily, into a slot. It is worth noting that it is generally more convenient to set up a closed mould which has its own alignment system than set up the mould halves separately and align them afterwards.

To save a considerable amount of machining on deep mould plates, the design shown at (c), in which local bolting recesses only are provided, is often used. The size of the recess, naturally, must be sufficient to permit the use of a suitable spanner for inserting and removing the bolt.

2.5.2 Indirect Bolting Method. With this design the attachment of the mould to the machine is by means of a clamp plate. It is used when it is not possible to use the direct bolting method which is generally to be preferred. For example it is often impracticable to incorporate direct bolting due to the inept positioning of tapped holes in the platen.

The indirect bolting assembly consists of three parts (Figure 2.38), namely the clamp plate, the bolt and the packing piece. Two alternative designs are given. At (a) the front plate (or back plate) incorporates a projection (or

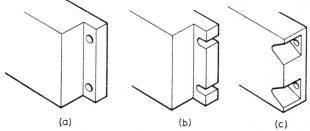

Figure 2.37—Attachment of mould to platen, direct bolting arrangement

61

GENERAL MOULD CONSTRUCTION

flange) and a clamping force is applied to this by the bolt via the clamp plate. Alternatively, a slot can be machined through the mould plate (as (b)) and the clamping force applied in identical manner.

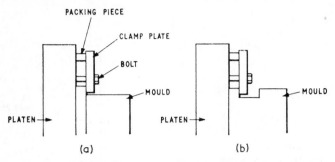

Figure 2.38—Attachment of mould to platen, indirect bolting arrangement

3
Ejection

3.1 GENERAL

The previous chapter deals with the basic two-part mould in which a moulding is formed by injecting a plastic melt, under pressure, into an impression via a feed system. The two parts by themselves, however, do not constitute an efficient design as no means are incorporated for removing the moulding once it is made. It must therefore be removed manually.

Furthermore, all thermoplastic materials contract as they solidify, which means that the moulding will shrink on to the core which forms it. This shrinkage makes the moulding difficult to remove.

It is normal practice, therefore, to provide some means by which the moulded part can be positively ejected from the core, and this chapter deals with the various methods which are used.

Facilities are provided on the injection machine for automatic actuation of an ejector system, and this is situated behind the moving platen. Because of this, the mould's ejector system will be most effectively operated if placed in the moving half of the mould, i.e. the half attached to the moving platen. We have stated previously that we need to eject the moulding from the core and it therefore follows that the core, too, will most satisfactorily be located in the moving half.

The ejector system in a mould will be discussed under three headings, namely: (i) the ejector grid; (ii) the ejector plate assembly; and (iii) the method of ejection.

3.2 EJECTOR GRID

The ejector grid is that part of the mould which supports the mould plate and provides a space into which the ejector plate assembly can be fitted and operated. The grid normally consists of a back plate on to which is mounted a number of conveniently shaped 'support blocks'.

There are three alternative designs:
  (i) The in-line ejector grid (Section 3.2.1)
  (ii) The frame-type ejector grid (Section 3.2.2)
  (iii) The circular support block grid (Section 3.2.3)

3.2.1 In-Line Ejector Grid (Figure 3.1). This consists of two rectangular support blocks mounted on a back plate. The ejector plate assembly, shown in chain-dotted lines, is accommodated in the parallel space between the two

EJECTION

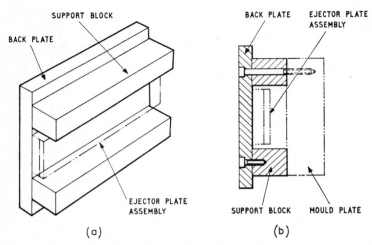

Figure 3.1—In-line ejector grid

support blocks. A cross-section through the ejector grid is shown in Figure 3.1b. The position of the mould plate is also indicated in chain-dotted lines for reference.

The design as illustrated is quite suitable for small types of mould where the overall size of the ejector plate assembly does not necessitate the support blocks being fitted a great distance apart. When this situation does arise, however, unless the mould plate is made reasonably thick there is the probability that the mould plate will be distorted by the injection force (see Figure 3.2a). To avoid the necessity of incorporating a thick, and therefore heavy, mould plate, extra support blocks are often added in the central region of the mould (Figure 3.2b).

The extra support can take the form of an additional rectangular support block (or blocks) fitted parallel to the outer pair (Figure 3.3). The ejector assembly used in conjunction with this type of ejector grid is shown in chain-dotted lines. It consists essentially of bars (rectangular cross-section) which extend completely across the mould and which are coupled together by a cross-bar at either end.

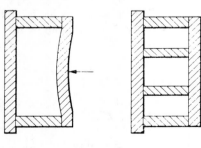

Figure 3.2—Mould plate distortion is likely when support blocks are far apart (a); extra support blocks fitted close to centre can avoid this hazard (b)

EJECTOR GRID

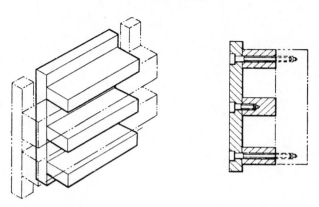

Figure 3.3—Multiple in-line ejector grid, used in conjunction with ejector-bar system

An alternative support arrangement is shown in Figure 3.4. In this system additional local support blocks (usually of a circular cross-section) are incorporated in judicious positions to provide the required additional support. These local support blocks are made from mild steel bar and are held in position by a single screw from the underside of the back plate. The ejector plate assembly (shown dotted) naturally must incorporate holes bored in positions corresponding to those of the local support blocks. Because of the last point, the positioning of these local support blocks is always delayed until after the position of the ejector element (i.e. ejector pin, ejector sleeve, etc.) has been decided upon.

3.2.2 Frame-Type Ejector Grid. Some frame-type ejector grid designs are

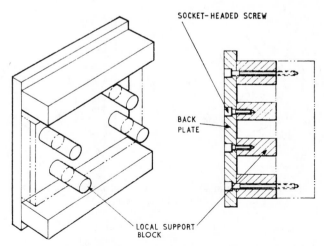

Figure 3.4—Extra support for mould plate can be obtained by judiciously positioning local support blocks

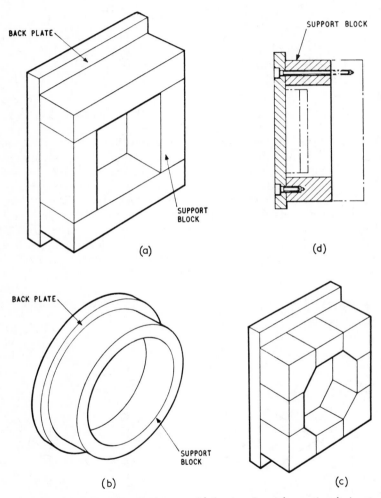

Figure 3.5— Frame-type ejector grid: (a-c) various alternative designs; (d) general cross-section

EJECTOR GRID

illustrated in Figure 3.5. The most common type encountered is the rectangular frame (a) constructed of four support blocks suitably mounted on a back plate. This design is favoured by many mould designers for the following reasons: (i) it is simple and cheap to manufacture; (ii) it provides good support to the mould plate on a small mould; (iii) it allows for the use of a conveniently shaped (rectangular) ejector plate assembly and (iv) the ejector plate assembly is completely enclosed, thereby preventing foreign bodies entering the system.

When the outside shape of the mould plate is circular it is often convenient to design a correspondingly shaped ejector grid. A typical design is illustrated at (b). It consists of a circular support frame mounted on to the back plate. The circular support frame, being machined from the solid block of steel, makes the design slightly more expensive to produce than the rectangular design.

We stated previously that the ejector grid must provide adequate support for the mould plate. Now, as the size of the mould plate increases (and assuming that the ejector plate assembly correspondingly increases in size), the effective support provided by either of the above ejector grid designs progressively decreases.

One method of improving this situation is to incorporate additional local support blocks (usually of circular cross-section) in judicious positions in a manner similar to that described for the in-line ejector grid system (Section 3.2.1; Figure 3.4).

It is often possible, however, to obtain additional support for the mould plate by designing the ejector grid of a shape other than the basic rectangular or circular. The precise shape is dependent primarily upon the positioning of the ejector elements, which in turn determines which part of the ejector plate assembly can be machined away to permit additional support to be incorporated in the design. One example is illustrated at (c). In this case greater support is achieved at each corner by a simple modification to the ejector plate design (i.e. the corners of a rectangular ejector plate are removed). Even more irregularly shaped frames are designed when warranted to give maximum possible support.

A general cross-section taken through any of the above frames (a, b or c) is shown at (d). The mould plate and the ejector assembly are shown in chain-dotted lines. Note that certain screws are used simply to attach the support block to the back plate, whereas other screws pass completely through the support block and are used to attach the mould plate to the ejector grid assembly. By undoing these latter screws the ejector grid can be removed from the mould as a unit. This feature facilitates repairs, etc.

3.2.3 Circular Support-Block Grid. In this design, circular support blocks are used to support the mould plate only, the rectangular outer support blocks of certain of the previous systems being dispensed with altogether. This system is used for large moulds when it is felt that no extra support would be gained by including rectangular blocks as well.

EJECTION

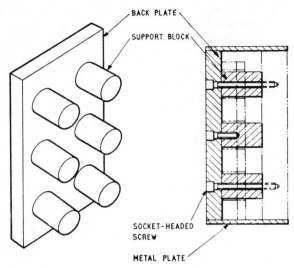

Figure 3.6—Circular support-block grid

A typical support-block grid system is shown in Figure 3.6. The design simply consists of a number of circular support blocks judiciously positioned on the back plate of the mould. The grid is attached to the mould plate by socket-headed screws. The ejector assembly (chain-dotted) can move freely as with previous designs, holes being bored through it to receive the circular support blocks. To prevent foreign matter getting into the ejector system it is desirable to attach thin metal plates to enclose the grid completely.

3.3 EJECTOR PLATE ASSEMBLY

The ejector plate assembly is that part of the mould to which the ejector element is attached. The assembly is contained in a pocket, formed by the ejector grid, directly behind the mould plate. This is illustrated in Figure 3.7. The

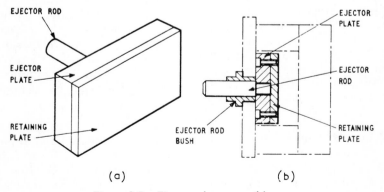

Figure 3.7—Ejector plate assembly

EJECTOR PLATE ASSEMBLY

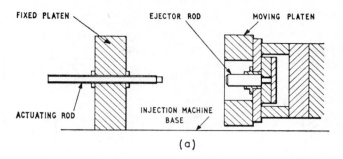

(a)

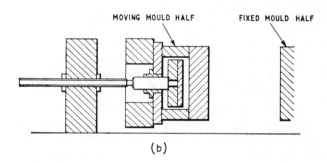

(b)

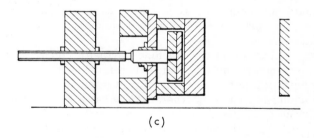

(c)

Figure 3.8—As moving platen moves left, ejector plate assembly is actuated by machine's actuator rod

assembly consists of an ejector plate, a retaining plate and an ejector rod. One end of this latter member is threaded and it is screwed into the ejector plate (see cross-section view (b)). In this particular design the ejector rod functions not only as an actuating member but also as a method of guiding the assembly. Note that the parallel portion of the ejector rod passes through an ejector rod bush fitted in the back plate of the mould.

Before proceeding to discuss the individual parts in more detail, let us consider how this assembly is actuated. A cross-section through the moving half of a typical mould is shown in Figure 3.8. (The core and ejector elements are

EJECTION

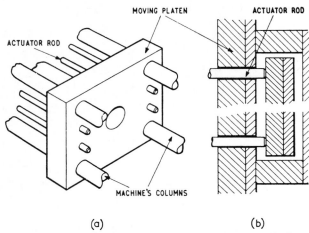

Figure 3.9—Direct actuation of ejector plate assembly by machine's actuator rods

excluded for clarity.) The mould is mounted on the moving platen of the injection machine. To the left of the moving platen is the machine's actuating rod. This member can be adjusted to allow for various alternative 'ejector strokes'. When the moving platen is caused to move to the left, and the mould opens, the mould's ejector rod at some point of the stroke strikes the actuating rod. The entire ejector plate assembly is arrested as shown at (b). The remainder of the moving half (i.e. the mould plate and the ejector grid) continues to move to the left until the opening stroke is complete (c). This relative movement between the ejector plate assembly and the mould plate is necessary to operate the ejector element.

In the above illustration the machine's actuator rod is shown passing through the centre of the moving platen. This is the normal arrangement for the smaller types of injection machine. However, on larger machines several actuator rods are normally incorporated so that a balanced force can be applied to the ejector plate. Such a system is illustrated in Figure 3.9. A view of the moving platen of the machine without the mould is shown at (a). In this example four actuator rods are incorporated and these pass through suitable clearance holes in the moving platen. The method of actuation is identical to that described above for central actuation except that in this case the actuator rods push directly on to the ejector plate as shown at (b). If the ejector rod and ejector rod bush are not incorporated (as in this design) then a separate method of guiding and supporting the ejector plate assembly must be incorporated (Section 3.3.3).

3.3.1 Ejector Plate. The purpose of this member is to transmit the ejector force from the actuating system of the injection machine to the moulding via an ejector element (Section 3.4).

The force required to eject a moulding is appreciable, particularly with

EJECTOR PLATE ASSEMBLY

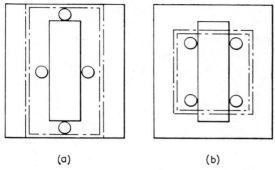

Figure 3.10—Alternative ways of arranging four ejector pin elements. Note that method (a) results in larger ejector plate assembly than does method (b)

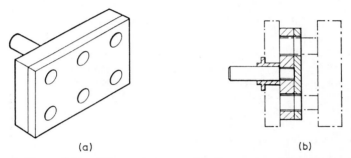

Figure 3.11—Ejector plate assembly for use in conjunction with circular support grid (see Figure 3.6)

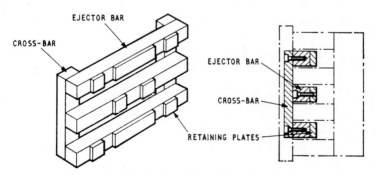

Figure 3.12—Ejector bar system

those mouldings which are deep and which incorporate little draft. Most ejector plates which fail in operation do so in fact because too thin a plate is specified in the design. The ejector plate must be sufficiently thick not to deflect to any significant extent. Deflection tends to occur at the beginning of the ejector stroke when there is maximum adhesion between the moulding and the core. The deflection of any beam is inversely proportional to the cube of its depth and, therefore, a relatively small increase in plate thickness will decrease deflection of the plate.

If an ejector plate does deflect to any extent, side forces are applied to the ejector elements which result in increased wear in the mould plate holes, bent ejector pins and, in extreme cases, in the complete seizure of the system.

During the injection part of the cycle, with certain pin and sleeve type ejector systems (see Section 3.4), the melt pressure acts directly on to the ejector element (Figures 3.22, 3.36). To prevent the ejector elements being hobbed into the ejector plate by the applied force, a reasonably tough steel must be specified for this member. A general purpose medium-carbon steel (BS 970-080 M40) is suitable.

The overall size in plan view of the ejector plate is dependent primarily upon the positioning of the ejector elements. For example, consider the plan view of the mould plate for a rectangular box (Figure 3.10). Suppose we decide that four ejector pins are sufficient to eject the moulding. They may be arranged either as at (a) in the figure or as at (b). The ejector plate must back up all the elements in either case, so it is apparent that method (b) permits a smaller ejector plate (shown by inner dotted lines) to be used. It must be remembered that the smaller the ejector plate the greater the support one can obtain from the ejector grid system. For example, compare the support (indicated in both drawings by the outer dotted line) provided by the in-line ejector grid at (a) with that provided by the frame-type ejector grid at (b).

A typical rectangular type of ejector plate which may be used in conjunction with either an in-line or a frame-type ejector grid is shown in Figure 3.7.

For the circular support block ejector grid system a similar rectangular ejector plate assembly design is used, but in this case holes are bored through the ejector plate (and retaining plate) to provide clearance for the columns. This type of ejector plate is illustrated in Figure 3.11a. The mould plate and ejector grid system is shown in chain-dotted lines in the cross-sectional view (b).

Finally we turn to the ejector bar system. This is used where relatively few ejector elements are incorporated in straight lines (or comparatively straight lines) on large moulds. In this system (Figure 3.12) individual bars are used instead of a plate, and the bars are joined together normally at the outer ends by cross-bars.

3.3.2 Retaining Plate. This member is securely attached to the ejector plate by screws (Figure 3.7). Its purpose is to retain the ejector element (or elements) and particular examples are illustrated in Figure 3.29 for pin-type ejection, and in Figure 3.36 for sleeve-type ejection.

The thickness of the plate is governed by the depth of the head of the ejector element it retains. In general, retaining plates are within the 7 mm (¼ in) to 13 mm (½ in) thickness range.

For small moulds the retaining plate is made to the same general dimensions (plan view) as the ejector plate (Figure 3.7). For larger moulds, how-

EJECTOR PLATE ASSEMBLY

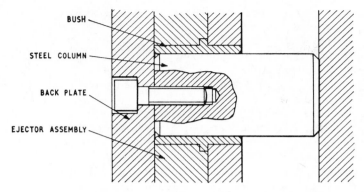

Figure 3.13—Guiding and supporting ejector plate assembly

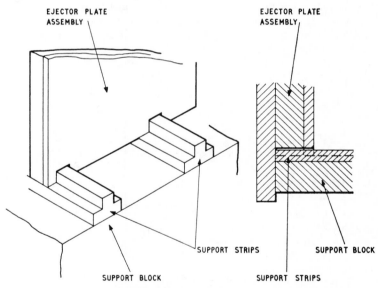

Figure 3.14—Method of guiding and supporting heavy type of ejector plate

ever, it is convenient from the mould making viewpoint to incorporate local retaining plates (i.e. small blocks of steel) in judicial positions to accommodate one or a small number of ejector elements only. For a similar reason, local retaining plates are normally fitted on ejector bar systems (Figure 3.12).

Retaining plates are normally made from a mild steel (BS 970-040 A15).

3.3.3 Guiding and Supporting Ejector Plate Assembly. This assembly must be guided and supported if there is any possibility of undue strain being

EJECTION

applied to any ejector element. The type of guide system used will depend largely upon the size of the mould.

We have previously noted that, for the smaller type of mould, the ejector plate incorporates an ejector rod which slides within an ejector rod bush which, in turn, is securely fitted into the back plate of the mould (Figure 3.7). This system very conveniently maintains alignment and provides support for the ejector plate assembly.

An alternative method for aligning and supporting the ejector assembly is shown in Figure 3.13. Bushes are incorporated within the ejector assembly and these slide on hardened steel columns attached to the back plate. These columns are normally also used as local support blocks.

For heavy types of ejector plate or bar assemblies, the plate (or bar) may be supported on its bottom edge as illustrated in Figure 3.14. In this design support strips are attached to the lower support block. The support strips are of either hardened steel or phosphor bronze. An alignment feature may be incorporated if desired in which case T-section support strips are used as illustrated. The projecting portion is a slide fit in a mating recess in the ejector plate assembly. It is common practice, however, on heavy moulds to use hardened steel columns for the main alignment, and incorporate strips purely for the purpose of supporting the member.

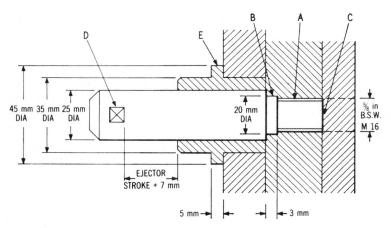

Figure 3.15—Ejector rod and ejector rod bush assembly (dimensions given are suitable for a small type of mould)

3.3.4 Ejector Rod and Ejector Rod Bush. The conventional design for these members is shown in Figure 3.15. The dimensions given apply to a relatively small mould.

The ejector rod is attached to the ejector plate by means of a thread as shown at A. To ensure concentricity a small parallel length of a slightly larger diameter than the thread is provided at B on both the ejector rod and the ejector plate. The threaded hole may either extend completely through the ejector

EJECTOR PLATE ASSEMBLY

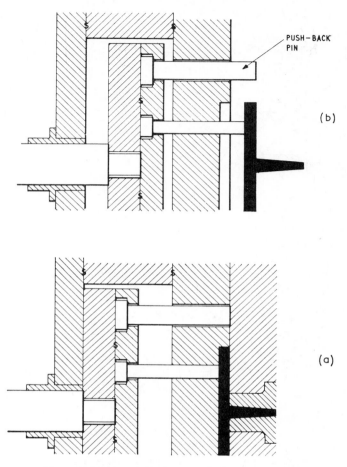

Figure 3.16—'Push-back' ejector plate return system

plate as shown, or it may be blind. If the former design is adopted it is normal practice to make the end of the ejector rod at C level with the ejector plate surface. This is particularly desirable when a central sprue puller is used.

Spanner flats are provided on the ejector rod at D. The position of these flats must be such that at no time do they enter the ejector rod bush. (If the top end of the ejector rod is damaged by unskilled use of the spanner, and this is then forced into the ejector rod bush, a seizure is likely.) The dimension specified for the distance between the end of the ejector rod bush and the ejector rod flats (see diagram) should therefore exceed the maximum ejector plate movement by at least 7 mm (¼ in).

The ejector rod bush is normally made a press fit into the back plate of the mould. Some designers, however, prefer to extend the flange (E) and positively secure the member to the back plate with screws.

EJECTION

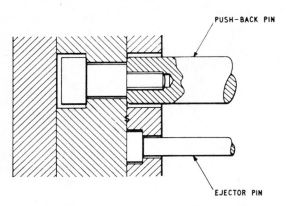

Figure 3.17—Alternative method of attaching push-back pin to ejector plate assembly

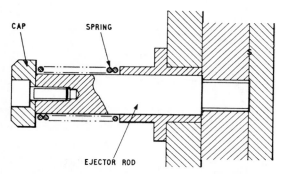

Figure 3.18—Spring return system for ejector plate assembly

Both the ejector rod and the ejector rod bush are normally made in a low-carbon steel (BS 970-080M15) and suitably case-hardened. This gives both members a wear resisting surface.

3.3.5 Ejector Plate Assembly Return Systems. We saw, earlier in this chapter, the mechanism by which the ejector plate assembly is moved forward relative to the remainder of the moving half (Figure 3.8). We must now consider how we are to return the ejector plate assembly to its rear position in preparation for the next shot, when the mould closes.

Certain ejection techniques provide for the positive return of the ejector assembly by virtue of the mould geometry. The stripper plate design is a good example of this (Figure 3.51). In this design the stripper plate is directly connected to the ejector plate by tie-rods. When the mould closes the stripper plate strikes the cavity mould plate thereby causing the stripper plate and the ejector plate to be returned to their rear positions.

However, the ejector pin and ejector sleeve ejection techniques do not have

EJECTION TECHNIQUES

a large surface contact with the fixed mould half and these techniques require, therefore, the use of a special system to return the ejector plate.

Two systems in common use are (i) the push-back return system and (ii) the spring return system.

(i) *Push-back return system.* 'Push-back pins' (return pins) are basically large-diameter ejector pins fitted close to the four corners of the ejector plate assembly. A cross-section through part of a mould which illustrates a push-back pin is shown in Figure 3.16. In the moulding position as shown at (a) the push-back pins are flush with the mould plate surface. In the ejected position the push-backs protrude beyond the mould plate surface (b). Thus, when the mould is in the process of being closed, the push-back pins strike the fixed mould plate and progressively return the ejector plate assembly to the rear position (a).

The shouldered head design illustrated is adopted for push-back pins under 13 mm (½ in) working diameter. For large pins an alternative design is sometimes adopted. In this, the push-back pin is secured to the ejector plate assembly by a shouldered screw (Figure 3.17).

(ii) *Spring return systems.* For small moulds, where the ejector assembly is of light construction, a spring or a stack of 'Belleville' washers can be used to return the ejector plate assembly. A typical arrangement of the former actuating method is illustrated in Figure 3.18. In this design the spring is fitted on the ejector rod. A cap is attached to the end of the ejector rod to hold the spring in position under slight compression.

In operation, when the ejector assembly is actuated, the spring is compressed further. Immediately the mould closing stroke commences, however, the spring applies a force to return the ejector assembly to its rear position.

An alternative design, used for heavier ejector assemblies, is to incorporate a multiple spring system between the retaining plate and the rear face of the mould plate. These springs are often fitted on local circular support blocks.

3.3.6 Stop Pins. With a large ejector plate or large ejector bar system, it is often preferable to incorporate stop pins on the underside of the ejector plate. This design drastically reduces the effective seating area. In so doing, it diminishes the possibility of the ejector elements remaining slightly proud of their correct position due to foreign matter being trapped behind the ejector plate.

These stop pins are normally fitted directly behind the ejector elements as illustrated (Figure 3.19). The heads of the stop pins should be of relatively large diameter to prevent the possibility of their being hobbed into the back plate.

3.4 EJECTION TECHNIQUES

When a moulding cools, it contracts by an amount depending on the material being processed. For a moulding which has no internal form, for example a solid rectangular block (Figure 3.20a), the moulding will shrink away from

EJECTION

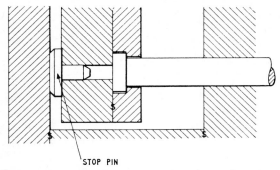

Figure 3.19—'Stop pins' incorporated to reduce seating area of ejector plate

the cavity walls as shown, thereby permitting a simple ejection technique to be adopted (for example, perhaps, a jet of air).

However, when the moulding has internal form, the moulding, as it cools, will shrink onto the core and some positive type of ejection is necessary (Figure 3.20b).

The designer has several ejection techniques from which to choose but, in general, the choice will be restricted depending upon the shape of the moulding. The basic ejection techniques are as follows: (i) pin ejection; (ii) sleeve ejection; (iii) bar ejection; (iv) blade ejection; (v) air ejection and (vi) stripper plate ejection. Certain of the ejector elements used in the above techniques are illustrated in Figure 3.21.

3.4.1 Pin Ejection. This is the most common type of ejection as, in general, it is the simplest to incorporate in a mould. With this particular technique the

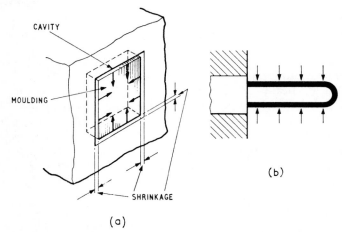

Figure 3.20—Effect of shrinkage on two different types of moulding: (a) moulding shrinks away from cavity; (b) moulding shrinks on to core

EJECTION TECHNIQUES

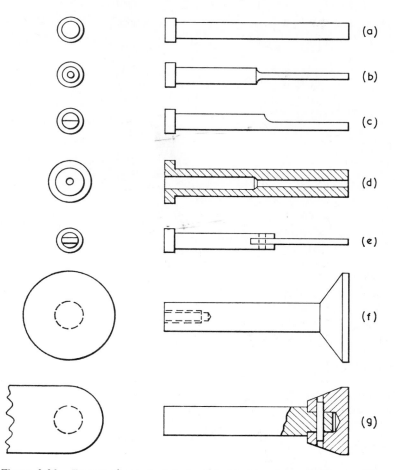

Figure 3.21—Ejector elements: (a) pin; (b) shouldered pin; (c) D-pin; (d) sleeve; (e) blade; (f) valve headed type; (g) bar

moulding is ejected by the application of a force by a circular steel rod, called an *ejector pin* (Figure 3.21a). The ejector pin is headed to facilitate its attachment to the ejector plate assembly.

Figure 3.22 shows the principle of operation. In the top drawing the ejector pin is in the rear (moulding) position, and it is held back by push-back pins (not shown). In operation, the ejector plate assembly, to which the ejector pin is attached, is moved forward relative to the mould plate. Thus the ejector pin pushes the moulding from the cavity (lower drawing).

The working diameter of the ejector pin must be a good slide fit in its mating hole in the mould plate. If it is not, then plastics material will creep through the clearance and a mass of material will progressively build up behind the mould plate. In the example taken here the most suitable position for the ejector pin is on the axis of the moulding, as shown.

EJECTION

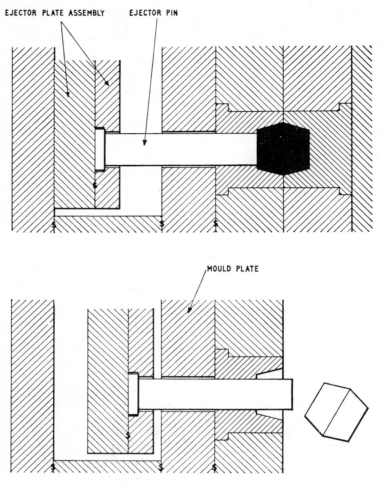

Figure 3.22—Pin ejection

The rear part of the ejector pin is fitted into a suitable hole which is bored and counterbored in the retaining plate. The rear face of the ejector pin head is backed up by the ejector plate. The accommodation so provided must allow the ejector pin to float. The reason why this feature is necessary is as follows.

As stated above, the ejector pin must be a good slide fit in the hole in the mould plate. The direction of movement of this ejector pin is, therefore, controlled by this hole. Should this not be bored absolutely at right angles to the mould plate face (an exaggerated example is shown in Figure 3.23 (top)) then when the ejector assembly is actuated (bottom) there will be relative lateral movement between the ejector pin and the retaining plate. Had the self-aligning feature not been provided the pin would have been subjected to consider-

EJECTION TECHNIQUES

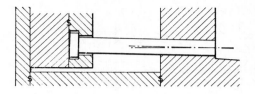

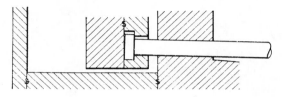

Figure 3.23—Illustrating transverse movement of misaligned ejector pin

able bending strain with the likelihood of breakage or seizure, or both.

Suitable diameters for the hole and the counterbore in the retaining plate are shown in Figure 3.24.

The plain diameter ejector pin may be used in one of two ways. It may be used as a moulding face pin, in which case the whole of the top surface of the pin is in contact with the moulding and we have seen an example of this in Figure 3.22. The alternative is to use the ejector pin as a *parting surface (butting face) pin*. In this case only a part of the top surface of the pin is in contact with the moulding; the rest abuts on to the fixed mould plate when the mould is closed.

The moulding face pin is normally specified in one of two circumstances. (i) Where the moulding has no internal form and must, therefore, be ejected from the cavity. An example of this is illustrated in Figure 3.25 which shows a moulding in the form of a Z-plate. In this case two moulding face pins are used to eject the moulding. (ii) In the alternative case where there is internal form, sometimes it is undesirable (from the point of view of appearance), or it is impracticable, to use parting surface pins. An example of this latter case is shown in Figure 3.26. The sides of the moulding terminate at a feather edge and the only simple positive ejection is by means of moulding face pins as illustrated.

However, in general, it is undersirable to use the moulding face pin for ejecting box-type mouldings (Figure 3.27a). When the ejection element is actuated, due to the adhesion between the moulding and the core, there is a tendency to bow the base of the box (Figure 3.27b). With certain of the softer plastics materials there is a tendency for the pins to push straight through the base of the moulding.

Moulding face pins are sometimes incorporated in particular positions in addition to other ejection methods, simply for the purpose of allowing air,

EJECTION

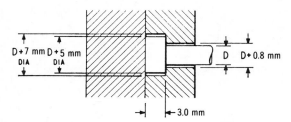

Figure 3.24—Float in retaining plate to allow for ejector pin misalignment

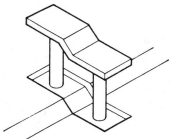

Figure 3.25—Ejection of Z-plate type moulding with moulding face pins

entrapped within the impression, to escape. This is called *venting* (Section 5.5).

When the moulding face pin is in its rear (moulding) position, the top face of the pin should be flush with the top of the core. If it is not, the moulding face pin will form either an indentation or a boss on the moulding depending whether the ejector pin is projecting above or is sunk below the core (Figure 3.28). The top surface of the moulding face pin must be given the same high finish as the rest of the impression as it forms part of the moulding surface.

The *parting surface pin* is the standard type of ejector element used for ejecting most types of box-shaped mouldings. The ejector pins are arranged to push on the bottom of the side walls of the moulding (Figure 3.29a).

The ejector pin is positioned with respect to the core so that there is a step of 0.13 mm (0.005 in) between the edge of the ejector pin hole and the side wall of the core (Figure 3.29b). Without this step, there is a probability that at some stage in the mould's life the ejector pin will score the side wall of the core.

Students beginning their study of mould design have the tendency to make the diameter of the parting surface pin approximately the same as the wall section of the moulding. The ineptitude of this is best illustrated by a diagram. Assume a moulding wall section of 3 mm (Figure 3.30). If a 2.5 mm diameter ejector pin is used then the actual ejection area is only 4.92 mm^2 (Figure 3.30a). As the diameter of the ejector pin is increased (Figure 3.30b, c) to 5

EJECTION TECHNIQUES

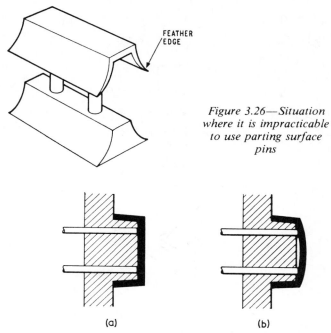

Figure 3.26—*Situation where it is impracticable to use parting surface pins*

Figure 3.27—*It is generally undesirable to eject box-type mouldings with moulding face pins*

mm and 10 mm, respectively, the ejection area increases similarly to 9.8 mm² and 15.3 mm², respectively. Thus the larger the parting surface pin used, the greater will be the ejection area.

Figure 3.31 is a graph on which is plotted the effective ejection area versus the moulding wall section for various sizes of ejection pin.

On certain large area mouldings it is often advantageous to use both parting surface pins and moulding face pins. The former are used around the periphery of the moulding and the latter are used to eject local bosses, etc., and to provide extra ejection area where it may be needed.

The location of the ejector pin elements, and the number used, is dependent on the component's size and shape. The aim of the designer must always be to eject the moulding with as little distortion as possible. The ejector pins should be located, therefore, so that the moulding is pushed off evenly from the core. Abrupt changes in shape (i.e. corners) tend to impede ejection; therefore an ejector pin or pins should always be located adjacent to these points. Once the size of the ejector pins has been decided upon, then the greater the number of ejector pins incorporated, the greater will be the effective ejection force and the less the likelihood of distortion occurring. For this reason it is better to err by having too many ejector pins than by having too few.

Pin ejection is the cheapest of the mechanical ejection methods. The close tolerance holes in the mould plate are made by a simple boring and reaming

EJECTION

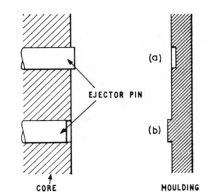

Figure 3.28—Unless moulding face pin is level with top face of core then either indentation (a) or boss (b) will be formed on moulding

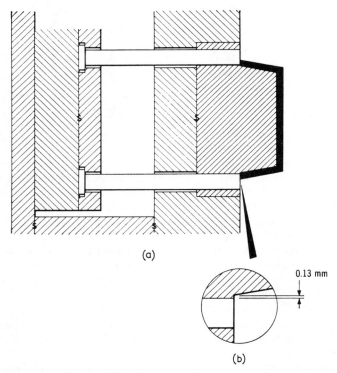

Figure 3.29—Parting surface ejector pins

operation and the ejector pins too are made by an equally simple turning and grinding operation. Ejector pins are manufactured by several companies who specialise in the standardisation of various mould parts. Either nickel - chromium steel (BS 970-835 M15) or chrome - vanadium steel may be specified for ejector pins; in either case heat treatment is necessary to obtain a wear-resisting external surface.

EJECTION TECHNIQUES

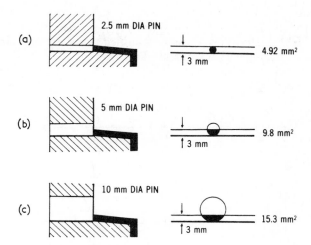

Figure 3.30—Showing effect of increasing diameter of parting surface pin

3.4.2 Stepped Ejector Pins. Next consider the case where small-diameter ejector pins (under 3 mm (⅛ in) diameter) are required for a particular design. Now slender, long length-to-diameter ratio ejector pins have the tendency to concertina in use. It is desirable therefore to keep the working length of such ejector pins to a minimum. This is achieved by designing the ejector pin as shown in Figure 3.21b. This is known as a stepped ejector pin.

The diagram illustrates a stepped ejector pin manufactured from a solid rod. Alternatively, it could have been made from two pieces of steel, the small diameter portion being fitted into a suitable hole machined in the large-diameter portion, the two parts being subsequently brazed together. This latter method has the advantage that should the ejector pin break only the small-diameter portion has to be remade.

The stepped ejector pin is normally used as a moulding face pin for the ejection of moulded bosses and ribs, etc. (Figure 3.32). Note that the main ejection is provided by standard plain type ejector pins.

The length of the small diameter portion of the stepped ejector pin should be kept as short as possible. This length need only equal the length in contact with the mould plate (i.e. length X) plus the ejector stroke (i.e. length Y) plus a small allowance of 5 mm (3/16 in). Note that the length of ejector pin contact with the mould plate (X) is kept to a minimum by incorporating a clearance diameter hole in the mould plate (at Z). A suitable contact length (X) for these small-diameter ejector pins is five to six times the diameter.

3.4.3 D-Shaped Ejector Pin. This is the name given to a flat-sided ejector pin. It is made quite simply by machining a flat on to a standard ejector pin (Figure 3.21c). It is used primarily for the ejection of thin-walled box-type mouldings.

The main advantage of this irregular-shaped ejector pin over the standard

85

EJECTION

parting surface pin is that, size for size, the former has a greatly increased effective ejection area. Figure 3.30 clearly shows that the ejection area obtained with the parting surface pin is confined to the extreme edge. Thus, because we are dealing with a segment of a circle, the effective ejection area rapidly diminishes as the wall section of the moulding is reduced. With the D-shaped ejector pin however, reduction of the moulding wall section does not

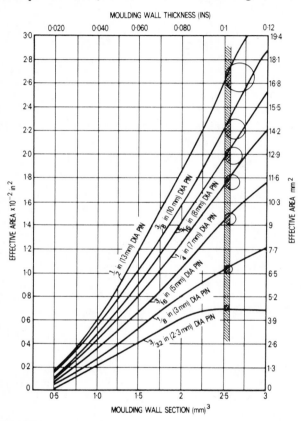

Figure 3.31—*Effective ejection area for various sizes of parting surface pins and for various moulding wall sections. The metric values are approximate*

have the same effect on effective ejection area because in this case this is situated at the centre of the ejector pin.

Now turn to the machining of the D-shaped hole. If we attempt to machine this complex shape through the mould plate or mould insert, the cost will be relatively high. However, by adopting a slightly modified insert - bolster assembly to that already described (Chapter 2) the incorporation of D-shaped ejector pins only increases mould cost slightly.

The D-shaped ejector pin is fitted into a complementary shaped hole in the mould plate, adjacent to the core, as shown in Figure 3.33. Note that the flat side of the ejector pin is parallel to the core insert face.

EJECTION TECHNIQUES

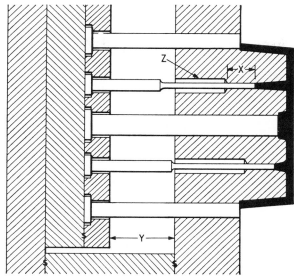

Figure 3.32—Moulding which necessitates use of parting surface, moulding face, and stepped ejector pins

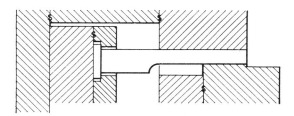

Figure 3.33—D-shaped ejector pin

The procedure adopted for producing the D-shaped hole is shown in Figure 3.34 (a part-plan view of the mould plate and a corresponding cross-section through the plate).

(a) Mark out the required position of the ejector pin hole.
(b) Bore and ream the required diameter hole in the bolster.
(c) Machine out the recess to accommodate the mould insert.
(d) Fit the insert and hold back with screws.

Thus the side wall of the mould insert forms the flat side of the D-shaped hole (Figure 3.33).

Another application for the D-shaped ejector pin is illustrated in Figure 3.35. This shows a part-section through a mould for a box lid. Note that the lid incorporates a projecting rim situated a short distance from the edge. By incorporating a part of the rim form on the ejector pin, as shown, it can be positively ejected. When the ejection is in the fully forward position a jet of air may be necessary to dislodge the moulding from the ejector pins.

EJECTION

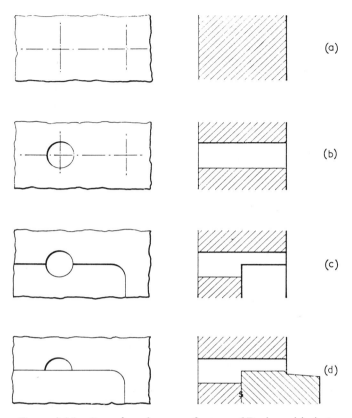

Figure 3.34—Procedure for manufacture of D-shaped hole in mould plate

3.4.4 Sleeve Ejection. With this method the moulding is ejected by means of a hollow ejector pin, termed a *sleeve* (Figure 3.21d). It is used in one of three circumstances:

(i) For the ejection of certain types of circular mouldings.
(ii) For the ejection (usually local) of circular bosses on a moulding of any shape.
(iii) To provide positive ejection around a local core pin forming a round hole in a moulding.

A part-section through a mould which incorporates sleeve ejection is shown in Figure 3.36. The sleeve, which is a sliding fit in the cavity insert and on the core pin, is fitted at its rear end to the ejector assembly. It is important for the beginner to note that the core pin extends completely through the sleeve, and is attached to the back plate. Beginners often attempt to anchor the core pin to the mould plate. This is impracticable because the core pin is completely surrounded by the sleeve.

The attachment of the core pin to the back plate can be accomplished in

EJECTION TECHNIQUES

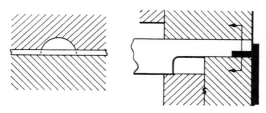

Figure 3.35—Ejecting a box lid (with rim) by means of D-*pins*

one of several ways. In the design illustrated, the core pin is anchored by a local core-retaining plate. An alternative method, to be adopted when several core-pins have to be held back, is to use one large retaining plate securely attached to the back plate by screws (Figure 3.37). Large-diameter (over 25 mm (1 in)) core pins are often held back directly onto the back plate by means of socket-headed screws.

Whichever method of attachment is used, because the outside diameter of the sleeve is a sliding fit in the cavity insert the sleeve and the core pin must be allowed to float in their respective anchorages.

When the ejector assembly is actuated, the sleeve is moved relative to the core (and to the cavity) and the moulding is ejected.

Ejection by means of a sleeve is a particularly efficient method because the ejection force is applied to a relatively large surface area. In the case of the example shown (Figure 3.36) the effective ejection area is practically the complete area at the base of the moulding.

While any shape of sleeve can be made, in practice the design is normally restricted to circular types, because of the high expense involved in the machining and the fitting of other shapes.

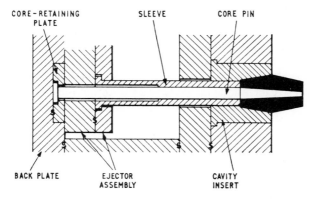

Figure 3.36—Sleeve ejection

Another sleeve design is shown in Figure 3.37. This is known as the stepped sleeve. It is used when a particular moulding necessitates a sleeve with a wall section of less than 2 mm (0.08 in). Long, thin wall-section sleeves are difficult

EJECTION

to manufacture and are prone to failure in operation. By enlarging the diameter of the sleeve at the lower end, the length of the thin wall section is shortened, thereby lessening the machining difficulties and producing a generally stronger sleeve. A recess may be provided in the rear side of the mould plate (directly below the cavity) to accommodate the larger-diameter portion of the sleeve when the ejector assembly is in the forward position. This permits the length of the thin wall section to be kept to a minimum.

It is undesirable to allow the sleeve to be in contact with the core pin over its entire length. To reduce frictional wear, to facilitate fitting, and to lessen the possibility of scoring, the surface contact between the two parts is kept to a minimum. There are two possible methods.

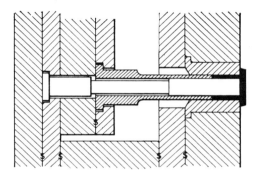

Figure 3.37—Stepped sleeve design

The first is to counterbore the hole in the rear side of the sleeve to a predetermined depth. This relief on the sleeve is only about 0.13 mm (0.005 in) all round (i.e. 0.26 mm (0.010 in) on diameter). This particular design is generally used with small diameter core pins. One point to note, however, is that as the ejector assembly is operated the length of the bearing contact between the sleeve and the core pin is reduced. Care must therefore be exercised in the design to ensure that when the sleeve is in the fully forward position a minimum effective bearing length of 7 mm (¼ in) is still maintained (Figure 3.36).

The alternative technique is to relieve a portion of the core pin by reducing its diameter locally by 0.26 mm (0.010 in). This offers the advantage (over the above technique) in that irrespective of the length of the relieved section the effective bearing area between the sleeve and the core pin remains constant over the complete ejector stroke. Thus a smaller length of contact bearing area is acceptable with this design. Note that this method is particularly desirable for use with stepped sleeve design in that it does not necessitate reducing the thin wall of the sleeve even further (Figure 3.37).

Typical examples of mouldings which can be ejected by the sleeve technique are now illustrated (Figure 3.38).

The first two components, the ring (a) and the bush (b), can also, in fact, be ejected by the alternative 'stripper plate' ejection (Section 3.4.9). Sleeve ejection is used for such components where a limited number of impressions only

EJECTION TECHNIQUES

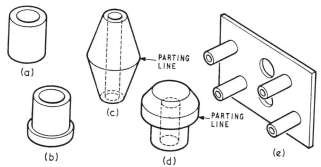

Figure 3.38—Typical mouldings which can be sleeve ejected

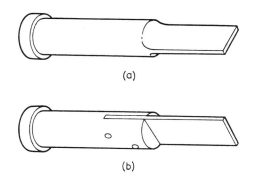

Figure 3.39—Alternative ejector blade designs

is required. For moulds containing more than two impressions the stripper plate design is more economical.

Now the second two components, the double cone reel (c) and the knob (d), illustrate examples which can only be sleeve ejected. Note that as the parting line occurs below the top surface of the moulding, the stripper plate technique is not an efficient ejection method, i.e. a part of the moulding will still be left in the stripper plate after ejection.

The final example is of a plate which incorporates a number of bosses and a couple of holes. The most effective ejection for this component is to incorporate a sleeve under each boss and a sleeve around the core pin forming each hole.

For components of a more complex nature, a combination of sleeve ejection with some other form of ejection is often desirable to achieve the optimum ejection results.

3.4.5 Blade Ejection. The main purpose of the blade ejector is for the ejection of very slender parts, such as ribs and other projections, which cannot satisfactorily be ejected by the standard type of ejector pin.

A blade is basically a rectangular ejector pin and examples are shown in Figure 3.39. While the blade ejector can be machined from a solid rod (a) it is

EJECTION

more usual to fabricate the element in which case a blade of steel is inserted into a slot machined into a standard type of ejector pin (b). The blade may be pinned (as shown) or, alternatively, it may be brazed. The advantage of the two-part construction is that the blade can easily be replaced should it become damaged.

The blade ejector element is fitted to the ejector assembly in an identical manner to that described for the standard ejector pin (Section 3.4.1). A cross-sectional view through part of a mould which incorporates blade ejectors is shown in Figure 3.40.

The rectangular blade ejector is accommodated in a complementary shaped hole in the mould plate. Now if this slot is to be machined into a solid mould plate, it presents a costly end-milling operation. It is desirable, therefore, to facilitate the machining of this aperture by adopting a built-up assembly.

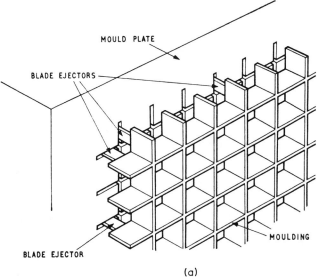

(a)

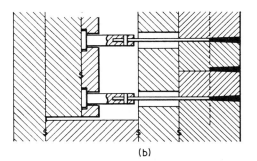

(b)

Figure 3.40—Blade ejection

EJECTION TECHNIQUES

For a moulding which incorporates a small projecting rib, it is usually possible to introduce a split bush into the design. Now this split bush (Figure 3.41) will carry the relevant part of the impression form, and by making the bush in two parts a slot can be milled to accommodate the blade ejector. The blade is fitted to the split bush before this latter member is fitted to the core plate. By adopting this technique, machining of both the local impression form and the rectangular aperture (for the blade) is simplified.

A somewhat similar method is adopted where a blade ejector is used on the edge of a moulding. In this case a shouldered core insert is adopted (Figure 3.42). A slot is machined completely through the shouldered portion, and when the core insert is fitted into a suitable recess in the mould plate a rectangular aperture is formed into which the ejector blade slides.

This latter method is adopted in the design shown in Figure 3.40. Assume that we wish to mould a lattice type component and that we want to have all the moulding form in one mould half (i.e. so that the other mould half will be perfectly plain). Pin ejection is not suitable here because only very small-diameter ejector pins can be fitted at the bottom of the slots in the mould plate. The effective ejection area of even a relatively large number of such pins would be insufficient to eject this type of moulding. Thus for such a moulding we would use blade ejection.

Now we come to the problem of machining the rectangular apertures for the blades. This problem is simplified by adopting the method discussed above. The cavity is built up with a number of local core blocks. Each alter-

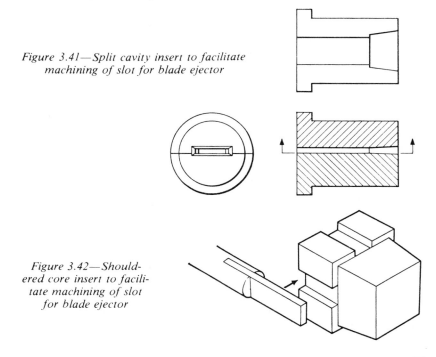

Figure 3.41—Split cavity insert to facilitate machining of slot for blade ejector

Figure 3.42—Shouldered core insert to facilitate machining of slot for blade ejector

EJECTION

nate core block incorporates a shoulder (Figure 3.42). Thus when these local core blocks are assembled in the cavity plate, rectangular apertures are formed into which the blade ejectors can be fitted. Note that a round clearance hole is provided in the mould plate directly below the rectangular aperture (Figure 3.40b).

3.4.6 Valve Ejection. The valve ejector element is basically a large-diameter ejector pin. Valve-type ejection is used, normally, for the ejection of relatively large components in situations where it is impracticable to use standard parting surface pins. It is also used as an alternative to stripper plate ejection in certain situations.

The valve-type ejector element is designed to apply the ejector force on to the inside surface of the moulding (Figure 3.43). Now we have previously

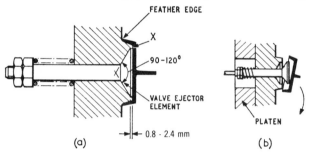

Figure 3.43—Valve ejection, single-impression mould: (a) moulding position; (b) ejection position

noted (for pin ejection) that this is a generally undesirable practice because of the possibility of distorting the moulding during ejection. However, with valve ejection, because the element has a large effective ejection area, the risk of the moulding being distorted is minimised.

The ejector element incorporates a valve-type head which is seated in a nest machined into the core as shown. To ensure a material leak-free joint, a small parallel portion is provided at the major diameter of the valve. This is a sliding fit in a complementary parallel recess in the core. The length of the parallel section is between 0.8 mm (1/32 in) and 2.4 mm (3/32 in).

Because the ejector element forms a relatively large part of the impression, a large force will be applied to the element by the melt tending to move it rearwards. The head must therefore be carefully bedded into its nest in the core plate. The bedding angle is usually between 90° and 120° inclusive.

Figure 3.43 illustrates a typical application of valve-type ejection. Note that in this example the moulding wall progressively reduces to a feather edge at its base. This makes the use of either the parting surface pin or the stripper plate-type ejection impracticable. One important point to note is that the major diameter of the head must not approach too close to the edge of this core (at X), otherwise a weakness at this point will occur.

For single-impression moulds, where the moulding is positioned on the cen-

EJECTION TECHNIQUES

tre line, the conventional ejector assembly is not required (Figure 3.43). The shank of the valve ejector (which must be of a suitable diameter to resist bending) passes completely through the mould plate and projects from the rear side. The valve is held in its nest in the mould plate by a heavy spring.

In operation, when the mould opens, the shank of the valve ejector element is arrested by the actuator rod of the injection machine and the moulding is ejected (Figure 3.43b). Immediately the mould closing stroke commences the spring causes the element to be returned to its seating (Figure 3.43a). For fully automatic operation it is normal practice to use a jet of air to blow the moulding off the valve as there is a tendency in certain instances for the moulding to adhere to the valve.

For a multi-impression mould, an ejector assembly and an ejector grid must be used (Figure 3.44). The valve element is attached directly to the ejector plate by screws (note that no retaining plate is used). A definite gap of 3 mm (⅛ in) minimum should be incorporated behind the ejector plate (at X) to ensure that both valves are seated in their respective nests. Push-backs, not shown, are incorporated to return the ejector plate and valve elements to their rear positions as the mould is closed.

An alternative method of actuating the valve element is by compressed air. A cross-section through a mould which uses this method is shown in Figure 3.45. It is, however, restricted to moulds which have a core insert large enough to contain a pneumatic actuator as shown. The shank of the valve ejector is threaded and screwed into the ram of the actuator. The power lines (not

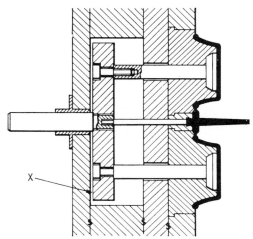

Figure 3.44—Valve ejection, multi-impression mould

shown) to the actuator are accommodated in suitable slots in the mould body. The main advantage of this system is that the ejection can be operated, forward and return, irrespective of the position of the moving half of the mould.

Valve-type ejector elements should be specified to be made from nickel - chromium steel (BS 970-835 M15) and suitably heat-treated.

EJECTION

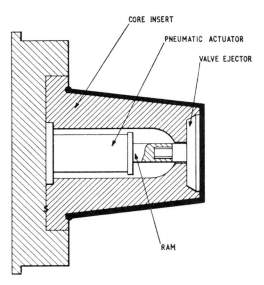

Figure 3.45—Valve ejector operated by compressed air

3.4.7 Air Ejection. With this ejection method the ejector force is provided by compressed air, which is introduced directly on to the moulding face via a small air ejector valve. For this method to operate efficiently the adhesion between the moulding wall and the core must be broken locally, to permit the compressed air to be introduced. This is achieved by causing the valve ejector to move forward slightly (by air pressure).

The effective ejector force is dependent upon the pressure of the compressed air and the area on which it acts. Therefore the larger the area of the component to be ejected the greater will be the ejector force. There is, however, one drawback in that immediately the compressed air finds an escape route the effective ejector force is rapidly diminished. From this it will be seen that this method of ejection is particularly suitable for box-type components where the side walls act as a seal during a major part of the ejection stroke, thus preventing the escape of the compressed air.

Now turn to the air ejector valve designs; Figure 3.46 shows a small valve ejector element similar to that discussed in the previous section. The head of the element nestles in a complementary shaped recess in the core plate, while the shank of the element is a sliding fit in a hole adjacent to the nest. To provide a passage for the compressed air, small slots are machined longitudinally on the periphery of this hole (at B). The element is held on to its seating by means of a compression spring. This latter member is held in position by a nut. To prevent the loss of compressed air from the rear side of the core plate, a sealing plate and associated O-ring are incorporated. A port (A) connects the air ejector valve to the external compressed air supply.

An alternative design of air ejector valve is illustrated in Figure 3.47. In this

EJECTION TECHNIQUES

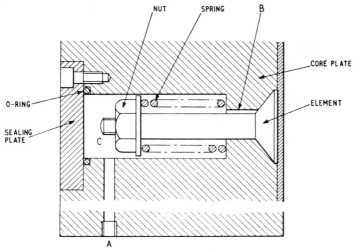

Figure 3.46—Compressed air ejection

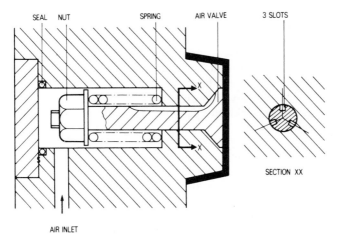

Figure 3.47—Compressed air ejection — alternative air valve design

design the air passageway is achieved by machining slots (usually three in number) into the shank of the valve as shown.

The introduction of the compressed air can be under either manual or machine control. Both cases are illustrated (Figure 3.48). With manual control (Figure 3.48a) the machine operator operates the two-way valve at any suitable point in the cycle. With machine control (Figure 3.48b) a cam is fitted on to the machine bed so that at the required point in the opening stroke, the valve spool is depressed and compressed air flows into the mould. Immediately the roller (attached to the spool) runs off the cam at either end the compressed air supply is cut off.

97

EJECTION

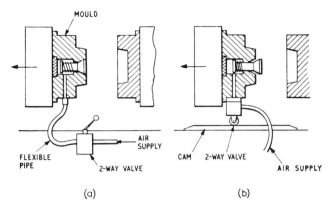

Figure 3.48—Methods of introducing compressed air to mould: (a) manual control; (b) machine control

The operation of this form of ejection is as follows. When the control valve is operated compressed air passes into chamber C (Figure 3.46). A force is applied to the air-valve ejector which overcomes the force exerted by the compression spring and the valve ejector leaves its seating. At this point the adhesion between the moulding and the core is broken in the area adjacent to the air-valve ejector. Compressed air is now free to flow from the chamber C through the slots (B), through the open valve and so into the space between the moulding and the core. The moulding is thereby ejected.

A few advantages of air ejection should be noted:
 (i) No ejector grid or ejector assembly is required. This reduces the cost of the mould.
 (ii) Air ejection can be fitted in either mould half. It is therefore a particularly suitable method of ejecting box-type components which are to be gated on the inside.
 (iii) The ejection can be operated at any time during the opening stroke of the machine.
 (iv) The air-valve ejector element acts as a vacuum breaker between the moulding and the core. This arrangement is often incorporated in addition to some other form of ejection simply to facilitate the ejection of the part without distortion. For this application, connection to an external compressed air supply is not necessary.

Against these advantages, the following limitations should be noted.
 (i) The method is only effective on certain types of component.
 (ii) A compressed air supply must be readily available.
 (iii) Air is an expensive service and if incorrectly used this system can be wasteful.

3.4.8 Stripper Bar Ejection. This method is an extension of the parting surface ejector pin principle, in which the ejector element is caused to push against the bottom edge of the moulding. However, a far greater effective ejec-

EJECTION TECHNIQUES

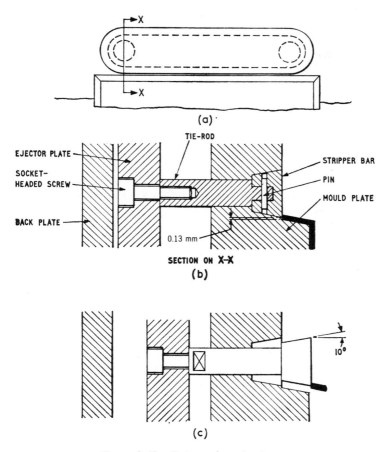

Figure 3.49—Stripper bar ejection

tion area is obtainable with the stripper bar method and, because of this characteristic, this method of ejection is particularly suitable for thin-wall box-type mouldings. As a single bar is used along each wall of the moulding, instead of, perhaps, three or four ejector pins, the marks left on the surface of the moulding by the ejection method are reduced to a minimum. However, against these advantages must be weighed the fact that the stripper bar is very much more expensive to manufacture and fit compared with the manufacture and fitting of an ejector pin.

The design of a typical stripper bar assembly is shown in Figure 3.49. At (a) the stripper bar is shown in plan view; at (b) a general cross-section through X-X is given; and (c) illustrates the stripper bar in the forward (ejection) position.

The stripper bar is fitted into a complementary shaped recess in the mould plate. A small angle of 10° is incorporated all around the periphery of the stripper bar to minimise wear. Note that if the sides were made parallel, each time the element was operated they would rub against the complementary face of the mould plate recess. A good fit is required between the stripper bar and the mould plate recess. If a good fit is not achieved, then plastics material will creep down the sides and probably under the stripper bar as well. This will prevent the stripper bar from seating correctly which may damage the opposite mould plate when the mould is closed.

To ensure that the working side of the stripper bar does not scrape the side face of the core, the ejector element should be set back from the core wall by at least 0.13 mm (0.005 in) (Figure 3.49).

The stripper bar is coupled to the ejector plate by a tie-rod. One end of the member is threaded and attached to the stripper bar. To prevent the tie-rod working loose it is desirable to incorporate a pin as shown. The rear end of the tie-rod is normally secured to the ejector plate by a socket-headed screw. (Note that a retaining plate is not necessary with this design.)

A gap of about 3 mm (⅛ in) should be provided between the ejector plate and the back plate to ensure that the stripper bar seats in its nest in the mould plate.

The stripper bar should be manufactured from a nickel - chromium steel (BS 970-835 M15) and suitably heat-treated. The tie-rods may be manufactured from a low-carbon steel (BS 970-080 M15), and these should be carburised to give a hard, wear-resisting surface.

3.4.9 Stripper Plate Ejection. This ejection technique is used mainly for the ejection of circular box-type mouldings. While the design is used for shapes other than circular, particularly for those which have thin wall sections, the mould cost which results is relatively high.

The principle of this ejection technique is illustrated in Figure 3.50. The stripper plate is mounted between the cavity plate and the core plate. The aperture in the stripper plate is a sliding fit on the core. The mould is shown closed at (a). When the mould starts to open the stripper plate moves back with the core plate (b). (Remember that the moulding adheres to the core because of the contraction on cooling.) Once the moulding is clear of the cavity, the movement of the stripper plate is arrested, while the core plate continues the rearward movement. The core is thereby withdrawn through the

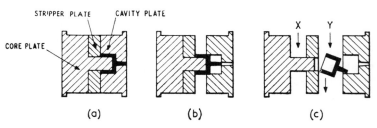

Figure 3.50—Principle of stripper plate ejection

EJECTION TECHNIQUES

stripper plate and the moulding ejected (c). The mould is then closed in preparation for the next cycle.

So far, discussion has been limited to moulds which consist essentially of two parts, namely: the cavity plate and the core plate, i.e. the fixed mould half and the moving mould half. From the diagram (c) it is apparent that this design consists of three parts, the stripper plate constituting a separate mould member. When the mould is open (as shown) there are two spaces (at X and Y) and these are termed *daylights*. This type of mould is therefore sometimes referred to as a *multi-daylight* mould, but, as there are several other multi-daylight designs (Chapter 12), the designation *stripper plate* design is preferred.

We now turn to the details of this design. A part-section through a typical mould is shown in Figure 3.51. The component is a round box with 1 mm

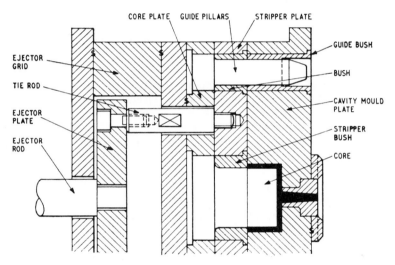

Figure 3.51—Part-section through a stripper plate mould

(0.040 in) walls. The stripper plate is mounted on guide pillars which are fitted in the core plate. The apertures in the stripper plate, through which the core and the guide pillars pass, are all bushed. The cavity mould plate is of conventional design.

On plan, the stripper plate is usually the same size as the other mould plates (except in the direct actuation design). This allows uniformity in the design. Stripper plate depth, however, must be relatively great: the depth dimension must be sufficient to resist the bending forces applied to the plate as the moulding is ejected. As mentioned previously, these forces can be considerable when ejecting deep mouldings which incorporate little draft.

The length of the working diameter of the guide pillar is generally longer than that specified for a corresponding two-part mould. The reason is that the guide pillar must support the stripper plate over the complete ejection

101

EJECTION

stroke and, in addition, it must still project enough to enter the fixed half guide bushes when the mould closes to ensure alignment between the cavity and the core.

Finally, the following methods are commonly used for actuating the stripper plate.

(*i*) *Tie-rod actuation—Method 1.* A conventional ejector plate and ejector grid system are adopted in this design but with the retaining plate removed (Figure 3.51). The stripper plate is coupled to the ejector plate by three or four tie-rods. In operation, as the moving half moves rearwards, the ejector rod strikes the actuating rod of the machine. The movement of the ejector plate and the stripper plate is arrested, thereby stripping the moulding from the core, as previously discussed. There are two main design points to note. First, the ejector plate does not seat on the back plate of the mould. This is because it is impracticable and unnecessary to have both the stripper plate and the ejector plate seating on their respective plates. Second, as the stripper plate is guided on the main guide pillars there is no point in including a separate ejector rod bush in this design.

(*ii*) *Operating pin method.* A standard two-part assembly is adopted for this design, and *'operating pins'* (basically ejector pins) are fitted (Figure 3.52). Note that a gap of 1 mm (0.040 in) is left between the top of the operating pin and the stripper plate.

When the ejector plate assembly is actuated, the operating pins push on the rear face of the stripper plate to perform the required function.

(*iii*) *Tie-rod actuation—Method 2.* This design (Figure 3.53) is very similar to the one discussed above in (*i*). The only difference is that the ejector grid system is dispensed with. For this method to be practicable, the aperture in the

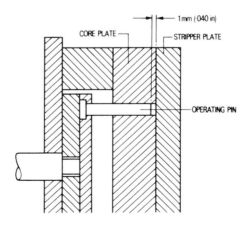

Figure 3.52—Stripper plate mould actuated via operating pins

EJECTION TECHNIQUES

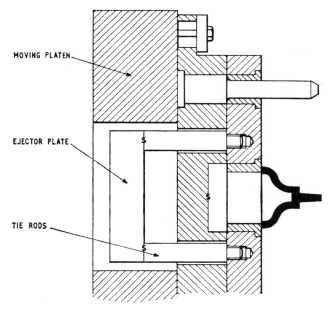

Figure 3.53—*Alternative method of actuating stripper plate*

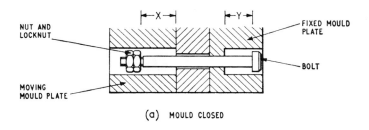

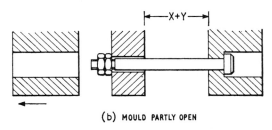

Figure 3.54—*Length-bolt actuation of stripper plate*

EJECTION

moving platen of the machine must be large enough to accommodate the ejector plate and the tie-rods.

(*iv*) *Length-bolt actuation.* In this design the ejector plate is arrested by length-bolts suitably situated within the mould. A part cross-section through relevant parts of the design is shown in Figure 3.54. The mould is shown in the closed position at (a). The fixed mould-plate is recessed to accommodate the head of the length-bolt and a clearance hole is provided in the moving mould plate to accommodate the nut and locknut. The amount of free movement that the stripper plate is permitted is the sum of the distances X and Y as shown at (b). (The mould is illustrated in the partly open position.)

With this design the maximum stripper plate movement is a function of the overall mould height, and this restricts the design to fairly deep moulds. However, by adopting telescopic length-bolts (Figure 3.55), the design can be used even on shallow moulds.

The main advantage of this design over the tie-rod method of actuation is that it is cheaper to produce (i.e. there is no ejector grid and ejector plate assembly) and the mould is lighter for handling purposes. However, against this advantage must be weighed the fact that the length-bolts (as they extend between the stripper plate and the fixed mould plate) tend to encase the moulding. Care must be exercised in this design therefore, to ensure that the operator has reasonable access to the impression and that the fall of the mouldings is not restricted by the length-bolts. Normally three length-bolts are used on circular-type moulds and four on rectangular-type moulds.

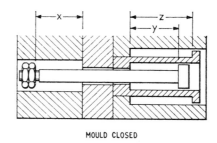

MOULD CLOSED

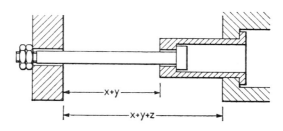

MOULD OPEN

Figure 3.55—Telescopic length-bolt actuation of stripper plate

EJECTION TECHNIQUES

(v) *Chain actuation.* This method of actuation is similar to the previous method except that chains are used to arrest the motion of the stripper plate instead of length-bolts. A typical arrangement is shown in Figure 3.56. One end of the chain is connected to the stripper plate, and the other end to the

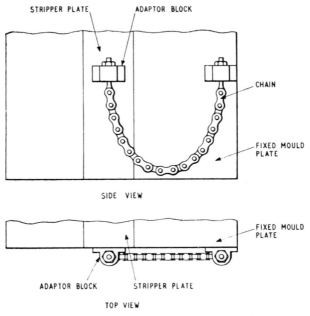

Figure 3.56—Chain actuation of stripper plate

fixed mould plate via adaptor blocks. When the mould is in the closed position (as shown) the chains hang down in a loop on either side of the mould. As the mould is opened the chains are progressively straightened until, finally, they arrest the movement of the stripper plate. One or two chains are used per side depending upon the size of the mould.

Again, while allowing for a cheaper mould than for the tie-rod design, there are two hazards to bear in mind.
(1) Should the chain swing in between the mould faces while it is being closed severe damage will occur.
(2) Should one of the chains break while the mould is opening, an offset load will be applied to the stripper plate which will tend to distort this member.

Both the above hazards can be avoided by careful design and choice of chain.

(vi) *Direct actuation.* On medium- and large-size injection machines, several actuator rods are normally incorporated for ejection purposes (Figure 3.9). Providing these actuator rods are positioned symmetrically about the centre line of the moving platen they can be used to actuate the strip-

EJECTION

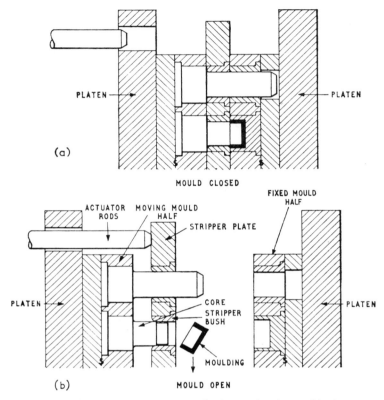

Figure 3.57—Direct actuation of stripper plate by machine's actuator rods

per plate directly. Figure 3.57a shows a part-section through the mould, suitably mounted on the platens of the machine. When the mould is opened (Figure 3.57b) the stripper plate moves back with the moving mould half until the stripper plate is arrested by the actuator rods. Further movement of the moving mould half causes the core to be withdrawn through the stripper bush and the moulding to be ejected.

The actuator rods either pass through suitable clearance holes in the moving half, or the ejector plate is extended beyond the edge of the mould plate, and the actuator rods contact the stripper plate on either side of the mould plate as illustrated.

This design, compared with the tie-rod actuation method, is less complex in that it does not require an ejector assembly; it is therefore less expensive to manufacture. Neither does this design suffer the disadvantages associated with the length bolt and chain actuation methods. The directly actuated design should therefore be adopted whenever the situation allows.

During the premoulding dry run operation there is just the possibility with this design (and with the chain operated design) that the stripper plate may

EJECTION FROM FIXED HALF

remain in contact with the fixed half while the moving mould plate is withdrawn. If this opening sequence occurs, the guide pillars may lose contact with the stripper plate, and the stripper plate would then fall between the two mould halves. To avoid this possible hazard, length bolts may be incorporated in the moving mould half to limit the travel of the stripper plate to a maximum consistent with the guide pillar length adopted.

If the mould opens in the wrong sequence when the length bolts are fitted, the stripper plate is pulled away from the fixed mould half when the free length of travel provided by the length bolt has been taken up. The mould setter must be instructed in setting this type of mould to ensure that the ejector stroke as set by the ejector rods is not greater than that permitted by the length bolt.

3.4.10 Stripper Ring Ejection. The *stripper ring* is basically a local stripper plate. It is used mainly for circular box and cup-type mouldings, and is generally restricted for use on moulds with one or two impressions, only. When there are multi-impressions the stripper plate design is more economic.

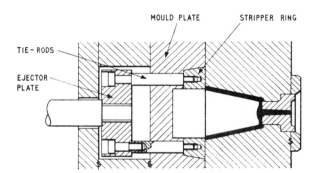

Figure 3.58—Stripper ring ejection

A stripper ring design is shown in Figure 3.58. The stripper ring is fitted into a recess in the mould plate. The internal bore of the stripper ring is a sliding fit on the major diameter of the core insert. This diameter should be at least 0.26 mm (0.010 in) greater than the core impression diameter, to avoid the possibility of the stripper ring scoring the side of the core during ejection. For the purposes of actuation the stripper ring is coupled to the ejector plate by tie-rods, as shown.

3.5 EJECTION FROM FIXED HALF

While it is generally desirable to situate the ejector system in the moving half of the mould this is not always practicable. Consider the case of a box-type component which must, for reasons of appearance, be gated from the inside (Figure 3.59). In this case the core and the ejector system are mounted on the fixed mould half. Now this presents certain complications.

107

EJECTION

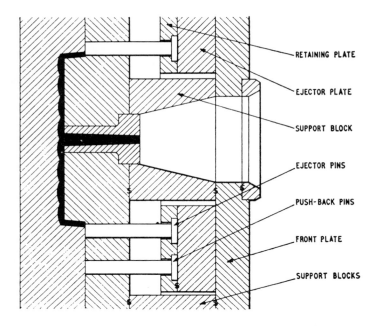

Figure 3.59—Ejection from fixed mould half

First, to incorporate an ejector grid and ejector assembly behind the fixed mould plate causes this mould half to be excessively deep. Because of this the melt has to travel a long distance from the machine's nozzle. This distance can be minimised, however, by sinking the sprue bush deep into the mould plate assembly as illustrated, though this necessitates using a special extension nozzle.

Second, facilities are not normally provided on the machine for actuating any type of ejector system from the fixed mould half side. This means that the actuating mechanism must be incorporated in the mould design.

While any of the ejection techniques discussed in Section 3.4 can be incorporated in the fixed half, certain of these techniques are more easily incorporated than others.

Air ejection. This ejection technique is the simplest to incorporate in the fixed mould half, as the complications mentioned above do not apply. There is no need, for example, to increase the depth of the mould plate unduly to accommodate the air ejector valve (Figure 3.46). Also, as the actuation is by the medium of compressed air, this is just as easily connected to the fixed half as to the moving half.

Stripper plate ejection. Again, this method can very easily be adapted for use on the fixed mould half. The stripper plate is mounted, as previously, between the cavity and the core plate. However, in this case, the stripper plate is coupled to the moving mould plate (not the fixed mould plate as previously) by either length-bolts or chains. When the mould is opened the stripper plate is retained on the fixed half side (the moulding having contracted on to the core) until the free movement allowed by the length-bolt (or chain) has been taken up. Further movement of the moving mould plate actuates the stripper plate and the moulding is ejected.

Pin, sleeve, bar ejection, etc. Any ejection method which necessitates the use of the ejector plate assembly is inconvenient, as noted above, for use on the fixed mould half. However, certain types of component necessitate the use of pin, sleeve or bar ejection, in which case a conventional ejector plate assembly must be incorporated.

Taking pin ejection as an example (Figure 3.59), the ejector plate assembly is contained in a pocket formed by the ejector grid which consists of support blocks and the front plate. The ejector pins and the push-back pins are attached to the ejector plate by the retaining plate. So this is a standard pin ejection arrangement—it is simply mounted on the moving half. We must, however, provide for the passage of the melt through the ejector assembly. A local circular support block is positioned on the centre-line of the mould and machined to accommodate the rear end of the sprue bush (Figure 3.59). The aperture, which also extends through the front plate, is of sufficient size to accommodate the machine's extension nozzle.

Next is actuation of the ejector plate. Where the overall depth of the mould is sufficient to accommodate hydraulic actuators these may be fitted, one on either side of the fixed half, to operate the ejector plate assembly directly. However, the normal practice is to couple the ejector assembly to the moving mould half by either length-bolts or chains. Thus, during the latter part of the opening stroke the ejector assembly is pulled, thereby ejecting the mouldings.

3.6 SPRUE PULLERS

When the mould opens it is essential that the sprue is pulled positively from the sprue bush. With single-impression moulds the sprue feeds directly into the base of the component (Figure 3.51) and the sprue is pulled at the same time as the moulding is pulled from the cavity.

For multi-impression moulds using a basic feed system (Figure 2.3) the sprue would probably be left in the sprue bush each time the mould was opened. This would necessitate a manual operation to remove the unwanted sprue. To avoid this undesirable feature, an arrangement for pulling the sprue should always be incorporated in the design.

The common sprue pulling methods utilise an undercut pin or an undercut recess situated directly opposite the sprue entry. The plastics material which flows into the undercut, upon solidifying, provides sufficient adhesion to pull the sprue as the mould is opened.

There are two basic designs of sprue puller. In one the undercut is produced within the cold slug well region, and is situated below the parting surface. In the second design, the undercut portion of the sprue pulling device is situated

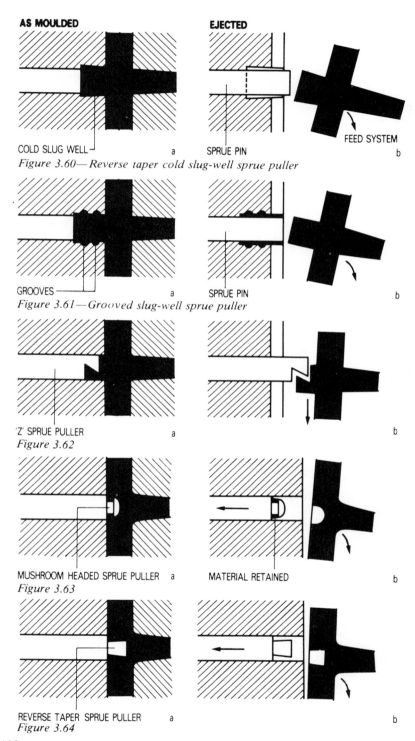

Figure 3.60—Reverse taper cold slug-well sprue puller

Figure 3.61—Grooved slug-well sprue puller

Figure 3.62

Figure 3.63

Figure 3.64

SPRUE PULLERS

above the parting surface. To differentiate between these two basic types, we designate them Type A and Type B respectively.

3.6.1 Type A Sprue Pullers. The simplest sprue puller design is the reverse taper cold slug well type, illustrated in Figure 3.60. The cold slug well walls are tapered inwards, as shown, creating an undercut in the in-line-of-draw direction. The *sprue pin* (which is identical in form to an ejector pin) is positioned behind the cold slug well so that when ejection occurs the slug is ejected with the feed system (Figure 3.60b).

In an alternative design, one or more grooves are cut into the cold slug well walls, thereby creating an undercut for sprue pulling purposes (Figure 3.61a). A sprue pin is again used to extract the slug, but in this case it shears through the plastic material, leaving solidified material in the grooves as shown in Figure 3.61b. During the next cycle of operations the incoming melt welds to this material.

Yet another design in the Type A classification, is the *'Z' type sprue puller*. This is possibly one of the earliest sprue pulling designs. The end of the sprue pin, now termed a *sprue puller,* is profiled in the form shown in Figure 3.62a. This 'Z' shape creates a positive undercut which pulls the sprue when the mould is opened. The sprue puller is moved forward during the ejection stroke as shown in Figure 3.62b, and the feed system removed in the direction indicated. Flash occasionally causes the feed system to adhere to this type of sprue puller and if positive automatic ejection is required, then some form of wiper device should be considered.

3.6.2 Type B Sprue Pullers. This type of sprue puller works on the principle of withdrawing the sprue puller through a plate, such as a stripper plate or an underfeed floating cavity plate, in order to eject the feed system.

One common design is the *mushroom headed sprue puller* illustrated in Figure 3.63a. The grooved head of the sprue puller creates an effective undercut which is used to pull the sprue, or secondary sprues in underfeed designs. During the mould opening sequence the sprue puller is effectively withdrawn with respect to the adjacent plate, causing the feed system to be sheared from the sprue puller. Note that a small ring of plastics material remains in the sprue puller undercut after the ejection phase is complete. During the next cycle of operations the incoming melt welds to this ring thereby recreating an effective sprue-pulling device.

A similar system, which is known as the *reverse taper sprue puller* design is illustrated in Figure 3.64a. The amount of undercut created by the reverse taper is not too severe, and while it is sufficient to pull the sprue and feed system, the elasticity of the plastics material allows the sprue puller to be withdrawn, without shearing the material during the extraction phase (Figure 3.64b).

4

Feed System

4.1 GENERAL

It is necessary to provide a flow-way in the injection mould to connect the nozzle (of the injection machine) to each impression. This flow-way is termed the *feed system*. Normally the feed system comprises a sprue, runner and gate. These terms apply equally to the flow-way itself, and to the moulded material which is removed from the flow-way in the process of extracting the moulding.

A typical feed system for a four-impression, two plate-type mould is shown in Figure 4.1. It is seen that the material passes through the sprue, main runner, branch runners and gate before entering the impression. It is desirable to keep the distance that the material has to travel down to a minimum to reduce pressure and heat losses. It is for this reason that careful consideration must be given to the impression layout.

The purpose of the cold slug well, shown opposite the sprue, is theoretically to receive the material that has chilled at the front of the nozzle during the cooling and ejection phase. Perhaps of greater importance is the fact that it provides positive means whereby the sprue can be pulled from the sprue bush for ejection purposes. (See sprue pullers, p. 109.)

4.2 RUNNER

The runner is a channel machined into the mould plate to connect the sprue with the entrance (gate) to the impression. In the basic two-plate mould the runner is positioned on the parting surface while on more complex designs the runner may be positioned below the parting surface (see Chapters 12 and 13).

The wall of the runner channel must be smooth to prevent any restriction to flow. Also, as the runner has to be removed with the moulding, there must be no machine marks left which would tend to retain the runner in the mould plate. To ensure that these points are met, it is desirable for the mould designer to specify that the runner (channel) is polished 'in line of draw'.

There are some other considerations for the designer to bear in mind: (*i*) the shape of the cross section of the runner, (*ii*) the size of the runner and (*iii*) the runner layout.

4.2.1 Runner Cross-Section Shape. The cross-sectional shape of the runner used in a mould is usually one of four forms (Figure 4.2): fully round (a),

RUNNER

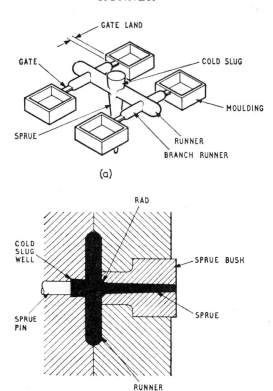

Figure 4.1—Feed system: (a) typical shot consisting of mouldings with sprue, runner and gates attached; (b) section through feed portion of mould

trapezoidal (b), modified trapezoidal (c) and hexagonal (d). The reasons why these particular forms are used in preference to others are outlined below.

The criterion of efficient runner design is that the runner should provide a maximum cross-sectional area from the standpoint of pressure transfer and a minimum contact on the periphery from the standpoint of heat transfer. The ratio of cross-sectional area to periphery will, therefore, give a direct indication of the efficiency of the runner design; the higher the value the greater the efficiency. Ratios for various types of runner section are given in Figure 4.3. As can be seen, the round and square types of runner are the two most satisfactory designs from this standpoint; whereas the ratios exhibited by the semicircular and rectangular types make their use generally undesirable.

Unfortunately, the square runner is not very satisfactory either, but for another reason: it is difficult to eject. In practice, because of this, an angle of 10° is incorporated on the runner wall, thus modifying the square to the trapezoidal section. The volume of the trapezoidal runner is approximately 25% greater than that of a round runner with corresponding dimensions ($W = D$,

Figure 4.2). To reduce this difference and still maintain corresponding dimensions, a modified trapezoidal form has been developed (Figure 4.2c) in which the volume is only 14% greater (approximately) than its round counterpart.

The hexagonal runner is basically a double trapezoidal runner, where the two halves of the trapezium meet at the parting surface. The cross-sectional area of this runner type is about 82% of that of the corresponding round runner. Naturally if similar cross-sectional areas are required, then the value for D (Figure 4.2d) must be increased accordingly. Some toolmakers feel that it is easier to match the two halves of the hexagonal runner compared with matching the two halves of a round runner. This point applies particularly to runners which are less than 3 mm ($\frac{1}{8}$ in) in width.

As the plastic melt progresses through the runner and mould system the melt adjacent to the cold mould surface will rapidly decrease in temperature and solidify. The material which follows will pass through the centre of this solidified material and, because of the low thermal conductivity that most thermoplastics possess, the solidified material acts as an insulation and maintains the temperature of the central melt flow region. Ideally, the gate should therefore be positioned in line with the centre of the runner to receive the material from the central flow stream. This condition may be achieved with the fully round runner (Figure 4.4a), and also with the hexagonal runner (Figure 4.2d).

The basic trapezoidal designs (Figure 4.2b and c) are not as satisfactory in this respect since the gate cannot normally be positioned in line with the central flow stream (Figure 4.4b).

The main objection to the fully round runner is that this runner is formed from two semicircular channels machined one in each of the mould plates. It is essential that these channels are accurately matched to prevent an undesirable and inefficient runner system being developed. A similar argument applies to the hexagonal runner system. The fact that these channels must be accurately matched means that the mould cost for a mould containing round or hexagonal runners will be greater than for one containing trapezoidal runners.

The choice of runner section is also influenced by the question whether positive ejection of the runner system is possible. Consider, for instance, the case of a two-plate mould in which a circular runner has been machined from both parting surfaces. In this case, as the mould opens, the runner is pulled from its channel in one mould half and it is then ejected from the other mould half either directly, by ejector pins, or by relying on its attachment to the mouldings by the gates (Figure 4.5).

For multi-plate moulds, however, positive ejection of the runner system is not practicable. Here the basic trapezoidal-type runner is always specified, the runner channel being machined into the injection half from which it is pulled as the mould opens. In this way the runner is free to fall under gravity between mould plates. If a circular runner had been specified, however, the runner system could well adhere to its channel and make its removal difficult (Figure 4.6).

RUNNER

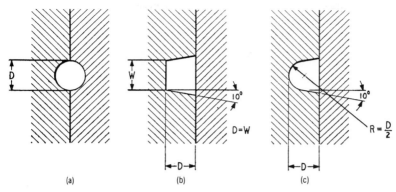

Figure 4.2—*Runner profiles: (a) round; (b) trapezoidal; (c) modified trapezoidal; (d) hexagonal*

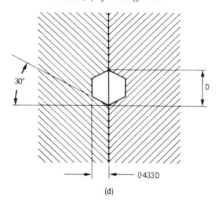

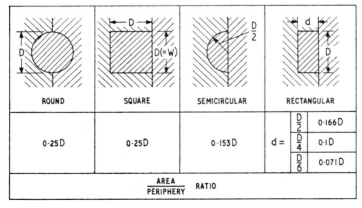

Figure 4.3—*Efficiency of various runner profiles (the higher the value the greater the efficiency)*

115

FEED SYSTEM

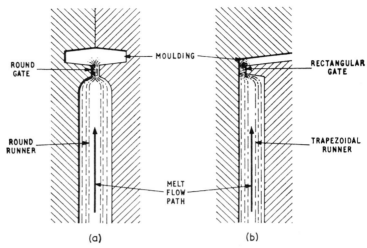

Figure 4.4—Flow of melt through round and trapezoidal runners, respectively

Summing up the points concerning cross-sectional shape, we can say that for simple two-plate moulds which have a flat parting surface the fully round runner or hexagonal runner is to be preferred, the increased mould cost being relatively small. For moulds which have complex parting surfaces, where it would be difficult to match accurately, the semicircular channels of the round runner or, for multi-plate moulds, the trapezoidal or modified trapezoidal section should be used.

4.2.2 Runner Size. When deciding the size of the runner the designer must consider the following factors: (*i*) the wall section and volume of the moulding, (*ii*) the distance of the impression from the main runner or sprue, (*iii*) runner cooling considerations, (*iv*) the range of mouldmaker's cutters available and (*v*) the plastics material to be used.

(*i*) The cross-sectional area of the runner must be sufficient to permit the melt to pass through and fill the impression before the runner freezes and for packing pressure to be applied for shrinkage compensation if required. Because of this, runners below 2 mm (3/32 in) diameter are seldom used and even this diameter is normally limited to branch runners under 25 mm (1 in) in length.

(*ii*) The further the plastic melt has to travel along the runner the greater is the resistance to flow. Hence the distance the impression is from the sprue has a direct bearing on the choice of cross-sectional size of the runner. For example, whereas a 5 mm (3/16 in) diameter runner may be suitable for a component weighing 60 g (2 oz) situated 25 mm (1 in) from the sprue, the same moulding 100 mm (4 in) from the sprue would require a 7 mm (¼ in) diameter runner.

(*iii*) The cross-sectional area of the runner should not be such that it controls the injection cycle, although this is sometimes unavoidable for very light

RUNNER

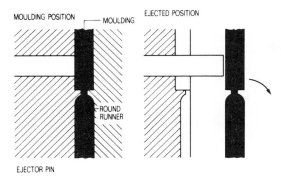

Figure 4.5—Ejection of round runner from two-part mould

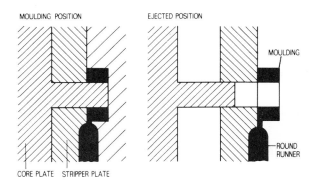

Figure 4.6—Incorrect design using a round runner for a stripper plate mould. Note that the runner may adhere to its channel, as shown, and thereby prevent extraction of the moulded shot

mouldings. The larger the cross-sectional area of the runner the greater is the bulk of material it contains and the longer the period it takes to cool sufficiently to enable the mould to be opened and the mouldings and runner ejected. For this reason it is undesirable to make the runner larger than 10 mm (⅜ in) diameter for most materials. However, the rigid PVCs and the acrylics are exceptions due to their high viscosity and diameters up to 13 mm (½ in) are used.

(*iv*) The size chosen for the runner should be in a range consistent with the mouldmaker's not having to carry in stock a multitude of different sizes of cutters. In practice the following are the more common sizes: 2 - 13 mm in 1 mm steps in the metric range and ⅛ - ½ in with 1/16 in steps in the imperial unit range. The following empirical formula is suggested as a guide to the size of the runner or branch runner for mouldings weighing up to 200 g (7 oz), and with wall sections less than 3 mm (0.125 in). For the rigid PVCs and the acrylics, increase the calculated diameter by 25%.

FEED SYSTEM

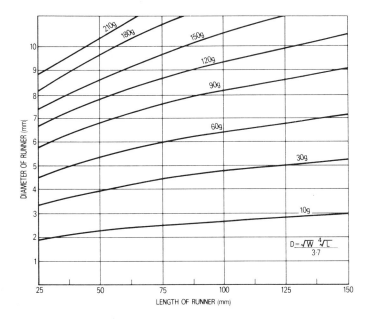

Figure 4.7—Guide to runner diameter for moulding section below 3 mm; for acrylic and PVC increase rated size by 25% (dimensions in metric units)

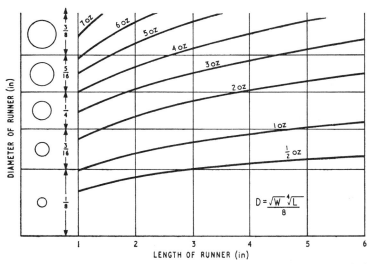

Figure 4.8—Guide to runner diameter for moulding wall sections below 0.125 in; for acrylic and PVC increase rated size by 25% (dimensions in Imperial units)

Metric system:
$$D = \frac{\sqrt{W}\sqrt[4]{L}}{3.7}$$
(4.1a)

Imperial system:
$$D = \frac{\sqrt{W}\sqrt[4]{L}}{8}$$
(4.1b)

where D = runner diameter (mm or in)
W = weight of moulding (g or oz)
L = height of runner (mm or in)
(A consistent system of dimensions must be used.)

The formula is used in conjunction with the notes given previously, i.e.:
(*i*) The runner should not be below 2 mm (3/32 in) diameter, nor above 10 mm (⅜ in) diameter (or 13 mm (½ in) diameter where applicable).

(*ii*) The calculated size should be increased to the next suitable cutter size.

Figure 4.7 shows a plot of diameter versus 'length of runner' for various weights of moulding, adopting the metric system of dimensioning. Figure 4.8 shows a corresponding plot using the Imperial dimensioning system. For example, a 120 g (4 oz) moulding in polyethylene being fed by a 50 mm (2 in) long runner will require a diameter of 7 mm (5/16 in).

Theoretically the cross-sectional area of the main runner should be equal to, or in excess of, the combined cross-sectional areas of the branch runners that it is feeding. This relationship is, however, ignored when the maximum suggested diameter is reached.

4.2.3 Runner Layout. The layout of the runner system will depend upon the following factors: (i) the number of impressions, (ii) the shape of the components, (iii) the type of mould (i.e., two-plate or multi-plate mould), (iv) the type of gate. There are two main considerations when designing a runner layout.

The runner length should always be kept to a minimum to reduce pressure losses, and the runner system should be balanced.

Runner balancing means that the distance the plastic material travels from the sprue to the gate should be the same for each moulding. This system ensures that all the impressions will fill uniformly and without interruption providing the gate lands and the gate areas are identical. Figure 4.9 shows examples of moulds all based on the balanced runner principle.

It is not always practicable, however, to have a balanced runner system and this particularly applies to moulds which incorporate a large number of differently shaped impressions (Figure 4.10). In these cases balanced filling of the impression can be achieved by varying the gate dimensions, that is by balanced gating (Section 4.3.2).

SINGLE-IMPRESSION MOULDS. Single-impression moulds are usually fed by a direct sprue feed into the impression (Figure 4.14) and hence no runner sys-

FEED SYSTEM

tem is required. However, it may be desirable to edge gate (for example when sprue marks must not appear on the main surface) in which case a short runner as shown in Figure 4.9a may be used. But note that by gating a single impression in this way the impression itself must be offset. This is undesirable, particularly with a large impression, as the injection pressure will exert an unbalanced force which will tend to open the mould one side and may result in flashed mouldings.

TWO-IMPRESSION MOULDS. The various alternatives for feeding two impressions are shown in Figure 4.9b, c and d. The simplest case (b) is where the runner takes the shortest path between the two impressions. Unfortunately, it is not always possible to adopt this short runner. This is because, as shown in the following discussion, the most desirable position for the gate may not be on the centre-line of the mould.

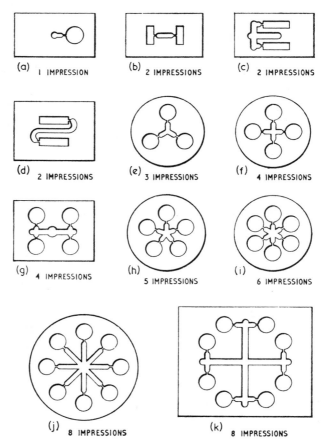

Figure 4.9—Balanced runner layouts; melt has same length of travel to all impressions

RUNNER

If we consider Figure 4.9b, which schematically shows the plan of a mould for two rectangular blocks, it is seen that solely from the viewpoint of mould layout it is desirable to have the impressions positioned as shown with short runners to the sides of the impressions, thus enabling the size of the mould to be kept to a minimum. However, there are other considerations, such as that of correct gating, and it may be desirable to gate at one end of the impression.

To achieve these end gates it is necessary to alter the design of the runner layout, so that either a T-shaped runner extends beyond the impressions and is then connected to the gates by short branch runners (Figure 4.9c), or the runner, in the form of an S, sweeps round to the gates (d), without the necessity for branch runners.

In general, providing that the impressions are approximately the same size and shape, no difficulty should be experienced in designing balanced runner systems for two-impression moulds.

THREE-IMPRESSION MOULDS. Figure 4.9e illustrates a balanced runner system for three similar impressions. In this case the impressions are placed on a pitch circle diameter 120° apart; this design allows the runners to be kept to a minimum length.

When, however, the impressions are of different shapes and sizes the layout may be as shown in Figure 4.10a. The large impression is shown being fed directly from the sprue via a short runner, while the smaller components are fed via a branch and main runner system. Balance in the feed system is ultimately attained by adjustment of gate size.

OTHER MULTI-IMPRESSION MOULDS. For four or more impression moulds the design of the runner layout is simply an extension of the previous discussion. For example, balanced runner systems for moulds containing four, five, six and eight impressions are shown in Figure 4.9. From the runner balancing standpoint it is simpler by far to situate the impressions on a pitch circle diameter and feed each impression directly from the sprue via a runner rather than to incorporate a main and branch runner system.

As the number of impressions increase, however, the pitch circle diameter design becomes progressively more impracticable as the runner length, which is a function of the pitch circle diameter, is also increased. This results in large-diameter runners being required which progressively lengthens the moulding cycle and more scrap (although reprocessable) is produced. When a large number of impressions have to be accommodated, or where the impressions are of greatly dissimilar shape, the alternative main and branch runner system is therefore usually adopted. Examples of moulds with this form of runner layout are shown in Figure 4.10. The main runner acts simply as a manifold which feeds a number of branch runners. The number and shape of the impressions will determine the precise layout but from the bolster machining standpoint it is convenient either to adopt a straight main runner (c) or design the main runner in the form of a cross (d). An eighty-impression mould is shown in Plate 2.

FEED SYSTEM

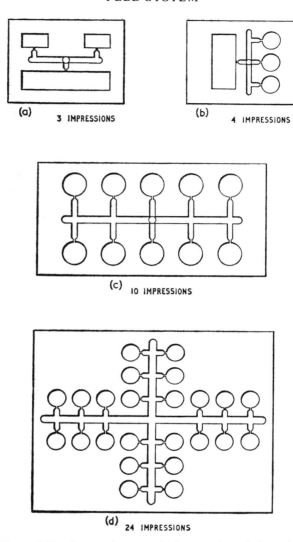

Figure 4.10—Runner layouts which necessitate balanced gating

4.3 GATES

The gate is a channel or orifice connecting the runner with the impression. It has a small cross-sectional area when compared with the rest of the feed system. This small cross-sectional area is necessary so that:
    (i) The gate freezes soon after the impression is filled so that the injection plunger can be withdrawn without the probability of void being created in the moulding by suck-back.

GATES

(ii) It allows for simple degating and in some moulds this degating can be automatic.
(iii) After degating only a small witness mark remains.
(iv) Better control of the filling of multi-impressions can be achieved.
(v) Packing the impression with material in excess of that required to compensate for shrinkage is minimised.

The size of the gate can be considered in terms of the gate cross-sectional area and the gate length, the latter being known as gate land. The optimum size for a gate will depend on a number of factors including (i) the flow characteristics of the material to be moulded, (ii) the wall section of the moulding, (iii) the volume of material to be injected into the impression, (iv) the temperature of the melt, and (v) the temperature of the mould.

No theoretical size exists for the ideal gate. The gate size chosen in practice for a particular component is normally based on past experience. However, the reader may not have this experience upon which to base a decision and, therefore, a guide to the dimensions for each gate type is given. The guide applies, except where otherwise stated, to the general case of a moulding with a wall section between 0.75 mm (0.030 in) and 4 mm (0.157 in).

4.3.1 Positioning of Gate. Ideally, the position of the gate should be such that there is an even flow of melt in the impression, so that it fills uniformly and the advancing melt front spreads out and reaches the various impression extremities at the same time. In this way two or more advancing fronts would rarely meet to form a weld line with consequent mechanical weakness and surface blemish in the moulding.

Such an ideal position for the gate is possible in certain shaped mouldings such as those with circular cross-sections, for example, a cup or a cone in which material is fed through the centre of the base or apex. The direction of melt flow of a centre feed compared with a side feed is illustrated in Figure 4.11.

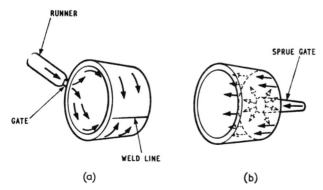

Figure 4.11—Position of gate: (a) edge gating of cup-shaped moulding—note resulting weld line where two flow paths meet; (b) same moulding with a sprue gate — note smooth even flow path for material

FEED SYSTEM

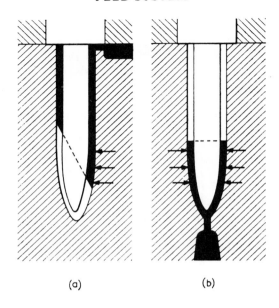

Figure 4.12—Position of gate for a pen cap: (a) edge gate — note deflection of core; (b) underfeed pin gate — even flow theoretically holds core central

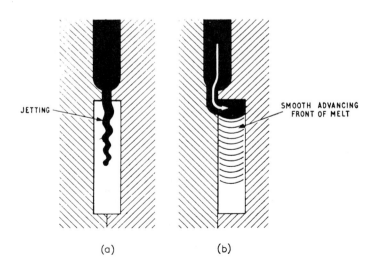

Figure 4.13—Position and choice of gate for solid block type of moulding: (a) if edge is used, jetting of material occurs; (b) overlap gating avoids this undesirable feature

GATES

Another strong reason for central gating at the apex of slender cone-like components such as pen caps is that side gating may cause deflection of the core. This can arise because side gating gives rise to a more rapid flow of material down one side of the impression, resulting in a differential pressure which can move the core out of position. This results in a thinner wall section on one side, thus adding another weakness to that of the weld line. The position of the melt front in the partly filled impression when the alternative gate positions are used is shown in Figure 4.12 at (a) and (b) respectively. From this it can easily be seen in (b) that a centre feed results in even flow and that the core is held more centrally. Central gating for this type of impression can, however, lead to complications in multi-impression moulds as either an underfeed mould or a hot runner mould will be required.

For rectangular mouldings the ideal position does not exist but even here the central position is often the one to be preferred. But in the case of thin-walled rectangular mouldings, particularly when the material used can exhibit differential shrinkage causing distortion, off-centre multi-point gating or film gating is advantageous.

When the edge gate is used, and the majority of mouldings are edge gated for reasons of mould economics, the gate should be positioned so that the melt flow immediately meets a restriction. An example of incorrect gating of a solid rectangular block type of moulding is shown in Figure 4.13a. The impression is fed in the centre at one end and the material, on entering at high velocity, 'jets' and quickly sets on reaching the cool mould walls. More material then enters and flows around the original jetted material. The resulting flow lines are often visible on the finished moulding.

The trouble in this particular case can be overcome by overlap feeding. The flow of material issuing from the gate is forced to impinge on an opposing face of the impression (Figure 4.13b) and this causes the material to form an advancing front which progressively fills the impression displacing the air in front of it, thus forming a moulding free of flow lines.

The box type of component, which necessitates a cavity and core, automatically provides opposition to jetting so that the edge gate here is quite permissible. However, some weld lines are to be expected where the two flows meet on the opposite side of the impression. In such cases, although flow lines cannot be prevented, their effects, such as mechanical weakness and surface blemish, can be largely overcome. This is accomplished by keeping the *cooling* medium away from the neighbourhood of weld lines; the mould temperature then increases slightly at these points, helping the two fronts to knit together more easily.

4.3.2 Balanced Gating. It is often necessary to balance the gates of a multi-impression mould to ensure that the impressions fill simultaneously. This method is adopted when the preferred balanced runner system cannot be used.

Consider the runner layout shown in Figure 4.10c. The melt will take the easiest path; hence, once the runner system is filled, those impressions closest to the sprue will tend to fill first and those at the greatest distance to fill last. In consequence, some impressions may be overpacked while others may be starved of material. To achieve balanced filling of these impressions it is necessary to cause the greater restriction to the flow of the melt to those impressions closest to the sprue and to progressively reduce the restriction as the distance from the sprue increases.

By adopting the method of balanced gating there are two ways of varying the restriction; (i) by varying the land length and (ii) by varying the cross-sectional area of the gate. In practice balanced gating is a matter of trial and error: the land length normally is kept constant; starting with a small gate width, the mould is tried out with a short injection stroke so that short mouldings are obtained. On inspection it will be obvious which impressions are filling first. The gate width can then be progressively enlarged and adjusted until balanced filling is achieved.

4.3.3 Types of Gate. To obtain the optimum filling conditions the type of gate must be carefully chosen. On most occasions, however, the choice will be obvious as only one type of gate will meet the particular requirements for the moulding on hand.

The types of gate commonly used are: sprue gate, edge gate, overlap gate, fan gate, diaphragm gate, ring gate, film gate, pin gate and subsurface gate.

SPRUE GATE. When the moulding is directly fed from a sprue or secondary sprue, the feed section is termed a sprue gate. The main disadvantage with this type of gate is that it leaves a large gate mark on the moulding. The size of this mark depends on: (i) the diameter at the small end of the sprue, (ii) the sprue angle, and (iii) the sprue length. Thus the gate mark can be minimised by keeping the dimensions of the above factors to a minimum. Note that as the sprue entry is controlled by the nozzle exit diameter, and, as it is undesirable to reduce the sprue angle below four degrees inclusive for withdrawing

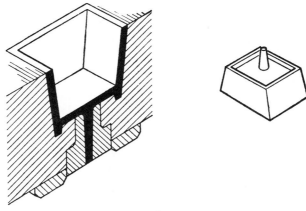

Figure 4.14—Sprue gate

GATES

purposes, the sprue length is the logical dimension for the designer to attempt to reduce. An extension nozzle can often be used to advantage, because it enters a recess in the mould and cuts down the overall sprue length. On basic two-plate moulds the sprue gate is used only for single-impression moulds. In this case the impression is positioned in the centre of the mould and the sprue is a direct feed into it. An example of the sprue gate is shown in Figure 4.14.

A modified form of sprue gate is also used on underfeed moulds and runnerless moulds. In both of these, any number of impressions can be accommodated and the sprue gates, (Figure 12.5) now termed secondary sprues, are fed from runner systems situated below the parting surface.

RECTANGULAR EDGE GATE. This is a general purpose gate and in its simplest form is merely a rectangular channel machined in one mould plate to connect the runner to the impression (Figure 4.15a). A section through the relevant parts of a typical mould is shown at (c).

This gate offers certain advantages over many other forms of gate:
(i) The cross-sectional form is simple and, therefore, cheap to machine.
(ii) Close accuracy in the gate dimensions can be achieved.
(iii) The gate dimensions can be easily and quickly modified. (In the wrong hands this is often a disadvantage.)

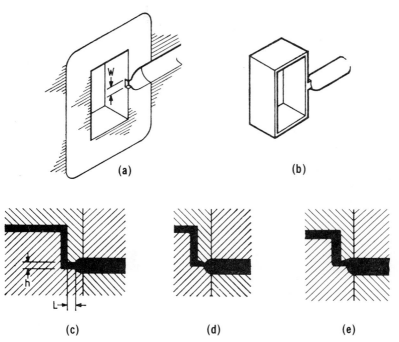

Figure 4.15—Rectangular edge gate: (a) gate machined into cavity plate; (b) moulding with gate attached; (c) cross-section through mould; (d) and (e) alternative shapes

FEED SYSTEM

(iv) The filling rate of the impression can be controlled relatively independently of the gate seal time.

(v) All common moulding materials can be moulded through this type of gate.

One disadvantage of this type of gate, however, is that after gate removal a *witness mark* is left on a visible surface of the moulding. This is more noticeable with certain materials (e.g. polystyrene, acetals, etc.), particularly if the mouldings are simply broken from the gates. To reduce the possibility of gate shatter marks on the moulding the modified form of gate shown at (d) may be used. For certain products, such as self assembly toy kits, a projecting gate is best avoided. In these cases adopt the alternate modified design (e).

Gate size: because of the rectangular shape, the dimensions of the gate are given by width (W), depth (h), and land length (L).

Now the pressure drop across the gate is approximately proportional to the land length and therefore this should be kept as small as possible consistent with the strength of the steel which remains between the runner and the impression. In practice a value of between 0.5 mm (0.02 in) and 0.75 mm (0.030 in) is satisfactory.

The minimum depth of gate controls the time for which the gate remains open. This *gate open time* must be sufficient for the material to reach the extremities of the impression. Now, providing that the wall section of the component has been correctly chosen with respect to the maximum length of flow required, it appears reasonable to expect a relationship between the gate depth dimension and the wall section of the component.

In practice the following empirical relationship for gate depth has been found useful:

$$h = nt \qquad (4.2)$$

where h = depth of gate mm (in)
t = wall section thickness mm (in)
n = material constant.

While theoretically it is probable that each material should have a different value for n, in practice it is convenient to group certain materials together and to use a group constant.

group 1: polythene, polystyrene	$n = 0.6$
group 2: polyacetal, polycarbonate, polypropylene	$n = 0.7$
group 3: cellulose acetate, polymethyl methacrylate, nylon	$n = 0.8$
group 4: PVC	$n = 0.9$

Figure 4.16 shows the above relationship plotted for the various groupings.

The cross-sectional area of the gate ($h \times W$) controls the rate at which the plastic material enters the impression. If the concept of the gate-depth - wall-section relationship is accepted then the gate depth is established first. This

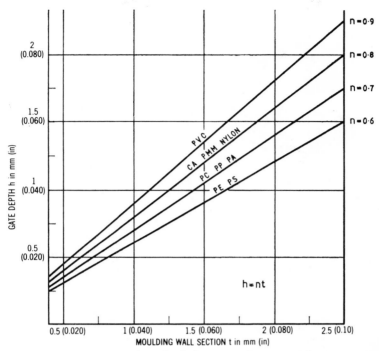

Figure 4.16—Guide to depth dimension for rectangular gates

means that the width of the gate becomes the controlling dimension for the flow rate. Very little has been published on the optimum flow rate for particular materials and even if this were known the relationship between optimum flow rate and gate width would probably be complex because of the large number of variables involved.

The gate width is, therefore, usually based upon experience gained when moulding components of similar shape and size. However, the beginner does not have this, and therefore as a guide to the student we suggest the following empirical relationship:

$$W = \frac{n\sqrt{A}}{30}$$

(4.3)

where W = gate width (mm or in)
A = surface area of cavity (mm^2 or in^2)
n = material constant
(A consistent system of dimensions must be used.)

A graph of the above equation is given in Figure 4.17.

Consider an example using the SI system of dimensions. Alternative Imperial dimensions are given in brackets, but note that these dimensions are not necessarily direct equivalents of the SI dimensions.

What gate dimensions are required for moulding a polythene box the dimensions of which are as follows:

FEED SYSTEM

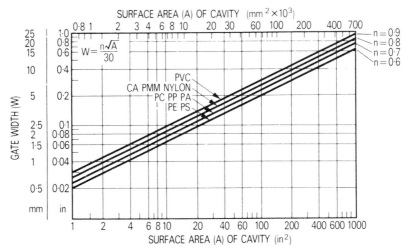

Figure 4.17—Guide to width dimension for rectangular gates

Length = 150 mm (6 in); width = 130 mm (5 in); depth = 50 mm (2 in); wall section thickness = 1.3 mm (0.050 in).

(i) Land length (L) = 0.5 mm (0.020 in). (Minimum recommended length.)
(ii) Gate depth (h) = 0.78 mm (0.030 in). (From Figure 4.16.)
(iii) Gate width (W).

First determine the approximate surface area of the cavity (i.e. the surface area of the outside of the component).

$$2(150 \times 50) + 2(130 \times 50) + (150 \times 130) = 47\,500 \text{ mm}^2$$
$$(\text{or} \quad 2(6 \times 2) + 2(5 \times 2) + (5 \times 6) = 74 \text{ in}^2)$$

Use equation (4.3):

$$W = \frac{n\sqrt{A}}{30} = \frac{0.6\sqrt{47\,500}}{30} = 4.36 \text{ mm } (0.172 \text{ in})$$

Should the value for W exceed the diameter of the runner then a fan type of gate can be used.

OVERLAP GATE. This gate can be considered as a variation of the basic rectangular type gate and is used to feed certain types of moulding. We have noted previously that the melt jets into an impression if it does not contact a restriction immediately (Figure 4.13a). Therefore, for block-type mouldings the rectangular gate is replaced by the overlap gate which, by virtue of its position, directs the melt flow against an opposite impression face (Figure 4.13b).

The overlap gate, which is of general rectangular form, is machined into the plain mould plate (Figure 4.18), in such a way that it bridges the gap between the end of the runner and the end wall of the impression (c). The general arrangement is shown at (d) which gives a local section through the mould at the relevant point.

The gate may be used for all the common moulding materials apart from rigid PVC. The gate, being attached to the moulding surface, does require

GATES

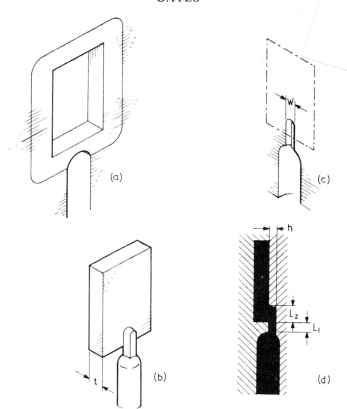

Figure 4.18—Overlap gate: (a) view of cavity plate; (b) moulding with gate attached; (c) overlap gate machined into plain plate; (d) cross-section through mould

more careful removal and finishing than for edge gates. The size of the gate can follow the general pattern suggested for the rectangular gate with the same limitations:

land length (distance between the end of the runner and the wall of the impression)

$$L_1 = 0.5 - 0.75 \text{ mm } (0.020 - 0.030 \text{ in})$$

gate width (see previous section)

$$W = \frac{n\sqrt{A}}{30}$$

gate height (see previous section)

$$h = nt$$

gate length (Figure 4.18)

$$L_2 = h + \frac{W}{2} \qquad (4.4)$$

FEED SYSTEM

FAN GATE. This is another edge-type gate but, unlike the rectangular gate which has a constant width and depth, the corresponding dimensions of the fan gate are not constant (Figure 4.19). The width (W) increases (normally from the basic runner diameter) while the depth (h) decreases so as to maintain a constant cross-sectional area throughout the length of the gate.

The width of the gate at the impression is relatively wide and, because of this, a large volume of material can be injected in a short time. This type of gate can, therefore, be used advantageously for large-area, thin-walled mouldings. The fan shape appears to spread the flow of the melt as it enters the impression and a more uniform filling is obtained with fewer flow lines and surface blemishes.

This gate may be used with all the conventional moulding materials apart from certain grades of rigid PVC.

The relevant gate sizes which must be decided upon are the land length (L), the gate width (W) and the gate depth (h).

The land length needs to be slightly longer than for the rectangular gate and a suggested size for this is 1.3 mm (0.050 in).

The dimension for gate width at the impression can be obtained using the formula suggested in the previous section, i.e.

$$W = \frac{n\sqrt{A}}{30}$$

However, in practice, to obtain the full benefits of the fan gate design a greater width is often used. For example 40 mm (1½ in) wide gates are not uncom-

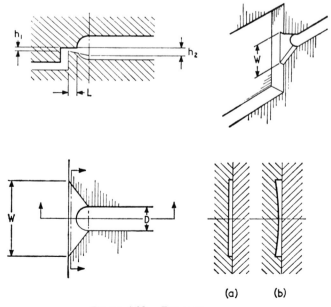

Figure 4.19—Fan gate

GATES

mon. The main disadvantage of the design is that a large witness mark is left on the moulding which must subsequently be trimmed and finished. It is therefore advantageous to design the gate relatively narrow and widen it only if necessary.

The height of the gate can be determined by the previously discussed empirical formula, i.e.

$$h_1 = nt$$

To maintain a constant cross-sectional area, because of the gate form, the depth of gate must be progressively increased back to the runner. The depth of gate at this point is given by

$$h_2 = \frac{Wh_1}{D} \qquad (4.5)$$

The effective length of the gate land between the runner and the impression progressively increases from a minimum at the centre line to a maximum at the outer gate wall. To compensate for the increased pressure drop which results at the wall it is common practice to increase the depth of the gate at either side to provide for more even flow through the gate. Figure 4.19 (a, b) shows a cross-section through the gate before and after correction.

TAB GATE. This is a particular gating technique for feeding solid block-type mouldings. A projection or *tab* is moulded on to the side of the component (Figure 4.20) and a conventional rectangular edge gate feeds this tab. The sharp right-angled turn which the melt must take prevents the undesirable 'jetting' which would otherwise occur. The melt is thereby caused to advance in a smooth steady flow and, providing the shape of the impression allows it, the impression will fill uniformly. Thus the tab gate is an alternative to the overlap-type gate. The choice of one gate design or the other will depend mainly upon whether the witness mark left by the gate is best, from the appearance

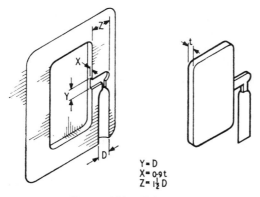

Figure 4.20—Tab gate

point of view, on the top or at the side. Both gates leave relatively large witness marks.

This gate, while being developed particularly for the acrylics, may be used for any of the common moulding materials. The size of the gate divides conveniently into two sections (i) the size of the rectangular gate and (ii) the size of the tab. For all practical purposes the rectangular gate size can be calculated using the empirical formula given above. The suggested dimensions for the tab are shown in Figure 4.20:

> tab width (Y) = D (where D is the diameter of the runner)
> tab depth (X) = $0.9t$ (where t is the wall thickness of the moulding)
> tab length (Z) = $1½D$ minimum

DIAPHRAGM GATE. This gate is used for single-impression tubular shaped mouldings on two-plate moulds. It may also be used in a similar manner for multi-impression tubular shaped mouldings on underfeed and runnerless moulds.

The sprue leads into a circular recess (Figure 4.21), slightly smaller than the inside diameter of the tube. This recess forms a disc of material and acts as a runner which allows material to flow radially from the sprue to the gate. The gate may be cut either on the core (inset (a)) or in the cavity (inset (b)). In both cases it connects the disc runner with the impression.

The choice of gate (a) or (b) will depend upon the use the tube is to be put to. If the internal bore is important then the gate should be cut in the cavity as shown (b). Thus by a simple machining operation on the face of the moulding the bore diameter is not disturbed. Alternatively, if the internal bore is not important (a) is the better choice as the gate is more easily removed by a blanking operation.

The gate dimensions which must be considered are, for method (a) land length (L) and depth of gate (h), and for method (b) land length (L), overlap length (L_1), depth of gate (h_1).

Consider first the side feed diaphragm gate (a). The land length again should be a minimum consistent with the strength of steel left between the circular runner and the impression. A value of between 0.75 mm (0.030 in) and 1 mm (0.040 in) is suggested.

The depth of the diaphragm gate is normally made slightly less than the values recommended for the rectangular gate as the corresponding value for W in this case (the inside surface of the moulding) is large.

The following relationship, with reference to section on the rectangular edge gate (above) has been found to be suitable.

$$h_1 = 0.7nt \tag{4.6}$$

For the overlap type of diaphragm gate the value for L can be as above. The overlap length L_1 should be at least equal to the depth of the gate (h_1) which may be computed from

$$h_1 = nt$$

GATES

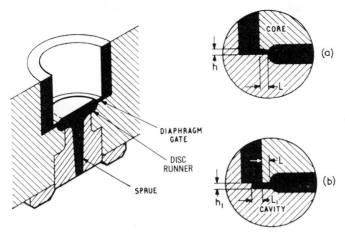

Figure 4.21—Diaphragm gate

The centre disc is sometimes tapered from the centre towards the gate to save material and to reduce the bulk for cooling reasons. This type of gate allows for constant filling of the moulding and minimises the formation of weld lines. It is recommended for use with all moulding materials.

RING GATE. The function of this gate is identical to that discussed in the previous section (for the diaphragm gate) and the same comments apply. This type of gate is used for tubular-type mouldings when more than one impression is required in a simple two-plate mould. The gate provides for a feed all around the external periphery of the moulding (as against the internal surface in the previous section) and permits the use of a conventional runner system to connect the impressions.

The runner, in the form of a trapezoidal annulus, is machined into the mould plate (Figure 4.22). The trapezoidal runner is normally used since this type of moulding would be ejected using a stripper plate.

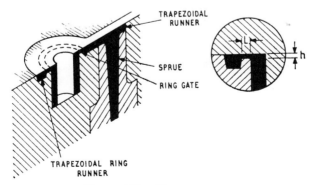

Figure 4.22—Ring gate

FEED SYSTEM

The gate is in the form of a concentric film between the runner and the impression. The dimensions of this gate are identical to those discussed for the diaphragm gate. That is $L = 0.75 - 1$ mm (0.030 - 0.040 in) and $h = 0.7nt$.

FILM GATE. This gate may be considered as a long rectangular-type edge gate and it is used for large, thin-walled components to assist in the production of warpage free products. The gate normally extends across the complete width (narrow side) of the moulding, although a smaller width may be used initially, which, if it proves satisfactory, will save some finishing time. The gate is similar in principle to the diaphragm and ring gates in that it provides for a large flow area and results in a quick fill time. Because of this feature, the gate depth may be somewhat less than for a corresponding rectangular gate. The same relationship as given above is suggested, i.e. $h = 0.7nt$.

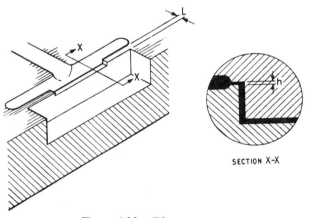

Figure 4.23—Film gate

A runner is provided parallel to the side face of the impression to feed the gate evenly with material as shown in Figure 4.23. This means that a very slender wall of steel exists between the runner and the impression, and to prevent this from collapsing in use, a minimum land length of 1.3 mm (0.050 in) is suggested for this particular gate. It is normal practice to extend the runner beyond the end of the impression, as shown, irrespective of the gate length.

The gate is used for all common moulding materials. It is particularly useful for those materials which exhibit differential shrinkage for which central feeding is impracticable.

PIN GATE. This is a circular gate used for feeding into the base of components and, because it is relatively small in diameter, it is often to be preferred to the sprue gate which necessitates a finishing operation. However, the pin gate may only be used in certain types of moulds and these are generally more complex in design than the moulds in which sprue gating or side gating techniques are used.

GATES

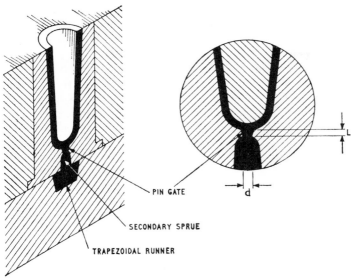

Figure 4.24—Pin gate

As an alternative to feeding into the centre of a component, this type of gate is often used for multi-point feeding and for single, off-centre, feeding. This technique is particularly desirable for use with materials which exhibit differential shrinkage characteristics, and it is often used as an alternative to film gating.

A section through a typical pin gate is shown in Figure 4.24. This shows a three-plate, underfeed-type mould. The gate is essentially a circular orifice which connects the secondary sprue to the impression. To prevent the gate fracturing the area of moulding adjacent to the gate, particularly with the more brittle moulding materials, it is desirable to taper the gate, as shown, so that the gate is caused to break at the junction with the secondary sprue. Certain material manufacturers advise increasing the wall section of the moulding locally opposite the gate to permit free flow into the impression.

To permit the use of a pin gate one of the following mould designs must be adopted.

(i) Three-plate, underfeed-type mould (Section 12.2). An extra plate is added behind the cavity plate to permit a runner system to be incorporated below the cavity or cavities. The pin gate connects the impression to the runner, either directly or via a secondary sprue.

(ii) Hot-runner moulds (Section 13.3). A heated insulated runner block is incorporated behind the cavity plate. The impression is connected to the secondary nozzle (of the hot-runner unit), again by a pin gate and secondary sprue. Both the above designs may be used for central or offset, single or multi-point gating, into one or more impressions.

FEED SYSTEM

(iii) Two-plate mould with special nozzles (Section 13.2.3). A relatively large recess is provided behind the cavity to permit a specially designed nozzle to protrude into the mould. The pin gate connects the impression to the nozzle usually via a hot well of plastics material. This is known as the antechamber design.

The gate dimensions which must be considered are the land length (L), and the gate diameter (d). To minimise the pressure losses, as for all other gates, the land length is kept to a minimum consistent with the strength of steel used. A land length of between 0.5 mm (0.02 in) and 0.75 mm (0.03 in) is suitable.

Because of its shape the round gate cannot be used independently to control the filling rate and gate seal time, so some compromise between these two requirements must be accepted.

A guide to the gate diameter is given by

$$d = nC\sqrt[4]{A} \qquad (4.7)$$

where d = diameter of gate (mm or in)
n = material constant (see above).
A = surface area of the cavity (mm^2 or in^2).
C = a function of the wall section thickness (t) and this value can be obtained from Table 4 for SI units and from Table 5 for Imperial units.

Note that the above equation applies to mouldings with wall section thicknesses in the range 0.75 - 2.5 mm (0.03 - 0.1 in). A consistent system of dimensions must be used.

TABLE 4 (SI UNITS)

$t(mm) =$	0.75	1	1.25	1.5	1.75	2	2.25	2.5
$c =$	0.178	0.206	0.230	0.242	0.272	0.294	0.309	0.326

TABLE 5 (IMPERIAL UNITS)

$t\ (in) =$	0.030	0.040	0.050	0.060	0.070	0.080	0.090	0.1
$c =$	0.036	0.041	0.047	0.051	0.055	0.058	0.062	0.065

For the antechamber design (see (iii) above) the calculated gate diameter may be reduced by about a third. This is because the nozzle is very close to the gate and the viscosity of the melt passing through will, therefore, be lower than with the alternative feed systems. Very small pin gates are possible with this design.

ROUND EDGE GATE. This gate is formed by machining a matching semi-circular channel in both mould plates between the runner and the impression

GATES

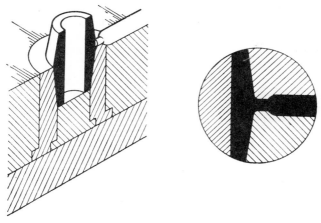

Figure 4.25—Round edge gate

(Figure 4.25). The slight radius incorporated on the entry into the impression is an advantage in preventing damage to the moulding when the gate is removed. However, because of its form, the round edge gate suffers many disadvantages as compared with the corresponding rectangular edge gate:
  (i) The matching form is more difficult to machine.
  (ii) Precise dimensions are more difficult to achieve.
  (iii) The filling rate of the impression cannot be controlled independently of the gate seal time.

Because of the above disadvantages this gate is seldom used for mouldings with wall thicknesses below 4 mm (0.150 in). The gate is used, however, for thicker wall sections. For this type of moulding a relatively large gate is an advantage to ensure that the gate remains open sufficiently long to allow follow-up pressure to be applied to prevent sinkage.

SUBSURFACE GATE. The subsurface gate is a circular or oval gate which submerges and 'feeds' into the impression below the parting surface of the mould. While similar to the round edge gate in that it is of similar or nearly similar shape and feeds into the side of the impression, it has several advantages over the round gate.
  (i) The form, being in one mould plate, has no matching problems and precise dimensions can be achieved.
  (ii) If the more oval form is used the filling rate of the impression can be controlled independently of the gate seal time.
  (iii) The gate is sheared from the moulding during its ejection.

The basic design is shown in Figure 4.26. The runner is terminated a distance X from the impression. A secondary runner, usually of conical form, is machined at an angle ϕ to the impression wall and is stopped short of the impression wall by a distance L. (Note that this corresponds to the gate land length.) The gate is then machined at the same angle ϕ to join the secondary runner to the impression.

FEED SYSTEM

The moulding and feed systems are removed separately from the mould and this means that a separate runner ejection is advantageous, particularly

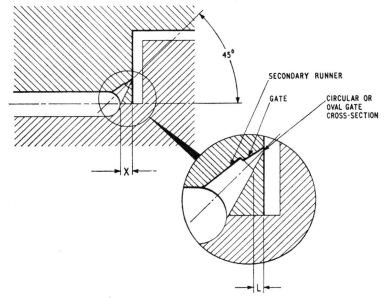

Figure 4.26—Subsurface gate

as a certain amount of deformation of the runner is necessary to remove the secondary runner from the mould.

The gate dimensions which must be considered are land length which, because of the form, needs to be L (minimum) = 1.9 mm (0.075 in). Φ is the angle subtended between the centre-line of the secondary runner and the impression wall. This angle is normally between 30° and 45°. The dimensions for the gate cross-section can be estimated from either the equation suggested for the rectangular gate (see above), if an oval gate is adopted, or the equation for the pin gate (see above), if the circular gate is used.

5

Parting Surface

5.1 GENERAL

The parting surfaces of a mould are those portions of both mould plates, adjacent to the impressions, which butt together to form a seal and prevent the loss of plastic material from the impression.

We can classify the parting surface as being either flat or non-flat. Examples of the latter type include stepped, profiled and angled parting surfaces.

In general, the flat parting surface is the simplest to manufacture and maintain. It can be surface-ground, and is easily bedded down.

To bed down a pair of mould plates is the process of marrying the two mould surfaces together. This is accomplished by blueing one surface, momentarily bringing the two plates together and subsequently removing any high spots which will be apparent on the non-blued surface. The plates are said to be bedded down when an even film of blue is transferred from one plate to the other.

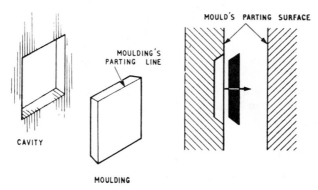

Figure 5.1—Parting line of moulding and parting surface of mould

5.2 FLAT PARTING SURFACE

The nature of the parting surface depends entirely on the shape of the component. For instance, consider the rectangular moulding shown in Figure 5.1. The cavity for this article can be die-sunk into one mould plate. The position of the parting surface will therefore be at the top of the moulding, the parting surface itself being perfectly flat. For appearance this is the ideal arrangement

PARTING SURFACE

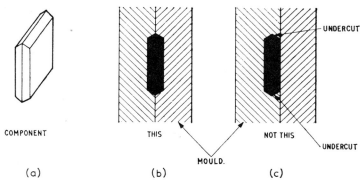

Figure 5.2—Practicable and impracticable choice of parting surface

as the *parting line* is not noticeable unless *flash* develops. Flash is the name given to the wafer of material which escapes from the impression if the two mould halves are not completely closed.

A further consideration is that the parting surface must be chosen so that the moulding can be removed from the mould. Let us consider an example. Figure 5.2 shows a flat rectangular moulding which incorporates a double-bevelled edge. Obviously the parting line for this component cannot be on its top surface (as for the previous case, Figure 5.1) as this will create an undercut in the mould (Figure 5.2c). The only suitable choice for the parting line is on the centre of the double-bevel which allows for half of the required form to be die-sunk into each of the two mould halves (Figure 5.2b).

To summarise the above two points we can say that the parting line must occur along the line round the position of maximum dimension when viewed in the draw direction. Then if this line lies on a plane the parting surface will be flat.

A number of typical mouldings which permit a flat parting surface to be adopted are shown in Figure 5.3. The arrows indicate the parting line on each moulding.

5.3 NON-FLAT PARTING SURFACE

Many mouldings are required which have a parting line which lies on a non-planar or curved surface. In these cases the mould's parting surface must either be stepped, profiled or angled.

5.3.1 Stepped Parting Surface. Consider the Z-plate component shown in Figure 3.25. The maximum dimension of this component when viewed in the draw direction occurs at the top of the Z-form. Thus, as this form is stepped, the mould's parting surface must likewise be stepped as shown. Note that as the edge of the component is square with the face (apart from moulding draft) the entire moulding form can be accommodated in one mould half. However, had the edge incorporated a radius, then in addition to the mould having a

*Plate 1   Three-dimensional machining of a cavity block on a 'Keller'
(Courtesy Alfred Herbert Ltd.)*

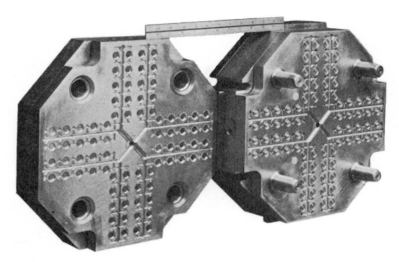

Plate 2   Eighty-impression mould for caps (Courtesy J. E. Snow (Plastics) Ltd.; mould manufactured by M.C.M. Tools Ltd.)

Plate 3   Two-impression mould for a handle. (Courtesy Bakelite Xylonite Ltd.)

*Plate 4 Single-impression mould for a child's training seat
(Courtesy Bakelite Xylonite Ltd.)*

Plate 5   Four-impression mould for dolls' arms and legs (Courtesy H. B. Sale Ltd.)

Plate 6   Single-impression mould for a café chair (Courtesy Tooling Products (Langrish) Limited: photo by Rowan Studios Limited)

NON-FLAT PARTING SURFACE

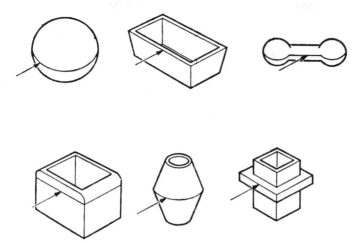

Figure 5.3—Typical mouldings which permit flat parting surface to be adopted

stepped parting surface, the required edge form would have to be die-sunk into each of the two mould halves.

A photograph of a mould which has a stepped parting surface is shown in Plate 3. This mould produces two handles for a refrigerator.

5.3.2 Profiled Parting Surface. An example of the profiled case is shown in Figure 5.4. The moulding is shown at (a). It will be noted that while in cross-section the moulding form is constant, the general form (side view) incorporates curves. As the edge of the component is square with the face (apart from moulding draft) the entire form can be die-sunk into one mould plate. Thus the general form of the parting surface will follow the inside surface of the moulding (b). To simplify the manufacture of a multi-impression mould it is

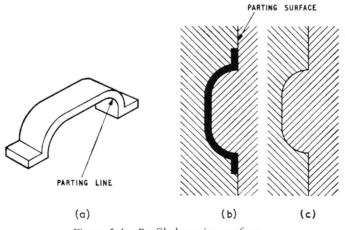

Figure 5.4—Profiled parting surface

143

PARTING SURFACE

often convenient to extend the profiled surface completely across the mould (c). The individual impressions can then be die-sunk as required.

A photograph of a mould which has a profiled parting surface is shown in Plate 4. This mould produces a child's training seat.

5.3.3 Angled Parting Surface. The designer is frequently confronted with a component which, while fairly regular in form, cannot be ejected from the mould if a flat parting surface is adopted. Figure 5.5a illustrates such a case (the component is shown at (c)). However, by adopting an angled parting surface (b) all parts of the moulding are in line of draw and it can therefore be ejected.

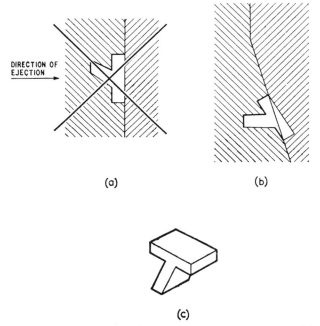

Figure 5.5—Angled parting surface: (a) impracticable design; (b) practicable design; (c) moulding

5.3.4 Complex Edge Forms. So far, we have considered only components which have a constant edge form (that is, either square, double-bevelled or incorporating a radius). Now we proceed to consider other components where the edge form is not constant. This often leads to quite complex parting surfaces but for the present consider the simple case illustrated in Figure 5.6.

The component (Figure 5.6a) in this example is a flat rectangular block, whose sides have a double-bevelled edge, but whose ends are square with the top surface. There are two alternative designs which may be used for this component. The simplest is to adopt a flat parting surface, in which case half of the component's form will be die-sunk into each of the two mould halves. A sketch of the relevant part of one mould plate is shown at (b). The parting line

NON-FLAT PARTING SURFACE

on the component will therefore occur down the middle of the double bevel, and also across the middle of the ends. This parting line (or witness mark) across the ends may not be acceptable to the customer, in which case the slightly more complex stepped parting surface must be adopted.

To obviate the parting line passing across the middle of the moulding it is necessary to raise the level of the mould surface at either end on one mould plate (Figure 5.6d). To accommodate these raised portions, the complementary form must be machined into the other mould plate (c). Now, as the raised portion follows the profile of the top of the component, a simple channel of the form shown is all that is required. The projecting male form (in the mould plate (d)) must, of course, be carefully bedded down into the complementary

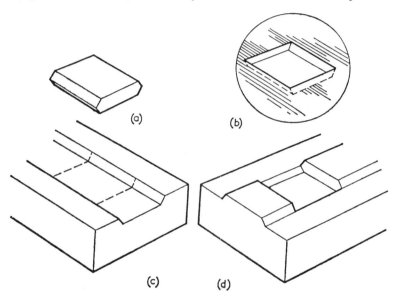

Figure 5.6—Complex edge form—example 1

female form (in the mould plate (c)); otherwise flash will result. The area enclosed by the dotted line (c) will require a higher degree of finish than the remainder of the channel as it forms part of the impression.

From the foregoing it may be concluded that whereas the flat parting surface is the simplest and therefore cheapest to produce, the stepped parting surface, in this case, allows the parting line to be positioned in the most inconspicuous place. Another factor to consider is that the stepped parting line allows for a slight longitudinal discrepancy between the two mould halves. A similar discrepancy with the flat parting surface would be very noticeable on the finished moulding.

The next case we consider is where the edge form continually changes. An example of this is a hairbrush stock (Figure 5.7).

To determine the parting line all we do is draw a number of cross-sections

PARTING SURFACE

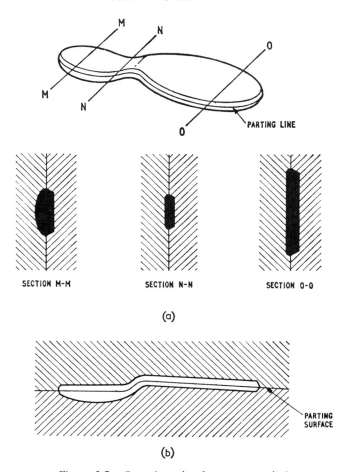

Figure 5.7—Complex edge form—example 2

through the brush stock and decide upon the maximum dimension of each when viewed in the draw direction (i.e. at M, N, O, etc.). The parting line will pass round all these points of maximum dimension. Once the parting line has been determined the mould's parting surface can be drawn as indicated at (b). This parting surface does not vary in the transverse direction and, if desired, it can be machined completely across the mould's surface. In general, however, this surface is confined to the area adjacent to the impression(s) to reduce the time spent in bedding-down, on the undulating surface, to a minimum.

On certain components it may be found that the change from one point of maximum dimension to another is quite abrupt. In these cases the parting line is stepped from one plane to the other, preferably on an angle. It is usually necessary to permit the mouldmaker a certain latitude in blending the impression form in the stepped region to ensure that local undercuts do not occur.

NON-FLAT PARTING SURFACE

Finally, we consider a component which requires a three-dimensional parting surface. A toy figure (Figure 5.8) provides an example of this. The parting line is shown. With such an intricately shaped original it is dangerous to rely on the positioning of a parting line simply from a drawing. It is far better to have a model of the component made and to specify the parting line on this. This will obviate possible costly errors. Photographs of moulds of this type are shown in Plates 5 and 6. These illustrate a four-impression mould for producing dolls' arms and legs, and a mould for a chair.

Figure 5.8—Moulding which necessitates three-dimensional parting surface

5.3.5 Local Stepped and Profiled Parting Surfaces. It is frequently necessary to incorporate a stepped or profiled parting surface to cater for one or two small irregularities in an otherwise regular form. Normally, this is best achieved by localising the change in parting surface to permit the major portion of the surface to be kept flat.

Consider for example the soap case shown in Figure 5.9a. This is basically a box type of component which we discussed in Chapter 2, and a typical mould design is given (Figure 2.4). However, for this particular component we have the complication of the finger-grip cutaways to surmount. This is quite simply accomplished by including a local profiled projection of semi-circular form on the core plate (Figure 5.9b). These projections are accommodated in complementary-shaped pockets die-sunk into the cavity plate.

A cross-section of the combined assembly is shown in Figure 5.9c. Note that the profiled projection is only bedded into the cavity recess for a relatively short distance and that clearance is provided at the ends (at X).

Another example, slightly more complex, is shown in Figure 5.10. This illustrates a box-shaped component with a lug on one of the side walls (a). This component can be moulded in two ways; either by locally stepping the parting surface or by designing a mould of the side cavity type (Chapter 9). The simpler of the two methods is to step the parting surface locally and it should be the one adopted where applicable. A sketch of the cavity and core is shown (b, c), together with a part cross-section through the mould (d). The lug is above the general parting surface (U), which necessitates a projection from the core side to raise the level locally to V. A complementary recess is die-sunk into the cavity plate (b) to accommodate this projection. The projection

PARTING SURFACE

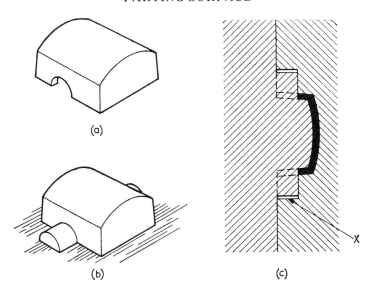

Figure 5.9—Local profiled parting surface

is bedded into the recess on a slight taper. An angle of between 2° and 5° per side, as shown, is suitable. While straight sides are simpler to produce and to bed down, in operation flash develops quickly owing to wear occurring between the two sliding members. It will be noted from the cross-sectional drawing (d) that a clearance is provided between the projection and the recess at the outer end to save an unnecessary fitting operation.

The recess in the cavity plate is locally deepened to level W (Figure 5.10b) to form the lower part of the impression for the lug. The top part is formed by the top surface of the projection V (Figure 5.10c).

A major disadvantage with this design is that a definite step (X) results on the side face of the moulding above the lug. This is due to the necessity of providing opposing draft angles within the impression for extraction purposes. The opposing draft angles are shown in Figure 5.10d as θ and ϕ, respectively.

5.3.6 Balancing of Mould Surfaces. When the parting surface is not flat, there is the question of unbalanced forces to consider in certain instances. This is best illustrated by an example. Figure 5.11 shows a mould with a stepped parting surface. The plastic material when under pressure within the impression, will exert a force which will tend to open the mould in the lateral direction (X). If this happens some flashing may occur on the angled face. The movement between the two mould halves will be resisted by the guide pillars, but even so, because of the large forces involved, it is desirable to balance the mould by reversing the step (Figure 5.11b) so that the parting surface continues across the mould as a mirror image of the section which includes the impression. It is often convenient to specify an even number of impressions

NON-FLAT PARTING SURFACE

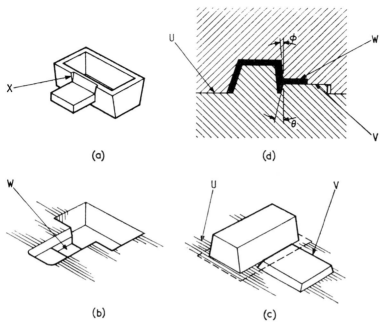

Figure 5.10—*Local stepped parting surface*

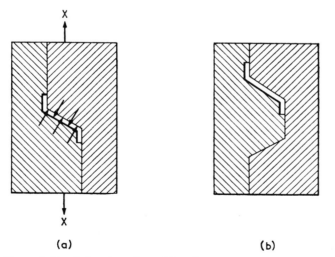

Figure 5.11—*Balancing of mould surfaces: (a) unbalanced; (b) balanced*

when considering this type of mould, as impressions positioned on opposite sides of the mould's centre-line serve to balance the mould (Plate 3).

When balancing is not practicable, due to size, then very sturdy guide pillars must be incorporated.

PARTING SURFACE

5.4 RELIEF OF PARTING SURFACES

We have, up to now, assumed that the parting surfaces of the mould are bedded down over the entire surface. However, this is not practicable, for not only would it be extremely expensive, but it would also affect the efficient functioning of the mould.

Consider the effect of injection pressure and locking force with respect to the area of contact between the two surfaces. The injection pressure, that is the pressure exerted by the plastics melt, is theoretically calculated using the basic hydrostatic formula

$$p = \frac{F}{A}$$

where p = the theoretical injection pressure (N/m² or lbf/in²)
  F = the applied force (N or lbf)
  A = the area of the injection ram (m² or in²)

The actual pressure exerted within the impression will be considerably less than this theoretical value for the following reasons:
  (i) The melt is non-Newtonian.
  (ii) The viscosity of the melt progressively increases as it passes through the mould due to cooling.
  (iii) The actual pressure within the impression depends on the length of the flow path, i.e. sprue, runners, etc.

The calculated value for the injection pressure can, therefore, only be used as a guide to the actual pressure within the cavity. In practice, a value for the effective injection pressure of between 25% and 45% of the theoretical is used, depending on the material and on the wall section of the moulding; i.e. a very fluid material, such as nylon, will transmit a higher pressure than the more viscous melts.

This effective injection pressure is transmitted to the projected area of the impressions, the runners and the gates (the latter being neglected for calculation purposes), and produces a force which tends to open the mould. This tendency to open is resisted by a locking or clamping force. It is generally desirable that the clamping force exceeds the opening force by at least 15%.

Bearing in mind the foregoing, a simple calculation enables the mould designer to calculate the maximum projected area of the impressions and the runners which he can accommodate within the mould. It will be apparent that this calculation depends on the parting surface being perfect for, should flash occur, the force tending to open the mould will be increased since the projected area of the flash must now be taken into consideration. To safeguard against very high opening forces being developed due to flash the parting surface adjacent to the impression and runner is bedded down on a relatively small area. The remainder of the surface in the vicinity is relieved to a depth of at least 2.4 mm (3/32 in).

This small area adjacent to the impression and runner is termed the *land*. The land width, i.e. the distance between the impression and the relief, is normally made between 5 mm (3/16 in) and 25 mm (1 in), depending on the shape

VENTING

and complexity of the impression (see examples in Figure 5.12). Note that a small width of land permits *venting* to be added where required very easily. It is simply a matter of scribing fine grooves across the surface of the land from the impression to the relief area.

Large clamping forces are provided by the machine manufacturers, and it is likely that if small land widths are adopted the effective land area will be insufficient to withstand the applied force and the relatively narrow steel projection will deform. To overcome this possible hazard the land area is increased by ensuring that other areas of the mould face are left proud in places unlikely to be affected by flash, for example at the corners of the mould (Figure 5.12).

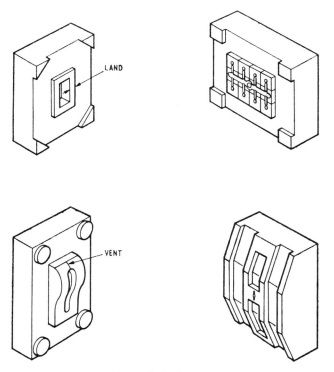

Figure 5.12—Relief of parting surfaces

5.5 VENTING

When plastics material enters an impression air is displaced. Normally the air can escape between the two mating mould plates. However, should the plates have a very fine lapped finish the air may be trapped within the impression so that moulding defects, such as discoloration, sinks, incomplete filling, etc., develop.

It is good design practice to provide vents in the mould to allow air (and other gases when present) to escape freely. It is not usually possible to pre-

PARTING SURFACE

determine where the vent will be required, so the vent is normally machined into the mould plate once the mould has been tried out.

The vent is normally a shallow slot, not more than 0.05 mm (0.002 in) deep by 3 mm (⅛) wide, machined in the land (see previous section). If a greater depth is adopted there is the likelihood of plastics material passing through the slot and the resultant undesirable flash mark being left on the moulding.

Positions where a vent is likely to be required are (i) at the point furthermost from the gate on symmetrical mouldings, (ii) at the point where flow paths are likely to meet and (iii) at the bottom of projections (i.e. blind recesses in cavities). Now the latter case cannot be vented by the surface vents discussed above. It is necessary in this case to provide a vent through the mould plate. This is most conveniently achieved by incorporating an ejector pin in the required position. The minute gap between the ejector pin and the mould plate hole is sufficient to allow the air to escape.

6

Mould Cooling

6.1 GENERAL

One fundamental principle of injection moulding is that hot material enters the mould, where it cools rapidly to a temperature at which it solidifies sufficiently to retain the shape of the impression. The temperature of the mould is therefore important as it governs a portion of the overall moulding cycle. While the melt flows more freely in a hot mould, a greater cooling period is required before the solidified moulding can be ejected. Alternatively, while the melt solidifies quickly in a cold mould it may not reach the extremities of the impression. A compromise between the two extremes must therefore be accepted to obtain the optimum moulding cycle.

The operating temperature for a particular mould will depend on a number of factors which include the following: type and grade of material to be moulded; length of flow within the impression; wall section of the moulding; length of the feed system, etc. It is often found advantageous to use a slightly higher temperature than is required just to fill the impression, as this tends to improve the surface finish of the moulding by minimising weld lines, flow marks and other blemishes.

To maintain the required temperature differential between the mould and plastic material, water (or other fluid) is circulated through holes or channels within the mould. These holes or channels are termed *flow ways* or *water ways* and the complete system of flow ways is termed the *circuit*.

During the impression filling stage the hottest material will be in the vicinity of the entry point, i.e. the gate, and the coolest material will be at the point farthest from the entry. The temperature of the coolant fluid, however, increases as it passes through the mould. Therefore to achieve an even cooling rate over the moulding surface it is necessary to locate the incoming coolant fluid adjacent to 'hot' moulding surfaces and to locate the channels containing 'heated' coolant fluid adjacent to 'cool' moulding surfaces. However, as will be seen from the following discussion, it is not always practicable to adopt the idealised approach and the designer must use a fair amount of common sense when laying out coolant circuits if unnecessarily expensive moulds are to be avoided.

Units for the circulation of water (and other fluids) are commercially available. These units are simply connected to the mould via flexible hoses. With these units the mould's temperature can be maintained within close limits. Close temperature control is not possible using the alternative system in which the mould is connected to a cold water supply.

MOULD COOLING

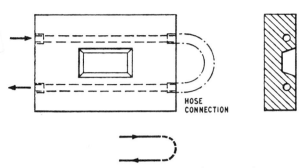

Figure 6.1—Integer cavity plate cooling, simplest circuit

It is the mould designer's responsibility to provide an adequate circulating system within the mould, and this chapter covers some of the different ways this can be achieved. In general, the simplest systems are those in which holes are bored longitudinally through the mould plates. However, this is not necessarily the most efficient method for a particular mould.

When using drillings for the circulation of the coolant, however, these must not be positioned too close to the impression (say closer than 16 mm (⅝ in)) as this is likely to cause a marked temperature variation across the impression, with resultant moulding problems.

The layout of a circuit is often complicated by the fact that flow ways must not be drilled too close to any other hole in the same mould plate. It will be recalled that the mould plate has a large number of holes or recesses, to accommodate ejector pins, guide pillars, guide bushes, sprue bush, inserts, etc. How close it is safe to position a flow way adjacent to another hole depends to a large extent on the depth of the flow way drilling required. When

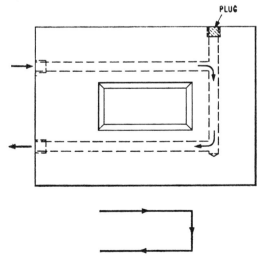

Figure 6.2—Integer cavity plate cooling, U-circuit

COOLING INTEGER-TYPE MOULD PLATES

drilling deep flow ways there is a tendency for the drill to wander off its prescribed course. A rule which is often applied is that for drillings up to 150 mm (6 in) deep the flow way should not be closer than 3 mm (⅛ in) to any other hole. For deeper flow ways this allowance is increased to 5 mm (3/16 in).

To obtain the best possible position for a circuit it is good practice to lay the circuit in at the earliest opportunity in the design. The other mould items such as ejector pins, guide bushes, etc., can then be positioned accordingly.

6.2 COOLING INTEGER-TYPE MOULD PLATES

The temperature of a mould plate of the integer type is controlled by circulating water through holes bored in the plate. The holes are normally interconnected to form a circuit. The circuit may be at one or more levels, the number of which will depend on the depth of the mould plate.

As the circuits for integer cavity plates and integer core plates are generally dissimilar, they will be treated separately.

6.2.1 Cooling Integer-Type Cavity Plate. Let us begin by considering the simplest case, that of a mould plate which incorporates a small, shallow cavity. The simplest approach to a circuit which we can adopt is to drill two flow ways, one on either side of the cavity, and to connect these at one end by means of a flexible hose, adaptors being fitted into the ends of the flow way as discussed in Section 6.5. The above circuit is illustrated in Figure 6.1.

To obviate any form of external connection, the two flow ways can be interconnected internally by means of an internal drilling (Figure 6.2). This forms a U-circuit and it is useful, in particular, for cooling long, narrow cavities.

Instead of using an internal cross drilling, a milled slot may be used in conjunction with a connecting plate. There are two basic designs. Figure 6.3 shows the two drilled flow-way holes interconnected by a channel milled in the side wall of the mould plate to provide a continuous flow path for the

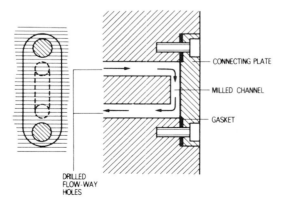

Figure 6.3—U-circuit utilising interconnecting milled channel (Design 1)

MOULD COOLING

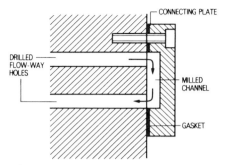

Figure 6.4—U-circuit utilising interconnecting milled channel (Design 2)

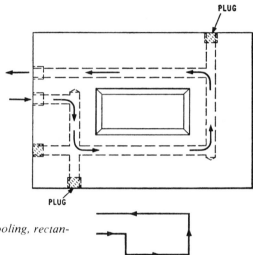

Figure 6.5—Integer cavity plate cooling, rectangular circuit

coolant. A connecting plate is sunk into the side wall and secured by screws, as shown. A gasket is incorporated to prevent leakage of the coolant.

The alternative design is shown in Figure 6.4 and it is similar to the previous design except that the connecting channel, in this case, is machined into the connecting plate, and the latter is attached directly to the side wall of the mould plate (i.e. the connecting plate is not sunk into the side wall). The latter design is, therefore, the cheaper of the two methods but it suffers from the disadvantage that the connecting plate may be disturbed during the mould setting operation and fluid leakage may occur.

The rectangular circuit (Figure 6.5) which incorporates further internal drillings, ensures that the flow ways are close to all four walls of the cavity, allowing a more even temperature control.

For a large-area, shallow cavity, the above circuits are not suitable. Now consider a circuit which can be positioned directly below the cavity. The mould plate is drilled, plugged and baffled (Figure 6.6). This forms a flow

COOLING INTEGER-TYPE MOULD PLATES

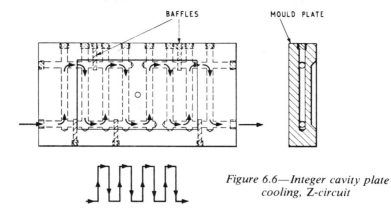

Figure 6.6—Integer cavity plate cooling, Z-circuit

path of a Z-configuration, through which the coolant is circulated. As mentioned previously, the temperature of a coolant progressively increases as it passes through the mould. Thus the cooling effect is going to be more marked on the left side of the mould plate than it will on the right side. This means that a temperature variation across the cavity face will occur with this flow-way design, with possible resulting problems.

A preferred design using basically the same circuit is shown in Figure 6.7. This is called the balanced Z-circuit. Note that the drillings for the left side of the mould plate are diametrically opposite to the drillings on the right side. The coolant inlets are 'A' and 'B' respectively and both drillings are adjacent to the vertical centre line and thus pass close to the sprue entry.

A modified approach is shown schematically in Figure 6.7c. In this circuit design all of the inlet and outlet ports are arranged on the same side to facilitate mould setting.

In the balanced Z-circuit, baffles are necessary to block certain flow ways to provide a continuous circuit without allowing sections to be bypassed and to become 'dead waters'. The baffles should be incorporated in such a manner that they are readily accessible should leakage past them occur.

A correctly and incorrectly fitted baffle is illustrated in Figure 6.8. Note that should leakage occur past the baffle fitted as in (b) the mould plate will have to be removed in order to permit a repair to be carried out. This is not necessary with method (a). In general, when a large number of cross drillings and baffles have to be incorporated in a design, it is good draughting practice to include a schematic drawing of the flow path to assist the mould-maker (see, for example, Figure 6.6).

Now consider the cooling of deep integer-type cavities. For these, a multi-level system is adopted. The circuit on each level is arranged to follow the contour of the cavity as far as possible. For a regular component, such as a box, this usually means that a number of identical circuits of rectangular type are arranged within the mould, one above the other, to provide for the transfer of heat from the walls of the cavity. (See circuits W and X, Figure 6.9.) The final circuit (Y) is normally of a Z-configuration to allow for the transfer of heat

MOULD COOLING

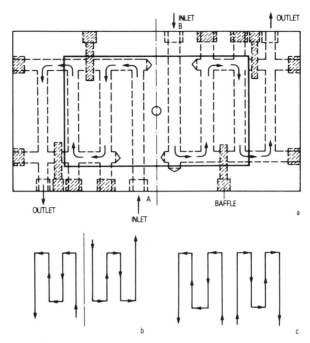

Figure 6.7—(a) Balanced Z-circuit; (b) schematic circuit; (c) alternative balanced Z-circuit

from the base of the cavity. In the example given, the individual circuits are interconnected by axial drillings so as to form one continuous circuit. The advantage of doing this is that the mould has only one inlet and one outlet which simplifies the connection to the supply.

The disadvantage, however, of adopting a continuous circuit for large moulds, is that close temperature control of the mould walls is impossible to achieve. The coolant fluid progressively gains heat during its passage through the mould, therefore opposite mould walls are likely to be at different temperatures. This feature may create moulding problems.

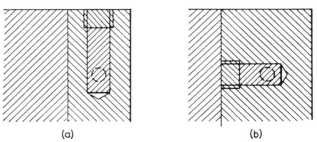

Figure 6.8—Water circuit baffles: (a) correctly fitted; (b) incorrectly fitted

COOLING INTEGER-TYPE MOULD PLATES

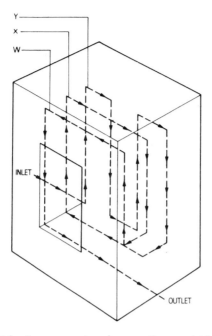

Figure 6.9—Integer cavity plate cooling; multi-level system

By adopting the individual circuit design, that is, each circuit is kept as a separate entity, the moulder can connect up the various circuits to achieve optimum results. To achieve a balancing effect it is desirable that the coolant flow through successive alternate circuit layers is reversed. This feature is illustrated in Figure 6.10.

Another method is to connect internally only certain of the circuits. This is most beneficially achieved in pairs, as shown in Figure 6.10b, which, while achieving balanced counterflow, reduces the number of external connections required.

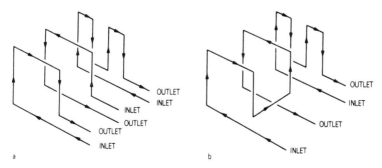

Figure 6.10—Integer cavity plate cooling; multi-level systems: (a) separate circuits; (b) partially interconnected

MOULD COOLING

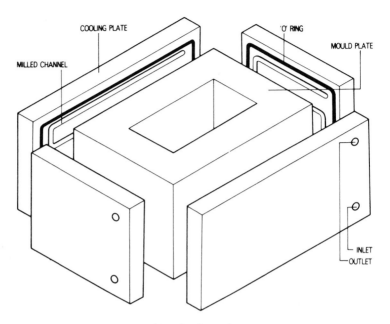

Figure 6.11—Cooling plate system

For irregularly shaped components, particularly those which require a differently shaped circuit on each level, internal connection of the various circuits is often impracticable. In these cases the circuits can be connected externally, using pipe fittings of the conventional type.

If control of the temperature of the individual walls of the cavity is required then the coolant plate method can be considered. In this design *coolant plates* are attached to the side faces of the mould plate and the coolant flows through channels machined into these plates. Drilled holes are therefore dispensed with, apart from the inlet and outlet apertures. An exploded view of the basic design is shown in Figure 6.11.

The individual coolant plates for the four sides are self contained and each has an inlet and an outlet. A typical configuration of the milled channel is shown in Figure 6.12. Note that while the channel cross-section may be kept relatively small to ensure turbulent flow, the length of the flow path can be extensive as required for the specific application. Turbulent flow is necessary to ensure good transmission of heat from the mould to the coolant fluid.

The coolant plate design suffers from the same problems that arise with drilled holes, namely corrosion and build-up of lime deposits. Thus, in operation, both designs become progressively less effective from the heat transfer viewpoint as time passes. It is common practice to clean the channels or holes at reasonable time intervals, and it is with respect to the removal of corrosion and lime deposit that the coolant plate method shows considerable savings in maintenance time. The coolant plates are relatively easy to remove, they are light and can be worked upon independently of the main mould structure.

COOLING INTEGER-TYPE MOULD PLATES

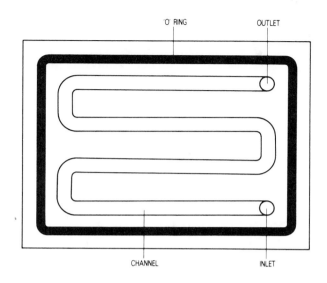

Figure 6.12—Typical configuration of milled channel for cooling plate

Problems associated with the removal of corroded plugs from drilled holes, and in the cleaning of corroded and blocked holes are thereby avoided.

Phosphating or similar protection of the channels or drillings is a sound investment to reduce the maintenance time spent on coolant systems.

Leakage of fluid is prevented either by a gasket or by an 'O' ring as shown in Figure 6.12.

6.2.2 Cooling Integer-Type Core Plate. Providing the depth of the core is fairly shallow (under 25 mm (1 in)) the Z-type single-level system can be adopted, the water ways being situated beneath the core in a manner similar to that for integer cavities discussed in the preceding section. For deeper cores, however, the single-level circuit is not sufficient to permit the coolant to transfer heat away from the core surface fast enough. Some arrangement must, therefore, be made to permit the circulation of coolant inside the core. There are several alternative ways of doing this and the method adopted will be determined to some extent by the actual shape of the core.

ANGLE HOLE SYSTEM (Figure 6.13). Flow ways are drilled at an angle from the underside of the core plate so that they interconnect at a point (X) relatively close to the surface. Each of these drillings is plugged. The inlet (Y) and outlet (Z) holes are drilled from either side of the mould and break into the angled waterways, as shown.

BAFFLED-STRAIGHT HOLE SYSTEM. In this system (Figure 6.14) holes are bored at right angles to the rear face of the core plate. The lower end of each

MOULD COOLING

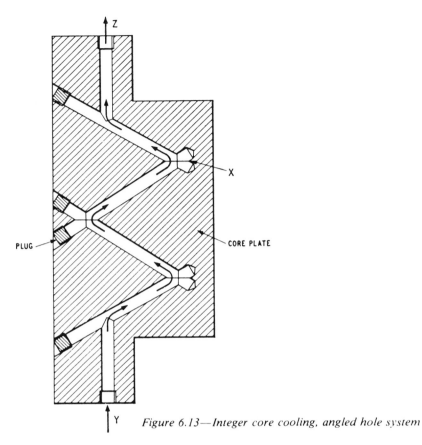

Figure 6.13—Integer core cooling, angled hole system

boring is plugged. The borings are interconnected by a hole (X) drilled from a side face. To ensure the coolant passes down each hole, baffles are fitted as shown.

STEPPED CIRCUIT. To obtain cooling channels which are positioned fairly close to the top surface of the core, the stepped circuit can be considered. In this system (shown in Figure 6.15) holes (X) are drilled through the side wall of the core, parallel to the core face. These holes must be very carefully plugged and finished as they form part of the impression. Badly fitted plugs on a moulding surface cause considerable moulding difficulties, and for this reason this particular design is not favoured by many designers. The stepped configuration of drillings, as shown, is necessary to provide a suitable inlet and outlet connection position.

The above three circuits can be used singly, as shown, for fairly narrow cores. For wider cores a number of identical circuits can be incorporated, positioned at suitable intervals along the core. The individual circuits can be internally or externally connected, whichever is the more convenient.

COOLING INSERT-BOLSTER ASSEMBLY

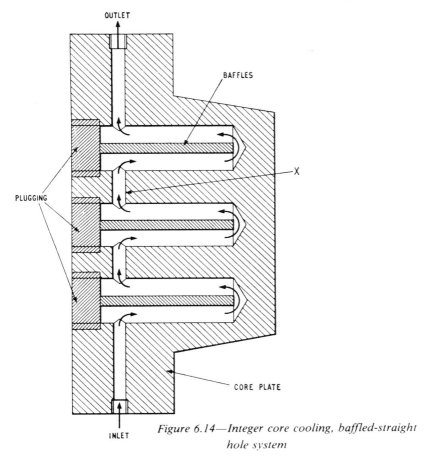

Figure 6.14—Integer core cooling, baffled-straight hole system

MULTIPLE CIRCUITS. Examples of typical multiple circuits are shown schematically in Figure 6.16. The first drawing (a) illustrates the angled hole system. Note that the six individual circuits have been connected as two sets of three. This feature permits a balanced temperature gradient across the core to be achieved. Figure 6.16b shows a pair of baffled hole circuits coupled together, while Figure 6.16c illustrates a multiple stepped circuit.

6.3 COOLING INSERT-BOLSTER ASSEMBLY

We will discuss cooling of the insert - bolster assembly under two headings, (i) cooling the bolster, and (ii) cooling the insert. The latter is further classified depending upon which type of insert is to be cooled (cavity or core).

6.3.1 Cooling Bolster. In moulds, constructed on the insert-bolster principle, where the depth of the impression is relatively small, the circulation of the coolant is often confined to the bolster. The designer relies on the

MOULD COOLING

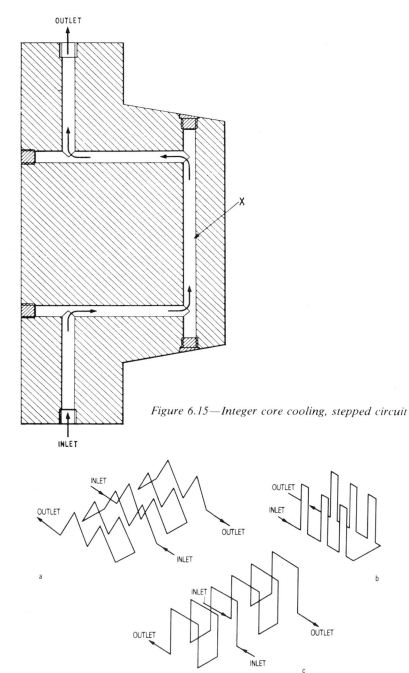

Figure 6.15—Integer core cooling, stepped circuit

Figure 6.16—Multiple circuits: (a) angled hole system; (b) baffled hole system; (c) stepped circuit

COOLING INSERT-BOLSTER ASSEMBLY

reasonably good thermal conductivity of steel to allow the heat to be rapidly transferred from the impression as required. Even better results can be achieved using a material with a higher thermal conductivity, such as beryllium-copper, for the insert.

The method adopted for cooling the bolster is identical to that described for cooling the integer cavity block (Section 6.2.1). That is, holes are drilled through the bolster and are interconnected, either externally or internally, to permit the circulation of a coolant.

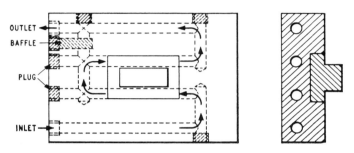

Figure 6.17—Bolster cooling, Z-circuit

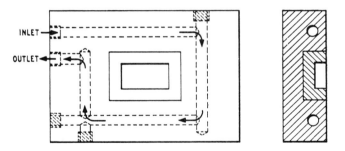

Figure 6.18—Bolster cooling, rectangular circuit

It is desirable that these flow ways are positioned as close to the insert as practicable. For a shallow depth of insert, the holes may be situated directly below the insert (Figure 6.17). A Z-type layout is normally adopted. The alternative method is to arrange holes close to the sides of the insert (Figure 6.18). In this case the rectangular type of circuit is used. For deeper inserts, a multi-level system is desirable. This is simply a combination of both the above layouts (Figure 6.10b).

6.3.2 Cooling Cavity Inserts. The method adopted for cooling cavity inserts depends, to some extent, upon the shape of the insert. This can broadly be classified as either rectangular or circular. (See Section 2.2.2.) The circulation of fluid within the insert is easily achieved, but a complication exists in that the flow way cannot be drilled into the insert from the bolster without incorporating some form of seal to prevent leakage.

MOULD COOLING

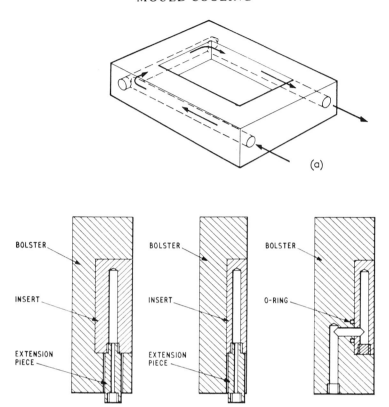

Figure 6.19—Insert cooling: (a) U-circuit; (b-d) alternative methods of connecting circuit to supply

COOLING RECTANGULAR INSERTS. A typical rectangular insert is shown (Figure 6.19). Now, while the simplest of circuits only is illustrated, more complicated single- and multi-level circuits can be adopted. This will be determined by the shape and the depth of the cavity. All drillings within the cavity insert should be interconnected, plugged, and baffled, so as to necessitate the minimum of external couplings. The designer should aim to have only one inlet and one outlet per insert.

The mould setter can more quickly set up a mould for production if the supply and return lines can be attached directly to adaptors which project from the side walls of the mould. This means that the designer should provide some positive connection between the insert and the outside of the mould. Three alternative methods are illustrated in Figure 6.19.

(i) An extension piece is screwed directly into the insert through a suitable clearance hole in the bolster (Figure 6.19b). This method has one disadvantage in that the connection between the extension piece and the insert must be broken each time the insert is removed from the bolster.

166

COOLING INSERT-BOLSTER ASSEMBLY

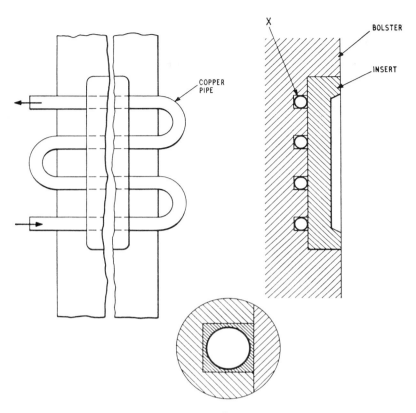

Figure 6.20—Insert cooling, copper pipe system

(ii) An extension piece is screwed directly into the insert through a suitable slot machined in the bolster (Figure 6.19c). This design overcomes the disadvantage stated in (i) but the slot does constitute an open crevice which may become a material trap if the mould flashes. Build-up of plastic material in this trap is difficult to remove and can damage the opposite face of the mould, if allowed to become excessive. However, providing the slot is well clear of the impression, this feature does not constitute an undue hazard.

(iii) Providing the depth of the bolster is sufficient, drillings can be made through the bolster below the insert and the two systems interconnected by cross-drillings (Figure 6.19d). Leakage between the insert and the bolster is prevented by an O-ring round each cross-drilling, suitably accommodated in a recess in the bolster as shown. This design does not suffer the disadvantages associated with the previous methods and is therefore the preferred design.

Another method of cooling inserts is by the use of copper pipes. This design is suitable when mould space is restricted. Pipes can be laid in channels which may be immediately adjacent to other holes in the mould plate without the possibility of leakage occurring.

A number of channels (X) are machined through the bolster (Figure 6.20).

MOULD COOLING

A copper pipe is bent to a suitable shape so that it can be placed into the channels (see plan view). The insert when fitted to the bolster is in direct contact with the pipe. To increase the contact area and to improve heat transfer, the space around the pipe can be filled with a low-melting-point alloy (see inset), or a square-section pipe can be used.

COOLING CIRCULAR INSERTS. The drilling methods discussed for rectangular inserts cannot normally be adopted for cooling circular inserts due to space limitations. However, because the insert has a circular form, an annular groove can be incorporated quite simply.

Most designs for direct cooling of circular cavity inserts are based upon this principle.

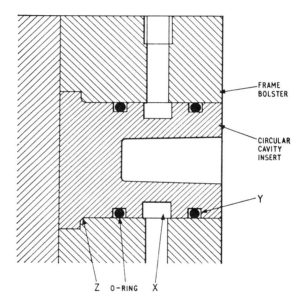

Figure 6.21—Insert cooling, coolant annulus method

Consider Figure 6.21. The circular cavity insert is shown fitted in a standard type of frame bolster. A coolant annulus (X) is machined on the periphery of the insert, and additional grooves (Y) provided above and below the coolant annulus to accommodate O-rings. When fitted to the bolster, these O-rings prevent leakage of fluid between the insert and the bolster. Some care must be exercised to prevent the O-ring being damaged when the insert is fitted. A lead-in on the bolster hole at Z facilitates this operation.

The annulus is connected to the supply and return line via drillings through the bolster. For multi-impression moulds, the inserts can be positioned in lines so that a vertical drilling interconnects each annulus to form a continuous circuit (Figure 6.22).

COOLING INSERT-BOLSTER ASSEMBLY

Instead of machining the coolant annulus into the periphery of the cavity insert, or coolant sleeve, the annulus can be incorporated as a groove machined into the mould plate as shown in Figure 6.23. The main object of this approach is to lay out the impressions so that the individual grooves interconnect and thereby avoid the necessity of additional drilled holes. While drilled holes are simple to incorporate for inserts arranged in line (see Figure 6.22), the holes are more difficult to incorporate to interconnect inserts which are arranged on a pitch circle diameter. It is for this latter layout that the interconnecting groove design is often adopted.

A plan view of a mould plate in which inserts are arranged on a pitch circle diameter, and located so that the individual grooves interconnect, is shown in Figure 6.24. To provide for a definite flow path for the coolant, a gap must be left between two of these grooves as shown at 'X'. These latter grooves are connected to the inlet and outlet apertures respectively.

To achieve this design of interconnecting grooves, a large amount of steel must be removed from the central region of the mould plate and this will have a considerable weakening effect. Therefore if this design is adopted the designer must ensure that the mould plate is adequately supported by the backing plate.

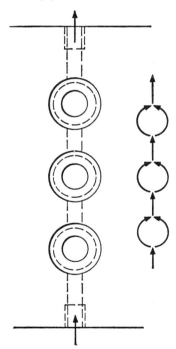

Figure 6.22—When inserts are arranged in line, one drilling can interconnect external cooling annulus of each insert

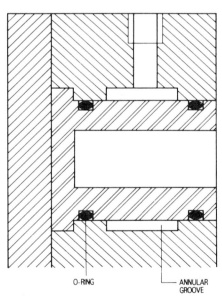

Figure 6.23—Insert cooling, interconnecting groove design

MOULD COOLING

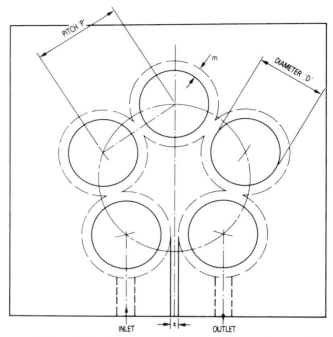

Figure 6.24—Interconnecting groove design, layout of impressions

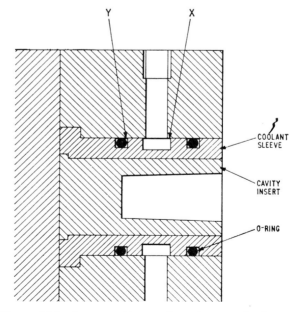

Figure 6.25—Insert cooling, coolant sleeve method

COOLING INSERT-BOLSTER ASSEMBLY

The relevant equations for determining the pitch circle diameter (*PCD*) and the pitch for the inserts are given below:

$$P = D + (3/2)m \qquad (6.1)$$

$$PCD = \frac{P + (X/n)}{\sin(180/n)} \qquad (6.2)$$

where P = pitch
D = diameter of inserts
m = depth of groove (see Figure 6.24)
n = number of impressions
X = space required between the grooves (see Figure 6.24).

Example. Determine the pitch and pitch circle diameter for the interconnecting groove design, given the following information:
diameter of insert, D = 25 mm (1 in)
gap between inlet and outlet grooves, X = 3 mm (1/8 in)
number of impressions, n = 6
depth of groove, m = 3 mm (1/8 in).

Solution. Use equations (6.1) and (6.2):

SI units	*Imperial units*
$P = 25 + \dfrac{3 \times 3}{2}$	$P = 1 + \dfrac{3 \times 0.125}{2}$
= 29.5 mm	= 1.187 in
$PCD = \dfrac{29.5 + (3/6)}{\sin(180/6)}$	$PCD = \dfrac{1.187 + (0.125/6)}{\sin(180/6)}$
= 60 mm	= 2.416 in

An alternative design is shown in Figure 6.25. A plain cavity insert is fitted into a coolant sleeve which incorporates both the coolant annulus (X) and the O-ring grooves (Y). This design has the advantage over the standard coolant annulus method (Figure 6.21) in that once the sleeves are fitted the O-rings need not be disturbed. The plain inserts can be removed and replaced without affecting the coolant system.

6.3.3 Cooling Core Inserts. Systems evolved for the efficient cooling of mouldings produced from the insert-bolster design, normally involve passing a coolant fluid directly through channels or holes incorporated within the body of the insert. The design adopted depends to a large extent upon the size and shape of the insert. Cooling the core for a large shallow box, for example, will require an approach quite different from that for cooling a long, small-diameter core for a pen barrel.

COOLING SHALLOW CORE INSERTS. Once the designer has decided not to rely on conducting the melt heat away from the core to rather remote holes drilled

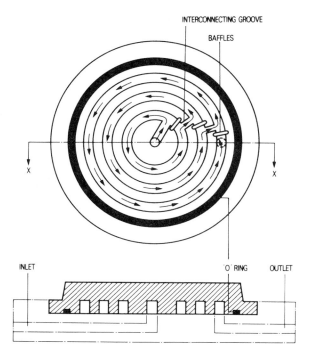

Figure 6.26—Cooling shallow inserts, circular milled groove design

in the bolster (see Section 6.3.1) he must then alternatively consider incorporating holes or channels directly into the core insert. One method for doing this was discussed in Section 6.3.2 under the heading 'cooling cavity inserts'. The method involved drilling holes using a basic 'U' circuit configuration. Useful variations on this approach use either the 'Z' or 'balanced Z' designs.

An alternative design which can be adopted for cooling shallow core inserts is the 'spiral circuit'. This design basically consists of a channel machined into the rear face of the core insert in the form of a spiral. Unfortunately in practice the spiral form is both difficult and expensive to produce, therefore compromise 'spirals' are normally adopted. Examples of these are shown in Figure 6.26, for cooling a large round insert, and in Figure 6.27, for cooling a large rectangular insert respectively.

In the first example a series of concentric grooves are machined, and these are interconnected via channels, as shown; the flow path is established by suitably positioned baffles. The inlet and outlet holes are machined into the bolster as indicated by the chain dotted lines. Note that the O-ring encompasses all grooves and thereby prevents leakage of the coolant fluid.

The second example, using a basic rectangular layout, is shown in Figure 6.27. The drawing shows the underside of the core insert only. In this design one continuous flow path is adopted without baffles. Alternatively a similar arrangement to that adopted for the round insert (Figure 6.26) may be used.

COOLING INSERT-BOLSTER ASSEMBLY

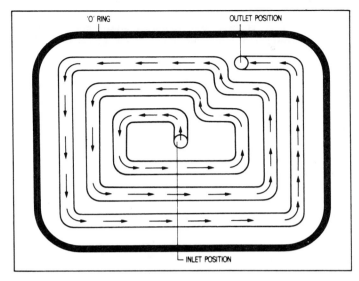

Figure 6.27—Cooling shallow inserts, rectangular milled groove design

COOLING DEEP CORE INSERTS. Only by circulating the coolant fluid deeply inside the core insert can efficient transfer of heat from the core surface be achieved. There are many alternative arrangements for cooling deep core inserts, and a number of these designs will be covered in the following discussion.

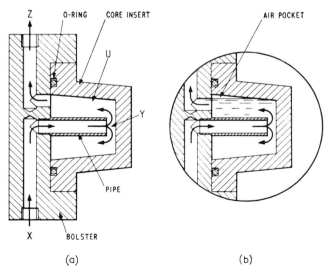

Figure 6.28—Core insert cooling, design 1: (a) deep chamber system; (b) incorrectly positioned outlet hole

173

MOULD COOLING

(i) *Deep chamber design.* In the first example shown in Figure 6.28 the rear face of the core insert is recessed to form a deep chamber (U). This chamber is normally circular for ease of machining. The insert is firmly held down on to a flat face at the base of a pocket machined in the bolster by screws (not shown). Leakage between the two surfaces is prevented by an O-ring fitted into a groove. In operation the chamber is completely full of water. The incoming coolant passes from the inlet (at X) through the internal drillings and pipe to impinge on the centre of the chamber (at Y). On single-impression moulds this is likely to be the hottest part of the core insert as it is directly opposite the sprue.

While being the cheapest of the deep cooling methods to incorporate, the deep chamber design suffers from two major disadvantages:

(a) The flow rate of the coolant drops markedly as it enters the chamber. This means that the required turbulent flow is not achieved in this region and that the transfer of heat to the coolant is less effective.

(b) It is possible by incorrect design or incorrect toolmaking for an air pocket to be formed at the top of the chamber as shown in Figure 6.28b. An uneven temperature profile, with associated moulding problems, will result. The air pocket is created by the incorrect positioning of the outlet port (Z) in relation to the chamber. It is essential that this port is always situated at the highest point of the chamber when the mould is mounted on the injection machine. (Compare Figures 6.28a and 6.28b.) It is for this reason that moulds incorporating the deep chamber design should be engraved with information as to which way the mould should be mounted on the machine.

The design can be usefully used for moulds in which the production rates are not important and where mould costs must be kept to a minimum. The deep chamber design forms the basis of more efficient designs, examples of which are discussed in Sections (ii) and (iii), which follow.

(ii) *The deep chamber design with central support.* This system is illustrated in Figure 6.29. The support feature is provided by a central column which can be integral with the bolster, as shown, or be a separate member. Obviously if the depth of the chamber necessitates a column which is relatively long, it is preferable to use the latter design.

The design has two primary objectives:

(a) To support the central region of the core against possible deflection.

(b) To have the central region solid to permit a valve type ejector element to be incorporated. Note that if the latter design is used in conjunction with a separate central column then an additional O-ring must be incorporated to avoid fluid leakage past the stem of the valve.

The disadvantages which applied to the deep chamber design, apply to this design as well. That is, the flow rate drops as it enters the annulus, and air pockets may be formed. If a large diameter ejector valve is incorporated, with its own coolant system (see Section 6.4.2) then the results of the above disadvantages are lessened. This is because an efficient coolant circulation system is incorporated at the point where it is required, that is, at the hottest part of the core, namely the front surface.

COOLING INSERT-BOLSTER ASSEMBLY

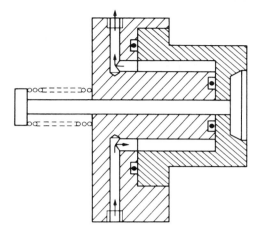

Figure 6.29—Core insert cooling, design 2; deep chamber with central support

(iii) *The helical channel design.* The more efficient design than either of the two previous systems is the helical channel design, which is illustrated in Figure 6.30. It is a more costly design, however, from the mould manufacturing viewpoint.

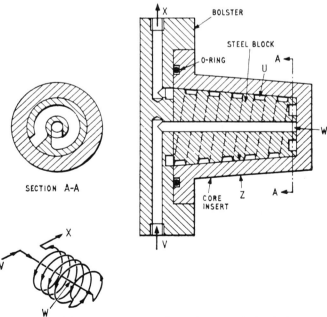

Figure 6.30—Core insert cooling, design 3; helical channel system

175

MOULD COOLING

In this design the deep chamber is plugged with a close fitting block of steel or brass (the latter being preferable from the corrosion resistance and block removal viewpoints). A helical channel (U) is machined into the outside of this plug. The internal drillings through the bolster and plug are arranged so that the coolant is directed from the inlet (V), through a central drilling (W), across the face of the plug (see cross-sectional drawing A-A) and so to the outlet port (X) via the helical channel. The core insert is fitted into a bolster of solid type and O-rings are fitted to prevent leakage.

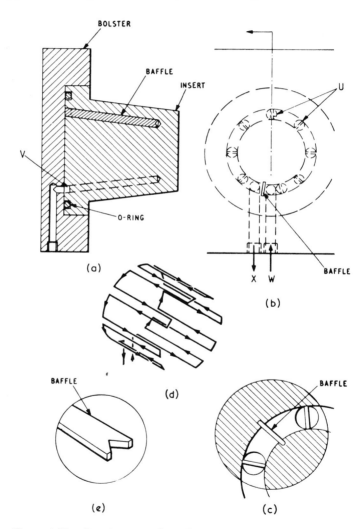

Figure 6.31—*Core insert cooling, design 4; baffled hole system: (a) cross-sectional view; (b) part plan view; (c) part view on underside of core insert; (d) schematic circuit; (e) top view of baffle*

COOLING INSERT-BOLSTER ASSEMBLY

This design ensures that the coolant follows a precise path and no 'dead waters' are possible. The plug forms a positive support to the insert shell and, therefore, the wall section of this shell (Z) can be less than for either of the previous designs. A more rapid transfer of heat from the moulding is, therefore, achieved as the coolant passes relatively close to the impression.

(iv) *Baffled hole system.* A completely different approach is illustrated in Figure 6.31. This design utilises a system of baffled holes (U). The holes, drilled into the rear face of the insert, may either be at right angles to the base or be parallel to the outside wall of the core. (The latter design is illustrated.) The diameter of the hole is normally in the range 13 mm (½ in) to 25 mm (1 in), depending on the size of the insert. To provide a flow path for the coolant the individual holes are interconnected by an annulus (V) which is machined into the base of the insert. A baffle is fitted into an end-milled slot which is machined at right angles to the annulus (Figure 6.31c). The inlet (W) and outlet (X) drillings through the bolster are situated on either side of this baffle (see plan view, Figure 6.31b). To ensure that the coolant circulates down each individual hole, baffles must be fitted into each. The baffle, the top end of which is shown in Figure 6.31e, is usually made of brass. The actual flow path of the coolant is shown in Figure 6.31d. The core insert is fitted into a bolster of solid type and an O-ring incorporated to prevent leakage. Each baffle must be flush with the rear face of the insert, to prevent the hole being bypassed.

The designs discussed so far in this section have been mainly applicable to relatively large core inserts. The following design examples offer alternative methods for cooling the smaller insert, particularly on multi-impression moulds.

(v) *Baffled hole system for small inserts.* In the design illustrated in Figure 6.32 the impressions are arranged in line(s), and the inserts, being circular, fitted into a frame type of bolster. Each insert incorporates a chamber (U) which is in alignment with a drilling in the bolster. To prevent leakage of coolant a small O-ring is fitted in a recess below each insert. The individual drillings are interconnected by a hole (V) drilled completely through the mould. The lower end of this hole is the inlet (W), and the top end becomes the outlet (X). To ensure the coolant passes down each chamber, baffles are necessary. The baffles are mounted in each insert chamber at right angles to the main drilling (see plan view). Note that the lower end of the baffle incorporates a radius to match that of the main drilling.

The flow path of the coolant is shown in Figure 6.32b. As the coolant progressively gains heat as it passes through the mould, this design is not efficient for cooling more than three or four impressions.

(vi) *The bubbler system.* The design is basically the same as the deep chamber design (i), suitably adapted for small inserts. A relatively small diameter hole is machined into the rear face of the insert as shown in Figure 6.33b. A

MOULD COOLING

'bubbler' pipe is fitted in the backing plate and protrudes into this hole, as shown, thereby forming an annulus. Suitable inlet and outlet holes are drilled in the backing plate.

One type of circuit for which this system can be used is illustrated. The

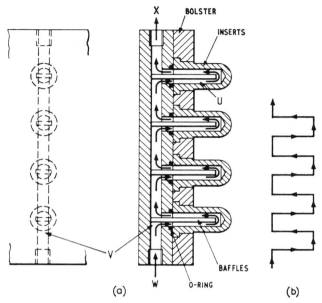

Figure 6.32—Baffled hole system for cooling small core inserts

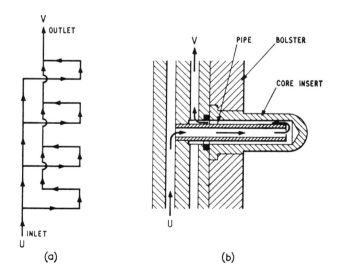

Figure 6.33—The bubbler system for small core inserts

COOLING OTHER MOULD PARTS

coolant passes from the inlet hole 'U' up the inside of the bubbler pipe, and then down the outside, into the outlet hole, 'V'. A schematic drawing of the complete circuit is shown in Figure 6.33a. Note that the temperature of the coolant is approximately the same in each insert as they are all connected in the same way.

(vii) *The coolant annulus system* (Section 6.3.2) for cooling cylindrical cavity inserts is also applicable for cooling cylindrical core inserts. The design details are identical and will therefore not be repeated.

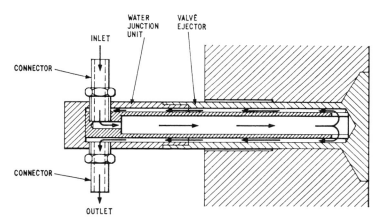

Figure 6.34—Cooling valve ejector elements; bubbler system

6.4 COOLING OTHER MOULD PARTS

6.4.1 Other Mould Plates. On multi-plate moulds it is necessary to consider the cooling of other mould plates in addition to that of the primary cavity and core plate. In particular, the stripper plate in a stripper plate mould, and the feed plate in a mould of the underfeed type. Separate control of the temperature of these plates is necessary to achieve the optimum production cycle.

In general, the maintenance of temperature of these plates is achieved in a manner identical to that described for cavity plates of integer type (Section 6.2.1). Flow ways are drilled and interconnected so that a coolant can be circulated through the plate.

6.4.2 Cooling Valve-Type Ejectors. The valve type of ejector normally forms a relatively large part of the surface of the impression. It is desirable, therefore, to provide facilities for the dissipation of heat from this component.

Several designs are illustrated. In the first (Figure 6.34) a bubbler system is adopted. The stem of the valve ejector is bored to accommodate a water junction unit. The connectors are coupled to the supply and return lines via flexi-

MOULD COOLING

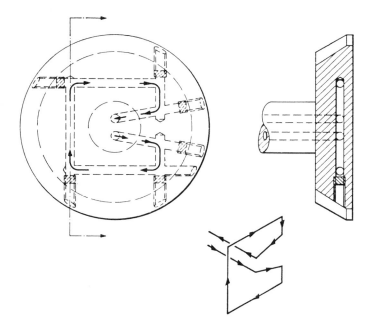

Figure 6.35—Cooling valve ejector elements; cooling circuit in valve head, design 1

ble hoses to allow for the ejector valve movement. The coolant passes via the inlet down the centre of the pipe, and back to the outlet via the outside of the pipe. This is the simplest method of cooling the valve-type ejector, particularly if a commercial water junction unit is used.

If it is desired to pass the coolant through the head of the valve ejector a more complex circuit is necessary. Figure 6.35 illustrates a valve, the head of which is drilled and plugged to form a square type of circuit. The coolant, to and from the head, passes through suitable drillings in the stem. Another design is shown in Figure 6.36. In this the head is designed as a two-part construction: channels are machined into one part as shown at (a). Baffles are fitted into end-milled slots. The coolant, to and from the head, passes through suitable drillings in the stem as in the previous case.

6.4.3 Cooling The Sprue Bush. A relatively large bulk of plastics material is contained in the sprue, which must be cooled during each cycle to a temperature at which it is sufficiently solid to allow for its removal from the mould. It is therefore desirable to incorporate a separate sprue bush cooling circuit so that heat can be transferred from this member as efficiently as possible.

The shape of the sprue bush is similar to that of circular inserts, so the methods illustrated for cooling circular inserts (Figure 6.21 and 6.25) are equally applicable for cooling the sprue bush.

WATER CONNECTIONS

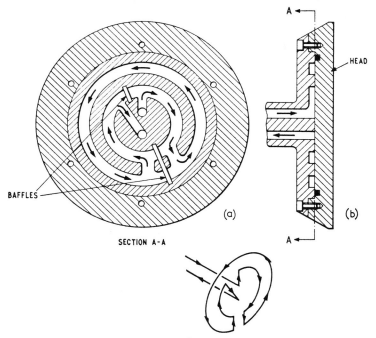

Figure 6.36—Cooling valve ejector elements; cooling circuit in valve head, design 2

6.5 WATER CONNECTIONS

6.5.1 Adaptors. As mentioned, the majority of moulds are drilled to provide a flow path through which a coolant can be circulated. These drillings are connected to the supply and return lines via adaptors. The adaptor is a standard pipe fitting which can be obtained in several sizes (Figure 6.37). At the time of writing the British Standard Pipe (BSP) thread has not been superseded by an SI equivalent for general use in the mould making industry. Two alternative methods of fitting the adaptor are shown. In method (a) the tapping hole size for the adaptor governs the diameter of water way used. This permits the minimum size of adaptor to be used for a particular diameter of water way. As the minimum spacing between two adjacent waterways depends on the space required to insert and remove the adaptor (using an appropriate spanner) the smaller adaptor is useful. However, this method has the disadvantage that the flow of the coolant is restricted by the relatively small-diameter hole in the adaptor.

In the second method (b) an adaptor is chosen which has an internal bore which closely matches that of the water way. A convenient size for a water way is 10 mm in diameter and from the chart it will be seen that either a ¼ BSP or a ⅜ BSP adaptor is applicable depending upon which method is adopted. For large mould plates, where the individual water ways are not positioned too close together, method (b) is the preferred design.

MOULD COOLING

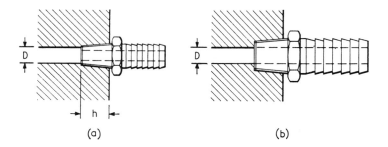

(a) (b)

Adaptor (BSP)	Tapping drill size (in)	Depth of threaded hole		Diameter of flow-way			
				Method A		Method B	
		(mm)	(in)	(mm)	(in)	(mm)	(in)
1/8	11/32	11	7/16	8	11/32	—	—
1/4	29/64	13	1/2	11	7/16	7	1/4
3/8	19/32	13	1/2	15	19/32	10	3/8
1/2	47/64	14	9/16	18	23/32	13	1/2

Figure 6.37—Adaptors for coupling mould water circuit to supply

An alternative design for an adaptor is to utilise a piece of copper pipe suitably threaded at one end. The pipe is screwed into the mould using a pipe fitting wrench. Here again it is necessary to ensure that there is sufficient space between the water ways to allow the pipe to be screwed in.

6.5.2 Quick Connection Adaptors. The disadvantages of the fixed adaptor design is two-fold.

(i) The rubber hose must be connected and disconnected each time the mould is set on the machine.

(ii) The adaptor projects a considerable distance from the side of the mould.

The first point results in an extension of the setting time, while the second point makes the mould setting just a little more difficult in that the projecting adaptors tend to get in the mould setter's way.

To overcome these disadvantages various quick connecting adaptors have been designed and are commercially available so as to minimise mould setting time. Those shown in Figure 6.38 are from the DME (USA and Europe) range. There are two types in this range called 'Jiffy-tite' and Jiffy-matic' respectively. Both types consist of two parts: the first part, shown on the left of the drawing, is the 'connecting plug' and it is screwed into the mould like a conventional adaptor. The second part, termed the 'socket', is attached to the rubber hose. By a suitable push action the two parts can be connected or disconnected very simply. The 'Jiffy-matic' has the additional advantage in that as soon as the socket is disconnected, the coolant flow ceases immediately. With the 'Jiffy-tite' system the coolant flow must be shut off at the source.

WATER CONNECTIONS

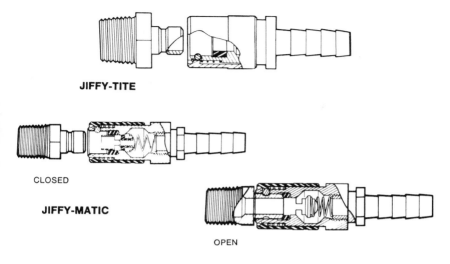

Figure 6.38—Quick connection adaptors (Courtesy DME (USA and Europe))

The connecting plug (with either type) may be sunk below the surface of the mould wall, so that it does not protrude above the relevant surface. This design feature prevents the plug from being damaged during the setting operation.

6.5.3 Position of Water Connections. The position of all inlet and outlet holes, should, whenever practicable, be positioned either at the base or at the rear of the mould. Connections at the front tend to get in the operator's way. With connections at the top of the mould, should leakage occur, water may get on to the polished face of the impression.

It is desirable that the position of all water connections (this includes holes which are interconnected externally) is such that the associated adaptors and hoses do not interfere with the bolting of the mould on to the machine.

6.5.4 Plugs. As discussed previously, plugging the ends of certain holes is necessary to form a continuous circuit. Commercial plugs are used and these are available in the tapered BSP range. The tapping size and length of thread h is the same as that indicated for adaptors (Figure 6.37). The plug has a square or hexagonal projection (or depression) so that a suitable key or wrench can be used to ensure a leakfree joint when the plug is screwed into the mould. By convention, plugs are shown with a dotted cross-hatch on the plan view of mould drawings (Figure 6.17). This simplifies the tracing of the required coolant flow path.

7

Procedure for Designing an Injection Mould

7.1 GENERAL

The major difficulty which beginners normally experience is the decision of where to start and the general procedure to adopt when designing a mould. This chapter is included as an introductory guide to permit the reader to follow a definite approach in order to establish a sequential technique of his own.

In practice a mould drawing normally comprises three views: that is, a plan view of the moving half; a plan view of the fixed half; and a side sectional view of both mould halves. This 'general arrangement' drawing should contain sufficient information to permit a draughtsman to make detailed drawings of each individual mould part. For the novice, however, it is sufficient to draw only two views, and the views normally chosen are the plan view of the moving half and the side sectional view of both mould halves.

In the following drawings the evolution of a typical design is given in seven separate stages consisting of 100 operations. The sequence starts at the primary positioning of the impression and finishes with the checking of the completed drawing. The operation number given in both the drawings and the notes indicate the procedural sequence of the operations.

STAGE A. *Primary positioning of inserts.* The object of the primary positioning stage is to determine the position of the impression with respect to the major horizontal and vertical centre-lines. During this stage, the size and position of the cavity and core inserts are established in both plan and sectional views.

STAGE B. *The ejector system.* During this design stage the size and position of the ejector pins and push-back pins are established. The overall sizes of the ejector plate and retaining plate are also determined. Before commencing to position the various ejector elements, however, it is prudent to consider the type of fluid circulation system that is most applicable for controlling the temperature of the mould. For this basic design, a parallel drilled hole system is adopted, with two holes being incorporated in each half, the holes passing on either side of the mould insert. This preliminary consideration regarding the position of the fluid circulation system is essential to prevent difficulties being experienced later in the design.

STAGE C. *The ejector grid.* During this design stage the drawing progresses

PROCEDURE FOR DESIGNING AN INJECTION MOULD

to include detail of the ejector grid in both plan and sectional views. Further details, such as the ejector rod and ejector rod bush, are added to the ejector system.

STAGE D. *Complete the top half.* The fluid circulation system and the guide pillars are added to the plan view which permits the final outside shape of the mould to be determined and drawn. In the sectional view the sprue bush drawing is completed. In addition, various items such as the feed system, sprue pin, bridge piece and register plate details are added.

STAGE E. *Complete the plan.* The lower half of the plan view can now be drawn as a mirror image of the top half. The runner and sprue pin details are added. Once the plan view is complete the *section cutting plane* can be decided upon. It is essential that the cross-sectional view illustrates each component part of the mould. Thus, as the cross-section already shows such items as support block, push-back pin, ejector pin and other centre-line details which have been incorporated in previous stages, the only remaining items to be included are the guide pillars and bushes, and the position of the fluid holes. Note that there is no advantage in taking a section through two identical parts. To do so is a complete waste of draughting time.

STAGE F. *Complete the cross-sectional view.* During this design stage the cross-sectional drawing is completed. Each part-section is considered independently in relation to the plan view to ensure that all relevant component parts of the mould are included and are drawn correctly. This will mean that the lower part of the cross-sectional drawing is completed in the process.

STAGE G. *Complete the drawing.* This final stage involves cleaning up the drawing, erasing unwanted lines, checking, cross-hatching and indicating those parts of the mould that are to be secured together.

In the following sequence of operations, the letter adjacent to the *operation number* indicates either plan view (P) or sectional view (S).

7.1.1 The Problem. Design a two-impression injection mould to produce the pin box shown in Figure 7.1. The box is to be manufactured in high-impact polystyrene. The drawing should include a plan view of the moving half and a side-sectional view of both halves.

7.2 STAGE A: PRIMARY POSITIONING OF INSERTS

Operation number	Operation
1	Divide the drawing paper into two parts. The left side is for the plan view of the moving half, the right side is for the cross-sectional view of both halves.
2(P&S)	Draw a horizontal centre-line across the paper.
3(S)	Draw a vertical line relatively close to the right side of the

STAGE A: PRIMARY POSITIONING OF INSERTS

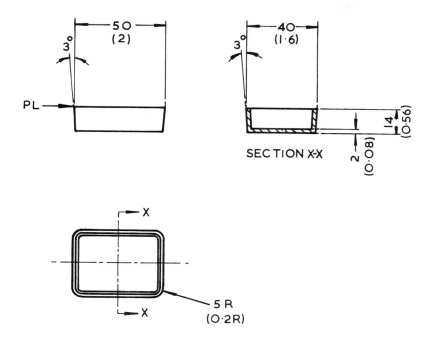

Figure 7.1—Pin box for which a two-impression mould is required

	paper. This is the *parting surface* of the mould.
4(S)	Draw two horizontal lines to represent the minor diameter of the sprue bush.
5(S)	Draw the horizontal centre-line for the impression. Note that the distance of this line from the major horizontal centre-line is computed from the following dimensions: (i) half the cavity length; (ii) wall thickness of the cavity insert; (iii) distance between the insert and the sprue bush; (iv) half the diameter of the sprue bush.
6(S)	Draw a sectional view of the impression positioned about the impression centre-line (Operation 5). This is basically a longitudinal view of the component. The parting line of the component is coplanar with the parting suface line (Operation 3).
7(S)	Draw the outside wall of the cavity insert. The depth of steel below the cavity impression must be sufficient to accommodate screws and dowels for securing purposes.
8(S)	Similarly, draw the outside wall of the core insert.
9(P)	Draw the vertical centre-line for the plan view.
10(P)	Draw the horizontal lines representing the *core* impression. The lines should be projected from the sectional view. (To avoid confusion, the draft angle is not included in this drawing.)

PROCEDURE FOR DESIGNING AN INJECTION MOULD

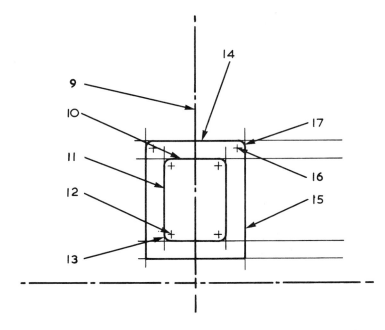

Figure 7.2—Procedure for designing an injection mould for a pin box. Stage A: plan view of moving half; primary positioning of the inserts

11(P)   Draw the vertical side walls of the core impression. The dimension for the distance between these two lines is obtained from the component drawing.
12(P)   Draw crossed centre-lines at each corner of the core impression for the specified component drawing radius.
13(P)   Draw the radius at each corner.
14(P)   Project the horizontal outside core insert wall lines from the sectional view.
15(P)   Draw two vertical lines to represent the outside core insert walls. The assumption is made that the wall thickness around the impression is constant.
16(P)   Draw crossed centre-lines at the top corners of the core insert.
17(P)   Draw a 4 mm radius at the interception of the above centre-lines (Operation 16). *N.B.* This radius is incorporated at each corner of the bolster recess to facilitate the machining operation.

7.3 STAGE B: THE EJECTOR SYSTEM

Operation Operation
number
18(P)   Decide upon the size, number and position of the ejector pins.

STAGE B: THE EJECTOR SYSTEM

STAGE A

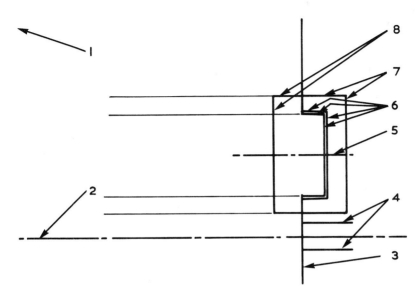

Figure 7.3—Stage A: cross-sectional view of both mould halves

(Three 5-mm ($\frac{3}{16}$ in) diameter ejector pins are suitable for ejecting this component in the positions indicated.) Draw crossed centre-lines.

19(P) Draw three circles to represent the ejector pins.
20(P) Similarly, decide upon the size, number and position of the push-back pins. (Four 10 mm ($\frac{3}{8}$ in) diameter push-back pins are suitable.) Draw crossed centre-lines as indicated.
21(P) Draw two circles to represent the top pair of push-back pins.
22(P) Draw two vertical lines to represent the width of the retaining plate in plan view. This plate lies below the parting surface (refer to Operation 30, sectional view) and it must be represented as *hidden detail*. *N.B.* The lines will be drawn as short dashes, as shown.
23(P) Draw a horizontal line (short dashes) to represent the top of the retaining plate.
24(S) Project the ejector pin centre-lines from plan to section.
25(S) Project the ejector pin details (two horizontal lines) from plan to section.
26(S) Project the push-back pin centre-line from plan to section.
27(S) Project the push-back pin details (two horizontal lines) from plan to section.
28(S) Project the top line of the retaining plate (and ejector plate)

189

PROCEDURE FOR DESIGNING AN INJECTION MOULD

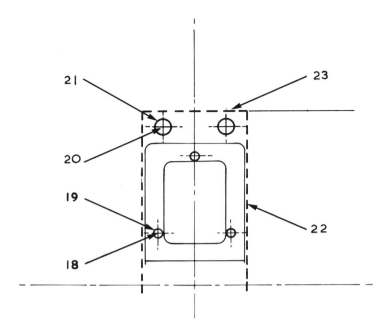

Figure 7.4—Stage B: plan view; incorporate the ejector system

29(S)   from plan to section.
Decide upon the thickness of the moving mould plate bolster. This bolster must be of sufficient depth to resist undue deflexion caused by the force applied to the core insert by the pressurised melt. A thickness of 35 mm (1.37 in) is suitable for this design.

30(S)   Draw the front face of the retaining plate. Note that a gap is required between the retaining plate and the rear face of the moving half bolster. (This is to allow for the movement of the ejector plate assembly during ejection.) For the majority of designs this distance may be taken as the height of the moulding plus 5 mm (0.2 in).

31(S)   Decide upon the thickness for the retaining plate and draw a vertical line to represent the rear face of this plate. (Note that this line will also represent the front face of the ejector plate.) A nominal thickness of 10 mm ($\frac{3}{8}$ in) is suitable for this design.

32(S)   Decide upon the thickness for the ejector plate, and draw a vertical line to represent the rear face of this plate. (Note that this line will also represent the front face of the back plate.) The ejector force is transmitted to the ejector elements via this ejector plate, therefore the thickness chosen must be sufficient to resist bending. A thickness of 19 mm ($\frac{3}{4}$ in) is suggested for this design.

STAGE C: THE EJECTOR GRID

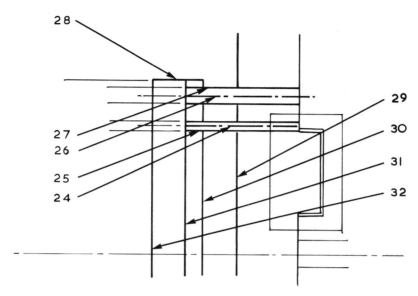

Figure 7.5—Stage B: cross-sectional view; incorporate the ejector system

7.4 STAGE C: THE EJECTOR GRID

Operation   Operation
number
33(P) Draw the two vertical lines representing the inside faces of the side support blocks (hidden detail). The position of these lines is obtained directly from the vertical retaining plate lines. A gap of 2 mm (0.08 in) is allowed between the two mould members.

34(P) Draw a horizontal line representing the inside face of the top support block (hidden detail). Other notes as in Operation 33.

35(S) Project the top horizontal ejector grid line from plan to section.

36(S) Extend the rear ejector plate line upwards to form the front face of the back plate.

37(S) Decide upon the thickness of the back plate and draw a vertical line. A thickness of 13 mm ($\frac{1}{2}$ in) is suitable.

38(S) Draw the ejector rod. This is drawn coaxial with the major horizontal centre-line. The front end of the ejector rod is threaded, and is attached to the ejector plate as shown.

39(S) The rectangle at the left side of the ejector rod represents spanner flats; that is, a machined flat surface on either side of

191

PROCEDURE FOR DESIGNING AN INJECTION MOULD

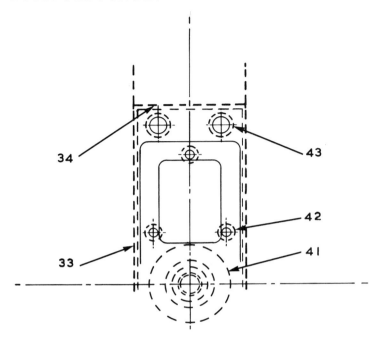

Figure 7.6—Stage C: plan view; incorporate the ejector grid

	the ejector rod to permit a spanner to be used for assembly purposes. Draw the spanner flats.
40(S)	Draw the ejector rod bush. This bush is coaxial with the major horizontal centre-line and circumscribes the ejector rod.
41(P)	Draw concentric circles to represent the plan view of both the ejector rod and ejector rod bush (hidden detail).
42(P)	Draw a circle concentric with each ejector pin to represent the shouldered head (hidden detail). The diameter can be 5 mm (0.2 in) larger than the ejector pins.
43(P)	Similarly, draw a circle concentric with each push-back pin to represent the shouldered head (hidden detail).
44(S)	Project the outside diameter of the ejector pin shouldered head from plan to section; that is, draw two short horizontal lines where indicated.
45(S)	Draw a vertical line to represent the front face of the ejector pin shouldered head.
46(S)	Project the outside diameter of the push-back pin shouldered head from plan to section; that is, draw two horizontal lines as shown.
47(S)	Draw a short vertical line to represent the front face of the push-back pin shouldered head.
48(S)	Decide upon a thickness for the fixed mould plate bolster.

STAGE D: COMPLETE THE TOP HALF OF THE DRAWING

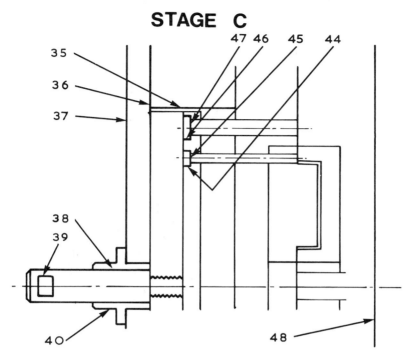

Figure 7.7—Stage C: cross-sectional view; incorporate the ejector grid

Note that this bolster accommodates the cavity insert. A dimension of 42 mm (1¾ in) is suitable for this design. Draw a vertical line to represent the front face of the bolster.

7.5 STAGE D: COMPLETE THE TOP HALF OF THE DRAWING

Operation number	Operation
49(P)	Draw two vertical centre-lines for the fluid circulation system. The position of these lines is dependent upon the relative position of the core insert, ejector pins, push-back pins, etc. Note that the controlling feature in this design is the push-back pin hole. The fluid circulation holes (flow-ways) must be well clear of this hole to avoid seepage.
50(P)	Draw the flow-ways. A 10 mm (⅜ in) diameter hole is suitable for this design. Note that as the holes are below the parting surface they are shown as hidden detail.
51(P)	Draw a second set of lines, of relatively short length (say 13 mm or ½ in), to represent threads for flow-way connector attachment purposes.
52(P)	Decide upon the number, size and position of the guide pillars.

PROCEDURE FOR DESIGNING AN INJECTION MOULD

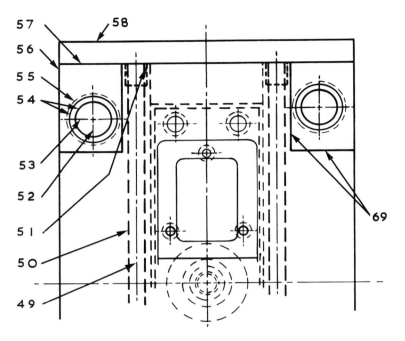

Figure 7.8—Stage D: plan view; complete the top half of the design

	Four 19 mm (¾ in) guide pillars are suitable for this size of mould. Bear in mind that the major diameter of the guide pillar must be clear of the flow-ways by at least 5 mm (0.2 in). Draw the vertical centre-line for the guide pillars.
53(P)	Draw the horizontal centre-lines for the guide pillars. (To prevent the two halves of the mould inadvertently being assembled incorrectly, it is normal design practice to offset one pair of guide pillars as shown.)
54(P)	Draw the major and minor diameters of the top guide pillars.
55(P)	Draw concentric circles (hidden detail) to represent the shouldered head of each guide pillar.
56(P)	Once the guide pillars have been incorporated in the design, the lines representing the vertical outside walls of the mould can be drawn.
57(P)	Draw the horizontal line to represent the top outside wall of the mould plate. A consideration here, in addition to the position of the guide pillars, is the thickness of the top support block. The thickness of the block should be about 20 mm (0.8 in) in order to accommodate securing screws.
58(P)	Draw the top horizontal line of the back plate. In practice the size of the back plate is often controlled by the position of the clamp holes in the injection machine platen. For this exercise extend the back plate beyond the top edge of the support block

STAGE D: COMPLETE THE TOP HALF OF THE DRAWING

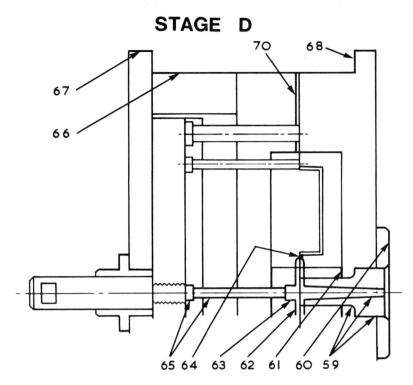

Figure 7.9—Stage D: cross-sectional view; complete the top half of the design

	by 13 mm ($\frac{1}{2}$ in) for clamping purposes.
59(S)	Complete the drawing of the sprue bush as indicated.
60(S)	Draw the register plate. In practice, the diameter chosen for this plate will depend upon the aperture size in the injection machine platen being used.
61(S)	Extend the vertical lines for the cavity and core inserts downwards to form a bridge piece in both mould halves.
62(S)	Draw the runner. A 5 mm (0.2 in) diameter runner is suitable which may be terminated 0.8 mm ($\frac{1}{32}$ in) from the cavity impression.
63(S)	Draw the cold slug well as indicated.
64(S)	Draw the gate. A rectangular-type gate has been chosen for this basic design with the following dimensions: width 1.5 mm ($\frac{1}{16}$ in); depth 0.8 mm ($\frac{1}{32}$ in); length (see Operation 62) 0.8 mm ($\frac{1}{32}$ in).
65(S)	Draw the sprue pin. A 5 mm ($\frac{3}{16}$ in) diameter pin is suitable. Include the shouldered head.
66(S)	Project the top outside wall lines of the mould plate from plan to section.

PROCEDURE FOR DESIGNING AN INJECTION MOULD

67(S)   Project the top outside line of the back plate from plan to section.

68(S)   Draw the 'clamping' ledge of the cavity plate. The dimensions for this ledge correspond to the complementary back plate extension.

69(P)   As noted in the text, it is normal practice to provide parting surface relief; that is, the surface contact between the two mould halves is restricted to the area adjacent to the impression plus additional supporting areas to prevent cavity plate collapse. In this design the surface of the moving mould plate is machined in order to leave four rectangular areas standing proud. Draw lines as indicated.

70(S)   Indicate in section the parting surface relief discussed above. A depth dimension of 1.5 mm ($\frac{1}{16}$ in) is suitable.

7.6 STAGE E: COMPLETE THE PLAN VIEW

Operation number   Operation

71(P)   Draw the bottom half of the plan as a mirror image of the top half. The vertical lines are continued downwards and the horizontal lines transferred using dividers.

72(P)   Draw the cold slug well: a circle and a dotted circle to correspond with Operation 63.

73(P)   Draw the runner. Refer to Operation 62.

74(P&S)   *Decide upon the section cutting plane.* Study both the plan view and the cross-sectional view drawn so far, and decide how many 'steps' are required to show all component parts of the mould. The first part-section (already drawn) from (i) to (ii) includes the support block and push-back pin details. The second part-section (also already drawn) necessitates a step ((ii)–(ii)) to the major vertical centre-line, to show ejector pin, core and cavity insert, sprue bush, sprue pin, cold slug well, bridge piece, register plate, ejector rod and ejector rod bush details. The second step ((iii)–(iii)) to the third part-section (iii)–(iv) is included solely to show the position of the flow-way relative to the parting surface. The third and final step ((iv)–(v)) is included to show details of the guide pillar, guide bush and parting surface relief. Draw the section plane line. This is the heavy chain-dotted line indicated in plan from (i) to (v) inclusive. The arrows at (i) and (v) indicate the viewing direction.

75(S)   Project the horizontal parts of the section plane line (i.e. the steps) to the section. These lines, too, are shown as thick chain-dotted lines. (Some designers follow normal drawing practice and do not show these lines. However, for the novice, these lines are essential.) The isometric view shown in Figure 7.16 will assist the novice in following this part of the drawing sequence.

STAGE F: COMPLETE THE CROSS-SECTION

7.7 STAGE F: COMPLETE THE CROSS-SECTION

Operation number	Operation
76(P&S)	Consider the plan view between section cutting plane line (i)–(ii). Check that all relevant parts are shown in the cross-sectional view. Work progressively downwards, along line (i)–(ii) from the top horizontal line which represents the back plate. Each time the line (i)–(ii) cuts a horizontal line (solid or dotted), check that a corresponding line is indicated in the cross-sectional drawing.
77(S)	Minor additions to be made. Add clearance holes in the mould plate and the retaining plate for the push-back pins. Note that, for reasons of clarity and to avoid confusing the novice, these clearance holes have not been indicated in the plan view.
78(S)	Consider the plan view between the section cutting plane line (ii)–(iii). Check that all relevant parts are shown in the cross-sectional view.
79(S)	Minor additions to be made. Add clearance holes in the mould plate and in the retaining plates. (See note in Operation 77.)
80(S)	Add clearance hole in the mould plate and the retaining plate for the sprue pin.
81(P&S)	Consider the plan view between section cutting plane lines (iii)–(iv). Check that all relevant parts are shown in the sectional drawing. In fact, nothing has been drawn so far in the space, and this subsection must be completed.
82(S)	Extend the vertical lines representing the back plate, support block, core mould plate and cavity mould plate downwards. Note that the section cutting plane line (iii)–(iv) passes through the support block in plan and not through the ejector plate as previous part-sectional view. Note, also, that while the register plate is not cut by the cutting plane line, it is shown because it is a normal side view. This point also applies to the bottom part of the ejector rod bush.
83(S)	The section cutting plane line (iii)–(iv) passes longitudinally through the flow-ways. Draw vertical lines to represent these holes in both mould plates. The centre-line distance to the parting surface may be taken as 13 mm ($\frac{1}{2}$ in).
84(S)	The above section cutting plane line also passes through the region of parting surface relief. Add a vertical line to correspond with Operation 70.
85(P&S)	Consider the plan view between the section cutting plane line (iv)–(v). Check that all relevant parts are shown in cross-section. Here, again, the part-sectional drawing needs to be completed.
86(S)	Extend the vertical lines, which represent the back plate, support block, cavity and core plates, downwards.
87(S)	Project the bottom horizontal line of the mould plate from plan to section.

PROCEDURE FOR DESIGNING AN INJECTION MOULD

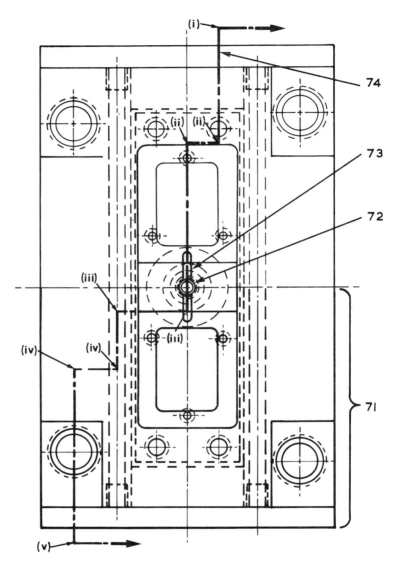

Figure 7.10—Stage E: plan view; complete the plan view and decide upon the section cutting plane

88(S) Project the bottom horizontal line of the back plate from plan to section.
89(S) Incorporate a clamping ledge at the base of the cavity plate. (See Operation 68.)
90(S) Project the centre-line for the guide pillar from plan to section.
91(S) Project lines representing the guide pillar on plan to the section.

STAGE G: COMPLETE THE DRAWING

STAGE E

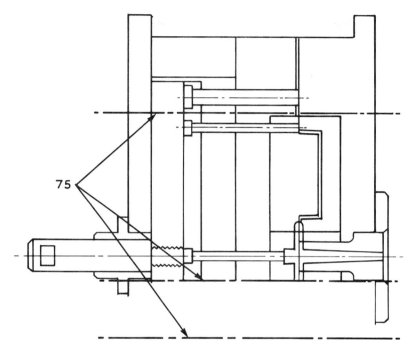

Figure 7.11—Stage E: cross-sectional view; show section cutting plane lines

92(S) Complete the drawing of the guide pillar, adding clearance for the shoulder in the core plate.

93(S) Complete the drawing of the guide bush, adding clearance for the shoulder in the cavity plate.

7.8 STAGE G: COMPLETE THE DRAWING

The following operation numbers are not indicated on the drawing.

Operation Operation
number
94(P&S) Erase all unwanted lines and clean up the drawing.
95(P&S) Re-check the parting plane line to ensure that the cross-sectional drawing is correct.
96(P&S) Make the outline of all component parts of the mould prominent. This entails redrawing all lines with a softer pencil or drawing pen.
97(S)   Cross-hatch the sectional drawing. All hatch lines should be drawn 45° to the major centre-lines. It is convenient to show the mould steels as a double cross-hatch line and use a single

PROCEDURE FOR DESIGNING AN INJECTION MOULD

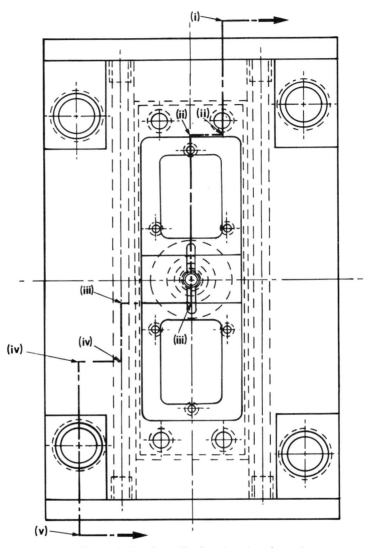

Figure 7.12—Stage F: plan view (no change)

cross-hatch line for the other steels. Note that the guide pillar, ejector rod and ejector pin follow the normal convention for round parts, and are not cross-hatched.

98(S) Indicate by a letter 's' those parts of the mould which are to be secured together with screws and dowels. (The precise location of screws and dowels is left to the detail draughtsman.) Add reference numbers and leader lines.

99(S) Show the feed and the impression in a solid colour.

100(P&S) Check the completed drawing. Add reference numbers.

STAGE G: COMPLETE THE DRAWING

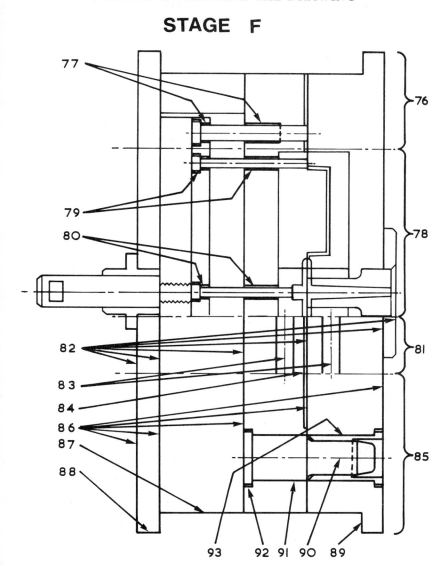

Figure 7.13—Stage F: cross-sectional view; complete the lower half of the drawing

PROCEDURE FOR DESIGNING AN INJECTION MOULD

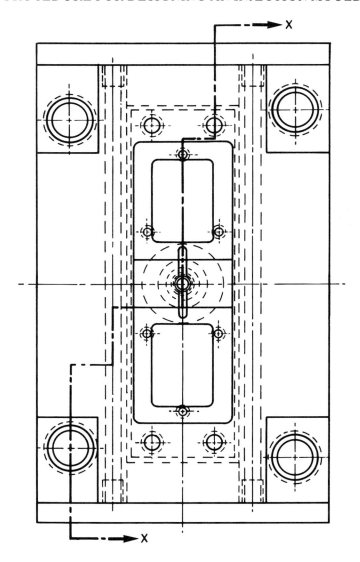

Figure 7.14—Stage G: plan view; check the drawing, erase all unwanted lines

STAGE G: COMPLETE THE DRAWING

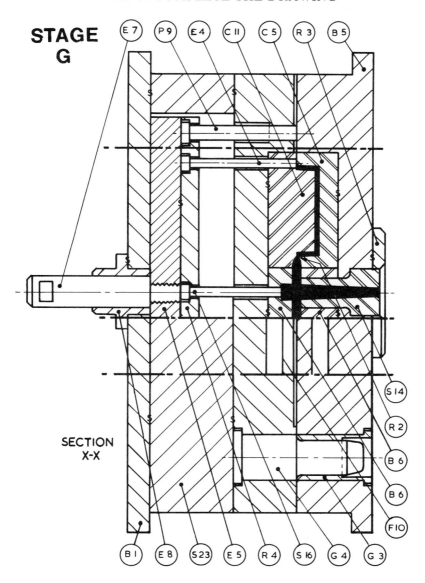

Figure 7.15—Stage G: cross-sectional view; add hatch and leader lines; check the drawing and erase all unwanted lines

PROCEDURE FOR DESIGNING AN INJECTION MOULD

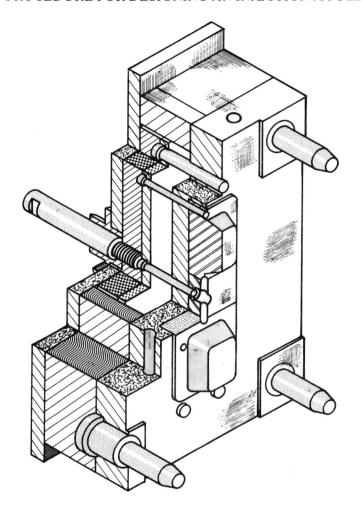

Figure 7.16—Isometric sketch of the moving mould half for reference

PART TWO

Intermediate Mould Design

8

Splits

8.1 GENERAL

The mould designer is frequently confronted with a component design that incorporates a recess or projection which prevents the simple removal of the moulding from the mould. The component designer, while endeavouring to produce in-line-of-draw component designs, has often to include a recess or a projection to perform a particular function or to satisfy an artistic requirement. A moulding which has a recess or projection is termed an *undercut moulding*. The mould design for this type of component is inevitably more complex than for the in-line-of-draw component, as it necessitates the removal of that part of the impression which forms the undercut prior to ejection.

8.1.1 External Undercut Components. Any recess or projection on the outside surface of the component which prevents its removal from the cavity is termed an *external undercut*. Various components which incorporate external undercuts are shown in Figure 8.1. The arrows on each drawing indicate the position of the undercut.

There are two forms of undercut to be considered.
  (i) The undercut may be local, in that the recess or projection occurs in one position only. The clip on a pen cap is an example of this (Figure 8.1j).
  (ii) The undercut may be a continuous recess or a projection on the periphery of the component. The water connector (Figure 8.1a) has a number of such undercuts. In either case, it is necessary to split the cavity insert into parts and open these, generally at right angles to the line of draw, to relieve the undercut before the moulding is removed.

Since the cavity is in two pieces, a *joint line* will be visible on the finished product. Now this joint line, on an undercut component, is comparable to the parting line on an in-line-of-draw component and the same careful consideration must be exercised in deciding its position before attempting to design the mould.

Let us now consider the choice of joint line for specific components. The symmetrical component will present no difficulties as the joint line can be positioned on any centre line (Figure 8.1a, b, e, f). Note, however, that as it is desirable to keep the *splits* movement to a minimum the joint line for regular rectangular components should be positioned on the longitudinal centre-line (for example, Figure 8.1g). For unsymmetrical components the choice of joint line is far more critical in that there is usually only one possible position.

SPLITS

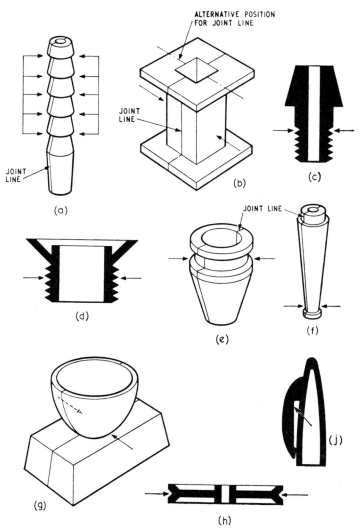

Figure 8.1—*External undercut components: (a) water connector; (b) spool; (c) threaded adaptor; (d) threaded ferrule; (e) connector; (f) pipe stem; (g) egg cup; (h) pulley; (j) pen cap
Arrows indicate position of undercut*

Consider, for example, the pen cap (Figure 8.1j). To extract this component from the mould, half the depth of the clip must be included in each split, therefore the joint line can only occur on the centre of this projection. For very complex components it is advantageous to have a model made to simplify the selection of the joint line. If an incorrect joint line is chosen, the moulding will tend to restrict the free opening movement of the splits which will result in scored or cracked mouldings.

GENERAL

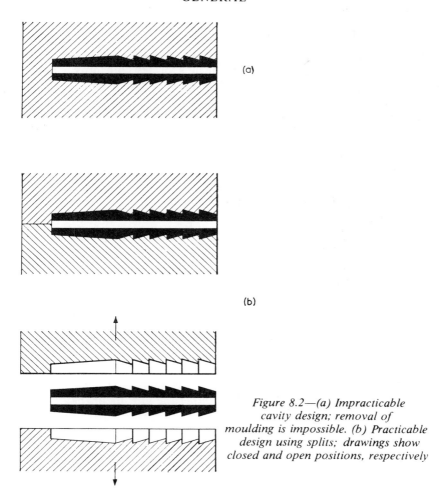

Figure 8.2—(a) Impracticable cavity design; removal of moulding is impossible. (b) Practicable design using splits; drawings show closed and open positions, respectively

Externally threaded components (Figure 8.1c, d) may be included under the general undercut component category but as there are alternative methods for dealing with these specific components they are discussed separately in Chapter 11.

The joint line should be as inconspicuous as possible on the moulding. To achieve this, the mould making must be excellent, the mould itself must be efficiently serviced during its life, and the correct conditions must be maintained during moulding.

8.1.2 Splits. Let us now consider why it is necessary to make the cavity insert in two parts for mouldings which incorporate an undercut. Figure 8.2a shows a part section through a two-plate mould for the water connector. In this design the cavity form is machined directly into the cavity plate. We note that when the mould is opened the core can be withdrawn, but the barbs on

SPLITS

the component form undercuts which make its removal from the cavity impossible. In the second design (b), the cavity form for the same component is shown machined into two separate blocks of steel. These split cavity blocks are called *splits*. One half of the component's form is sunk into each split and, providing the splits can be opened (Figure 8.2b, lower drawing), the moulding can be extracted.

The splits can be incorporated in the mould design in several ways. The simplest is by fitting the splits into a chase bolster. This method has a major disadvantage in that after each moulding operation the splits must be removed and opened prior to the mouldings being extracted. This manual operation, of necessity, lengthens the moulding cycle and therefore this design should be avoided except for moulds for prototype components.

Let us now consider more complex systems where the splits are retained on the mould plate and actuated automatically. There are two basic designs: sliding splits and angled-lift splits. In both designs there are moving parts and it is necessary to arrange for (i) guiding the splits in the desired direction, (ii) actuating the splits, and (iii) securely locking the splits in position prior to the material being injected into the mould.

The two basic designs will now be discussed in detail.

8.2 SLIDING SPLITS

In this design, the splits are mounted in guides on a flat mould plate and they are actuated in one plane by mechanical or hydraulic means. The splits are positively locked in their closed position by heels which project from the other mould half.

Sliding splits may be mounted on either the moving or the fixed mould plate, but as the former is the more general case, to prevent confusion, we will confine the discussion to this type only.

The principle of the sliding action is illustrated in Figure 8.3. This shows the splits in their closed moulding position (a) and in their fully open position (b).

8.2.1 Guiding and Retention of Splits. There are three main factors in the design of the guiding and retention system for a sliding splits type mould.
 (i) Side movement must be prevented to ensure that the split halves always come together in the same place.
 (ii) All parts of the guiding system must be of adequate strength to support the weight of the splits and to withstand the force applied to the splits by the operating mechanism.
 (iii) The two split halves must have a smooth, unimpeded movement. Although this point appears obvious, it is easy for a beginner to position guide pillars, for example, in the path of the splits.

In most designs, the guiding function is accomplished by providing an accurately machined slot in the mould plate in which the splits can slide. Close tolerances on both members are essential to prevent side play.

The splits retaining system usually adopted is based upon a T-design. Each

SLIDING SPLITS

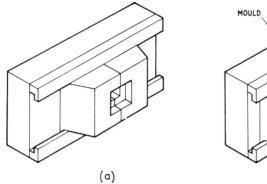
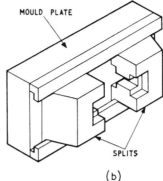

(a) (b)

Figure 8.3—Basic sliding split mould. Splits are shown in closed (a) and open (b) positions, respectively

split incorporates shoulders which are caused to slide in the T-shaped slot that extends across the mould plate (Figure 8.3). The depth of the shoulder and of its female counterpart in the mould plate must be carefully dimensioned to ensure the split has a smooth, judder-free movement.

By utilising the side walls of the T-slot for the guiding function as well, the mould design can often be simplified.

MOULD PLATE DESIGNS. There are several ways of producing the T-shaped slot in the mould plate, and three examples are shown in Figure 8.4.

(a) The form may be machined from a solid steel plate using a T-type of milling cutter. This system is seldom used, as it is comparatively difficult to grind the T-form to accurate dimensions.

(b) Separate *guide strips* may be attached to a flat mould plate as shown. The guide strips must be screwed and dowelled to the mould plate to ensure rigidity.

(c) In this design the required form is produced by machining a U-shaped slot across the face of the mould plate and then two flat steel strips are securely attached to the top surface.

No grinding difficulties are associated with either of the last two designs.

It is undesirable for the split to be in sliding contact with more than one pair of side faces of the T-form, as this makes the mould making unnecessarily difficult. Thus the actual guiding contact surface should either be as shown at X (Figure 8.4d) or as at Y (Figure 8.4e). A definite clearance of 0.75 mm (0.030 in) should be specified, as shown.

SPLIT DESIGN TYPE 1. The basic shouldered split design is shown in Figure 8.5. The only difference between its two variants is that (a) is made by machining the form from the solid block, while (b) is of two-part construction. While the economic advantage of the two-part construction for small splits is insignificant, some manufacturing time can be saved on larger splits. Note that if the latter alternative is adopted positive alignment between the two parts is essen-

SPLITS

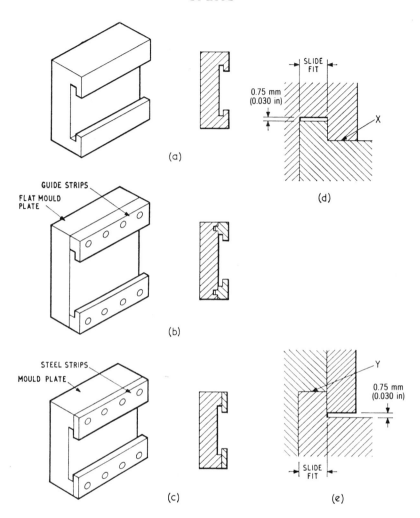

Figure 8.4—Mould plate designs for use in conjunction with sliding splits: (a-c) various designs; (d, e) fitting details

tial and in the example given (b) a rectangular key is fitted for this purpose.

The depth of the split's shoulder must be dimensioned to be a slide fit in the T-form of the guide (Figure 8.4d, e).

SPLIT DESIGN TYPE 2. This design incorporates a male T-shaped projection below the base of the split (Figure 8.6a). The guiding arrangement can thereby be positioned directly below the split, which permits a narrower mould than is possible with the previous design. A general arrangement scrap section assembly drawing is shown (Figure 8.6b) which gives details of the fit-

SLIDING SPLITS

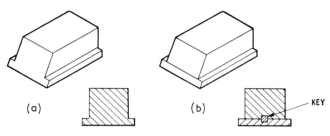

Figure 8.5—Split, design 1: (a) basic and (b) alternative methods of manufacture

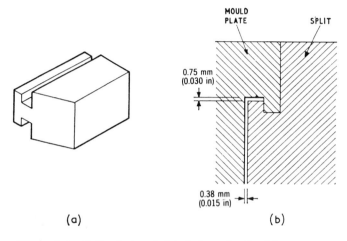

Figure 8.6—Split, design 2: (a) design of split; (b) part cross-sectional view through split and mould plate

ting arrangements. Note particularly where clearances are specified. As the split's side face and the mould plate edge are in general alignment, a cam plate can be conveniently fitted close to the split for actuation purposes. A pictorial view of the splits and mould plate assembly is shown in Figure 8.14.

SPLIT DESIGN TYPE 3. In this design (Figure 8.7), the splits are guided by a projection from the base of the split (b) which slides in a slot that extends across the face of the mould plate. The direct alignment between the split and the mould plate is advantageous when heavy splits must be guided and supported. General splits retention, as for the other designs, is by the split's shoulders, which fit into a T-shaped slot in the mould plate. An enlarged cross-sectional drawing (c) shows where specific clearances are required.

8.2.2 Methods of Operation. We will now consider the various methods which can be used to actuate the splits in relation to the mould plate.

The most frequently used designs are based on various types of cam. Under

SPLITS

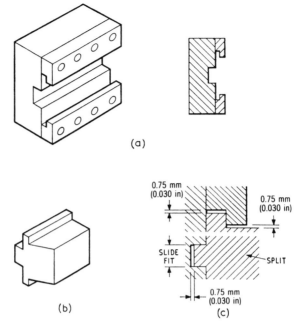

Figure 8.7—Split, design 3: (a) design of mould plate; (b) design of split; (c) part cross-sectional view through split and mould plate

this heading we will discuss finger cam, dog-leg cam, and cam track methods of actuation.

The basic operation with cam actuation is as follows. As the mould is opened, the cams attached to the fixed mould half cause the splits to slide across the moving mould plate. Conversely, when the mould halves are brought together, the splits are progressively closed. The cams generally lose contact with the splits as the mould opens and should either split be moved out of position prior to the mould closing serious damage will occur. To obviate this danger some safety features are incorporated in the design.

Another method of actuating the splits is by the use of compression springs, but as these can only be used to open the splits, the locking heels on the mould plate are utilised to fulfil the closing function. This system, while simple and cheap, has limited application as only relatively small split movement can be obtained.

Most machine manufacturers incorporate facilities in the hydraulic circuit for the operation of additional actuators if required. Thus, providing an actuator can be accommodated in the design, this method of splits operation should not be overlooked. The actuators can be mounted on opposite sides of the mould and the rams coupled directly to the splits. The main advantage of this system is that large splits movements are practicable. The various actuating methods will now be discussed in detail.

SLIDING SPLITS

FINGER CAM ACTUATION. In this system, hardened circular steel pins, termed *finger cams*, are mounted at an angle in the fixed mould plate (see qualifying note in Section 8.2). The splits, mounted in guides on the moving mould plate, have corresponding angled circular holes to accommodate these finger cams.

A typical design is given in Figure 8.8, which is of a mould for a spool component, the splits being shown in the closed position at (a). As the mould opens, the finger cam forces the split to move outwards, sliding on the mould plate (b). Once contact with the finger cam is lost, the split's movement ceases immediately. Continued movement of the moving half causes the ejector system to operate and the moulding to be ejected (c). On closing the reverse action occurs. The finger cam re-enters the hole in the split and forces the split to move inwards. The final closing nip on the splits is achieved by the locking heels and not by the finger cams.

The distance traversed by each split across the face of the mould plate is determined by the length and angle of the finger cam. The movement (Figure 8.9) can be computed by the formula

$$M = (L \sin \phi) - (c/\cos \phi) \tag{8.1}$$

As the required movement is known from the amount of component undercut, the following rearranged formula to determine the finger cam length is of greater use, apart from checking purposes

$$L = (M/\sin \phi) + (2c/\sin 2\phi) \tag{8.2}$$

where M = splits movement,
ϕ = angle of finger cam,
L = working length of finger cam,
c = clearance.

The designer must aim to keep the splits' movement down to a minimum, at the same time ensuring that the moulded part can be easily and quickly removed from the mould. The clearance c (Figure 8.9) serves a dual purpose. (i) It ensures that the force which is applied to the split during the injection phase is not transferred to the relatively weak cam. (ii) It permits the mould to open a predetermined amount before the splits are actuated; this movement can in certain circumstances be used to withdraw the core from the moulding. The amount of delay movement D before the splits are actuated is determined by the following relationship

$$D = c/\sin \phi \tag{8.3}$$

10° is a suitable angle for ϕ, but if the mould height has to be increased unduly to accommodate excessively long finger cams it is permissible to increase this angle up to a maximum of 25°. For actuating small splits, a finger cam diameter of 13 mm (0.5 in) is suitable, but for large splits or where a greater than 10° angle is specified the diameter should be increased accordingly. Note the lead-in angle at the front end of the finger cam (Figure 8.9). This facilitates the re-

SPLITS

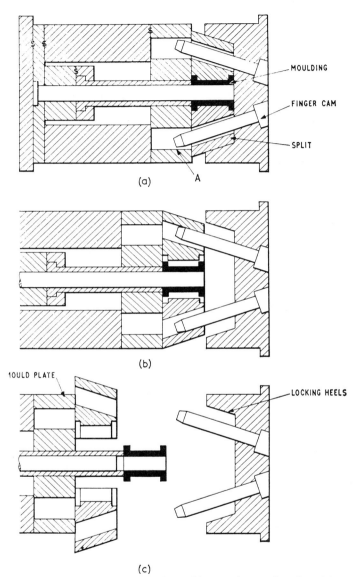

Figure 8.8—Finger-pin actuated mould: (a) splits in closed position; (b) intermediate position; (c) splits fully open; moulding ejected by sleeve

entry of the finger cam into the split as the mould is being closed. This angle is normally $(\phi + 5)°$.

One or two finger cams are used to operate each split. When two are used (i.e. for splits greater than 76 mm (3 in) in width) it is essential that both function in unison.

SLIDING SPLITS

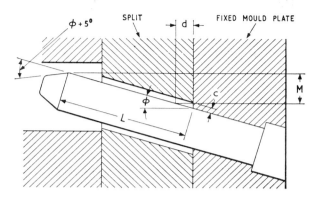

Figure 8.9—Split movement calculations (finger-pin actuation)

One method of fitting the finger cam into the mould plate is illustrated (Figure 8.8). A backing plate is not essential as the finger cam is supported by the machine platen when the mould is fitted to the injection machine. The injection platen must be checked, however, to ensure there are not any holes or depressions in the proposed finger cam position.

The hole in the moving mould plate, positioned below the split to accommodate the projecting end of the finger cam, may be circular or link shaped in the plan view. It is important to note that the side of this hole (see A, Figure 8.8) is straight, i.e. the angled hole of the split is not continued through the mould plate. A typical finger-cam actuated design is shown in Plate 7.

DOG-LEG CAM ACTUATION. This method of actuation is used where a greater splits delay is required than can be achieved by the finger cam method. The dog-leg cam (Figure 8.10), which is of a general rectangular section, is mounted in the fixed mould plate. Each split incorporates a rectangular hole, the operating face of which has a corresponding angle to that of the cam.

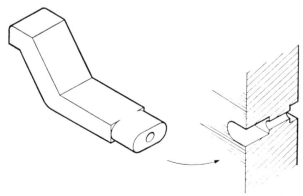

Figure 8.10—Method of attaching dog-leg cam to mould plate

SPLITS

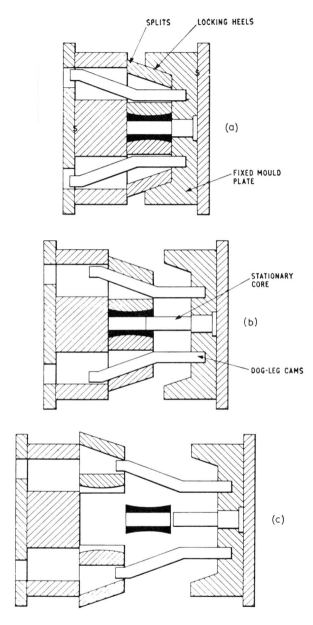

Figure 8.11—Dog-leg cam actuated mould: (a) splits closed; (b) intermediate position (note splits still closed, moulding pulled from core); (c) splits opened by cam

SLIDING SPLITS

The sequence of operation of this type of mould is shown in Figure 8.11. At (a) the mould is closed and the splits are locked together by the locking heels of the fixed mould plate. The feed to the impression is not shown. The splits do not immediately start to open when the mould halves are parted (b) because of the straight portion of the dog-leg cams. The moulding, which is encased within the splits, will thus be pulled from the stationary core. Further movement of the moving mould half causes actuation of the splits by the dog-leg cams, thereby releasing the moulding (c). The reverse action occurs when the mould is closed.

The simple form of the component in Figure 8.11 permitted the mould to be designed without an ejector system. There is a tendency for the moulding to be left in one of the split halves and it is desirable, therefore, to arrange for the splits to start opening before the moulding completely leaves the core; this holds the moulding central.

Another example is shown in Figure 8.12. This design is for a circular component which necessitates positive ejection. To make the problem a little more difficult let us assume that no joint lines are permissible on the major diameter (top and bottom). This means that we can only use the splits for the central region, the remainder of the moulding's exterior being formed in cavity inserts. The component design necessitates two cores which meet at Z. Now, in this example, it is desirable to pull that part of the moulding which is formed in the fixed mould plate completely clear of the impression before the

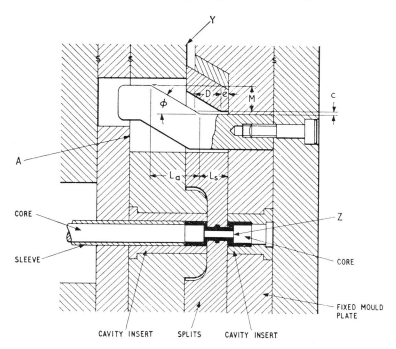

Figure 8.12—Typical dog-leg cam actuated mould

219

SPLITS

splits are opened. Thus the delay dimension D of the cam must be sufficient to accomplish this. After actuation of the splits by the dog-leg cams the component is ejected by the sleeve. (Note—it would be necessary with this design to limit the forward movement of the sleeve to prevent fouling by the splits as they are closed, or to provide a mechanism to ensure the ejector system is returned before the splits are actuated).

The relevant formula for calculating the opening movement, the length of cam, and the delay period (Figure 8.12) are given by

$$M = L_a \tan \phi - c \qquad (8.4)$$

$$L_a = (M + c)/\tan \phi \qquad (8.5)$$

$$D = (L_s - e) + (c/\tan \phi) \qquad (8.6)$$

where M = movement of each split,
L_a = angled length of cam,
L_s = straight length of cam,
ϕ = cam angle, c = clearance, D = delay,
e = length of straight portion of the hole.

Typical cross-section dimensions of a dog-leg cam for a small mould are 13 mm (½ in) by 18 (¾ in). The angle ϕ, ideally, is 10°, but here again this may be increased to 25° if by so doing the overall height of the mould can be reduced. A lead-in at the front end of the cam should be provided to facilitate the re-entry of the cam into the split as the mould closes. This lead-in may either take the form of a 10° taper or a generous radius.

One method of attaching the dog-leg cam to the mould plate is shown in Figure 8.10. The link shape at the lower end of the cam allows the complementary hole in the mould plate to be machined by a simple end-milling operation. The cam is held back on to the mould plate by a socket-headed screw. When a back plate is incorporated in the design (Figure 8.12), the link-shaped portion of the cam may extend completely through the mould plate and be secured as shown. For splits under 100 mm (4 in) in width, one cam is used, for splits of greater width two cams are preferable.

The clearance hole in the moving mould plate below the split (A, Figure 8.12) can be either circular or rectangular with radiused corners. The latter design is generally preferred as it results in less of the mould plate, which supports the split, being machined away.

CAM TRACK ACTUATION. This method of actuation utilises a cam track machined into a steel plate attached to the fixed mould half. A boss fitted to both sides of the split runs in this track. The movement of the splits can thus be accurately controlled by specific cam track design.

Figure 8.13 illustrates two typical cam track plates; they differ only in respect of whether the cam track is machined part way through (b) or completely through the plate (a). Design (b) is the stronger but to maintain the

SLIDING SPLITS

same depth of boss contact as (a) the plate must of necessity be thicker. To ensure smooth operation a generous radius should be incorporated at each point where the cam track form changes. A radius or taper should also be included at the entrance to form a lead-in for the boss as it re-enters the track.

Figure 8.14 shows a simplified pictorial view of a moving mould half incorporating splits used in conjunction with this actuating arrangement. It will be noted that the width of the splits in this design (Section 8.2.1) corresponds to the width of the mould plate. This is necessary so that the cam plates can be in the close proximity to the splits in order to actuate them. The drawing also shows the bosses which protrude from the side faces of the splits. Corresponding bosses protrude from the lower faces but are not visible.

Details of the operation and of the mould assembly are given in Figure

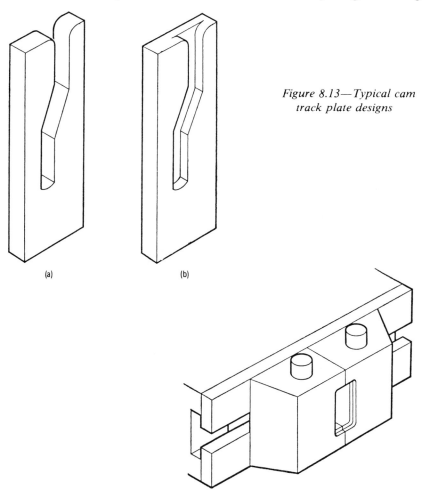

Figure 8.13—Typical cam track plate designs

Figure 8.14—Sliding split mould (cam track plate actuated)

221

SPLITS

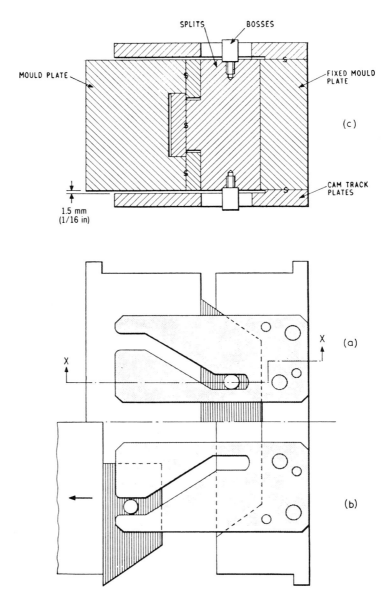

Figure 8.15—Cam track actuation: (a) mould closed; (b) intermediate position; (c) section X-X

SLIDING SPLITS

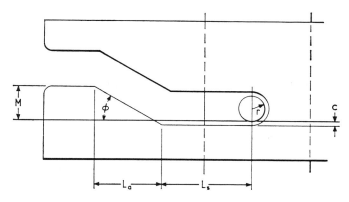

Figure 8.16—Split movement calculations (cam track actuation)

8.15. The top drawing shows a typical cross-section through this type of mould. The splits are mounted on a mould plate. The bosses, screwed into the side faces of the split, protrude into the cam track plates. The latter are securely attached to the side faces of the fixed mould plate. Note that a small clearance of 1.5 mm (1/16 in) is provided between the cam track plate and the moving mould half. The mould operation is illustrated in the lower drawing. At (a) the mould is shown in the closed position. As the mould opens the bosses follow the cam track and thereby cause the splits to open. An intermediate mould open position is illustrated at (b), which shows the boss on the point of leaving the cam track. Note that at this point the splits are fully open. When the mould is closed the boss re-enters the cam track and the splits are progressively closed.

The relevant formulae for calculating the distance traversed by each split, the length of cam track, and the delay period (Figure 8.16) are as follows:

$$M = L_a \tan \phi - c \qquad (8.7)$$

$$L_a = \frac{M+c}{\tan \phi} \qquad (8.8)$$

$$D = L_s + \frac{c}{\tan \phi} + r\left(\frac{1}{\tan \phi} - \frac{1}{\sin \phi}\right) \qquad (8.9)$$

where M = movement of each split,
L_a = angled length of cam track,
L_s = straight length of cam track,
ϕ = cam track angle,
c = clearance,
D = delay,
r = radius of boss.

There is a greater range of permissible operating angle with this design than with other methods because the cam plate is more rigid. The angle used is normally between 10° and 40°.

SPLITS

SPRING ACTUATION. This design, which obviates the use of cams altogether, incorporates compression springs to force the splits apart and utilises the angled faces of the chase bolster to close them. The outward splits movement must therefore be limited so that they will re-enter the chase bolster as the mould is closed. This design is limited to mouldings which incorporate relatively shallow undercuts.

A typical basic design is shown in Figure 8.17. The splits are mounted on the mould plate and retained by guide strips. Studs project from the base of the splits into a slot machined in the mould plate. The outward movement of each split is therefore controlled by the length of this slot. A compression spring is fitted between the studs in a link-shaped pocket situated in the lower mould plate. The splits are held closed by the chase bolster. The sequence of operations is shown in Figure 8.18.

(a) The splits are held closed by the chase bolster during the injection phase.

(b) Immediately the mould begins to open, the compression spring exerts a force to part the split halves.

(c) The splits movement is stopped by the stud reaching the end of the slot in the mould plate. Continued movement of the moving mould half operates the ejector system (not shown). During the closing stroke the splits re-enter the chase bolster and are progressively closed.

Calculations for this arrangement are limited to those for the splits opening movement (Figure 8.19).

$$M = \tfrac{1}{2}H \tan \phi \tag{8.10}$$

where M = movement of each split,
H = height of locking heel.
ϕ = angle of locking heel.

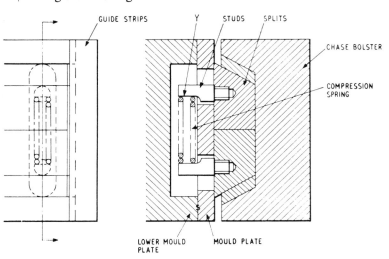

Figure 8.17—Spring actuation

SLIDING SPLITS

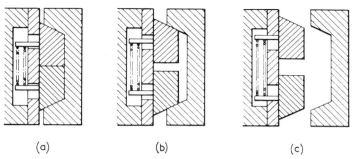

Figure 8.18—*Mould opening sequence of spring actuated split mould*

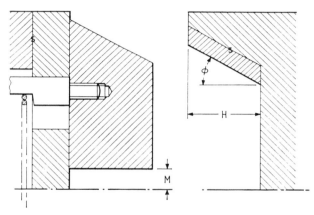

Figure 8.19—*Split movement calculation (spring actuation)*

A suitable angle for the locking heel is between 20° and 25°, and therefore approximately

$$M = 0.2 H \qquad (8.11)$$

The splits open immediately the mould parts and no delay is possible; the moulding must therefore be biased to remain in the moving half so that it can be positively ejected.

For splits under 76 mm (3 in) in width one stud per split is suitable, for those over this width two studs should be used. The studs can be flattened at Y (Figure 8.17) to provide a seating for the spring.

In operation, spring failure may result in the splits being damaged by the ejector system. Careful selection of the spring and regular mould maintenance are therefore essential.

HYDRAULIC ACTUATION. In this design the splits are actuated hydraulically and, unlike the previous systems, it is not dependent on the opening movement of the mould. The splits can be operated automatically at any specific time by including this function in the operating programme of the machine.

SPLITS

On machines which do not programme for auxiliary cylinder control it is necessary to add a separate hydraulic operating circuit to the existing system.

The designer does not generally rely on the locking force which can be applied hydraulically to keep the splits closed during the injection phase, as this would mean fitting large-diameter and, therefore, heavy cylinders. However, if the total projected area of the mouldings is relatively small this method should not be overlooked as the mould design is considerably simplified. (See Chapter 9, which deals with hydraulically-actuated side cores.) In

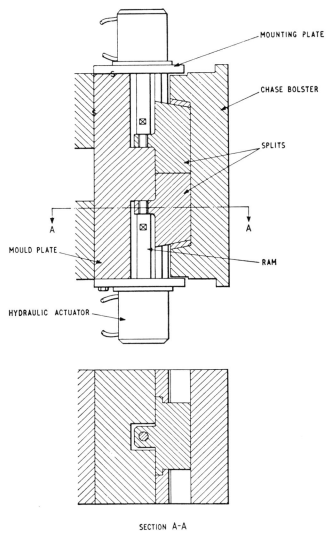

SECTION A-A

Figure 8.20—Hydraulic actuation of small splits

SLIDING SPLITS

all other cases, the conventional locking heel system (Section 8.2.3) is used.

A general design for shallow splits is shown in Figure 8.20. The splits incorporate a projection on the underside to which the ram of the hydraulic actuator is attached. The projection is free to move in a slot which extends partway through the mould plate (see also section A-A). The hydraulic actuator is fitted on a mounting plate which is securely attached to the side wall of the mould plate. The splits are shown in the closed position, and are held by the locking heels of the chase bolster.

For deeper splits, a similar arrangement is adopted, except that the ram is attached to the angled face of the split (Figure 8.21). In this design it is necessary to provide a slot through each locking heel (see part-section X-X) to clear the ram. A foot-mounted hydraulic actuator is shown fitted to an angle-plate, which in turn is securely attached to the mould plate wall.

In either design, the splits can be actuated at any time after they are clear of the locking heels, even in the extreme case when the mould is fully open. How-

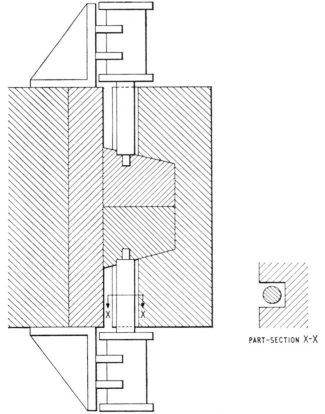

Figure 8.21—Hydraulic actuation of large splits. (Guide strips are not shown in this drawing, for reasons of clarity)

SPLITS

ever, to reduce the cycle time to a minimum it is desirable to operate the splits while the mould is opening. On the return stroke it is essential the splits are closed before they re-enter the chase bolster. From the above comments it will be apparent that large delay movements and large split movements can be achieved with this design. Against these advantages, the fact that the mould is more bulky as compared with the other designs makes the mould setting more difficult and, of course, the hydraulic system has to be connected each time the mould is set up.

8.2.3 Splits Locking Method. It is essential that the splits are held rigidly during the actual injection phase, as high pressures developed within the impression will tend to force them apart. To resist these pressures some means of locking the splits must be provided. Sliding splits can be conveniently locked in position by the use of a chase-bolster. When the mould is closed the chase-bolster encloses and clamps the splits (Figure 8.8).

Each split will have a sloping or angled face accurately matching the complementary angled face of the chase-bolster. Where the cam method of actuation is adopted this locking angle must be at least 5° greater than the cam operating angle to ensure the splits do not foul the chase bolster heels as the splits open.

There are two basic designs of chase-bolster, namely the *open-channel* type and the *enclosed-channel* type. Both types are used for *sliding* and *angle-lift* split designs. However, as the basic mould assembly is different for each method the following discussion refers specifically to the sliding split design.

OPEN-CHANNEL CHASE-BOLSTER. The basic design of the chase-bolster is illustrated in Figure 8.22. It is made by machining a channel with angled side across the width of a steel plate. The resultant projections on either side of the plate are termed *locking heels*. To resist wear the locking heels are faced with *wear plates* which are secured with socket headed screws as shown.

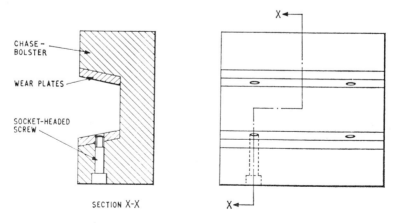

Figure 8.22—Open-channel chase-bolster

SLIDING SPLITS

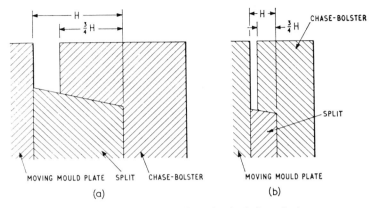

Figure 8.23—Depth of locking heel of chase-bolster

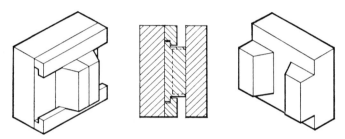

Figure 8.24—Open-channel chase-bolster design for shallow splits

The height of the locking heel (depth of the channel) should be at least ¾ the depth of the split (Figure 8.23). Thus the gap between the chase-bolster and the moving mould plate depends on the height of the split (compare drawings (a) and (b)). If we now refer back to Figure 8.3 we will note that with this particular design of split mould plate assembly the guides extend down either side of the mould plate and are accommodated in the gap between the mould plate and the chase-bolster. Therefore, if the depth of the split is less than a specified minimum value (e.g. 16 mm (⅝ in) for a 11 mm (7/16 in) deep guide strip) the chase-bolster design must be modified. Figure 8.24 shows the alternative design. The locking heels are reduced in width to permit the guides to straddle the locking wheels when the mould is closed. The centre sectional view of the closed mould illustrates the position of the locking heel (shown dotted) in relation to the guides.

This alternative design is not necessary when the guiding arrangement is mounted below the split.

The open-channel design can be made as an assembly. The *heel blocks* are individually attached to a flat plate (Figure 8.25). Projections are incorporated on the underside of the heel blocks to resist applied forces. These projections accurately fit into recesses machined in the mould plate. This method of construction economises in steel though it is not as strong as the solid designs.

SPLITS

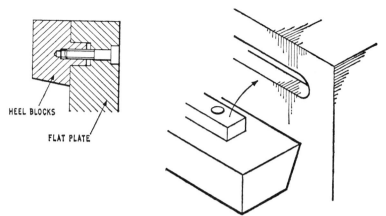

Figure 8.25—Open-channel design made as an assembly. Note method of fitting heel blocks

ENCLOSED CHASE-BOLSTER. For deep splits, the enclosed chase-bolster design is preferred as it results in a more rigid structure than can be obtained with the open-channel design. The chase-bolster is made by machining a pocket, which may be of tapered circular or tapered rectangular form, into a solid steel block (Figures 8.26 and 8.27 respectively).

An example of the tapered circular design is illustrated in Plate 8 which shows a sliding split mould for a jug top (the moulding is in the foreground). The left-hand view illustrates the fixed mould half which consists of the enclosed chase-bolster, the core (fixed half), dog-leg cams, guide pillars and lifting eyebolt. The moving half (right-hand view) consists of the tapered circular splits mounted on the moving mould plate being retained by the guides. Also visible are the core (moving half) and the guide bushes. The sprocket and chain are part of a split balancing mechanism. The circular chase-bolster design is adopted in this case as the component is of a general circular form and a single impression only is required. If two impressions are required, or if

Figure 8.26—Tapered cylindrical enclosed chase-bolster

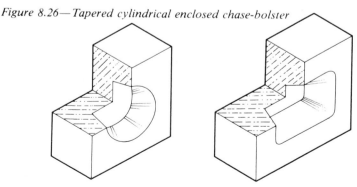

Figure 8.27—Tapered rectangular enclosed chase-bolster

SLIDING SPLITS

the component is more rectangular in shape, rectangular splits fitting into a complementary-shaped pocket will be used.

From the standpoint of cost, the circular form is preferred as this design allows for the tapered pocket and for the splits (external form) to be produced by a simple turning operation. The alternative rectangular design necessitates expensive milling operations. It should be noted, however, that with the rectangular pocket design wear strips can be fitted, thereby allowing a medium carbon steel (BS 970-080 M40) to be used, whereas a more expensive case-hardening nickel - chrome steel (BS 970-835 M15) is required for the circular design as wear strips are not easy to fit.

8.2.4 Splits Safety Arrangements. It is necessary to incorporate certain safety features in moulds which utilise the cam method of actuation. As noted previously, the splits are not in contact with the cams when the mould is fully open. This means that they may be accidentally moved out of alignment by

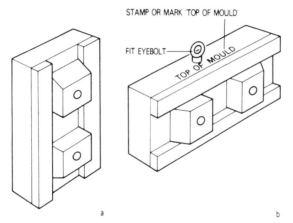

Figure 8.28—Splits safety arrangements. The splits should operate horizontally as shown (b) otherwise they will be affected by gravity (a)

shock, by vibration or even by gravity. If this occurs, the splits and cams will certainly be damaged as the mould is closing.

To safeguard against being opened by gravity, the splits should operate horizontally with respect to the machine. It is advisable to stipulate that the top of the mould is so marked to prevent the mould being fitted incorrectly on the machine (Figure 8.28). To prevent movement of the splits by shock or vibration the splits must be retained nominally in the open position. There are several methods from which to choose, two of which are now given.

SPRING-DETENT METHOD. A spring-loaded plunger is fitted below the surface of the split (Figure 8.29). When the split is open the plunger is engaged in a shallow conical depression which retains the split nominally in that position.

SPLITS

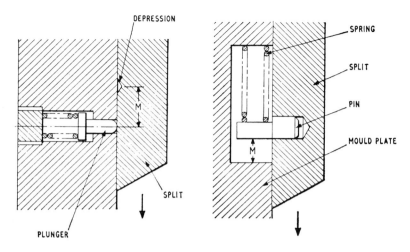

Figure 8.29—*Spring-detent method*   Figure 8.30—*Spring-loaded method (1)*

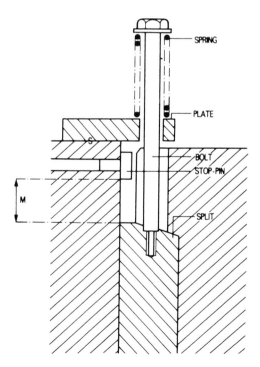

Figure 8.31—*Spring-loaded method (2)*

SPLITS SAFETY ARRANGEMENTS

SLIDING SPLITS

The distance M between the plunger and the depression must equal the calculated movement of the split. The plunger is located on, or as close as possible to, the centre-line of the split (plan view); the exact position depends on whether one or two cams are used.

SPRING-LOADED METHOD (1). The splits may be individually spring loaded so that after they are actuated they remain in the open position. The design is shown in Figure 8.30. A stud, fitted to the underside of the split, is free to move in a recess in the mould plate. A spring is fitted, between the stud and the end of the slot. The strength of the spring must be such that it nominally holds the splits open. Too strong a spring will put unnecessary loading on the cams during closing. To ensure that the cam re-enters the split correctly, the distance M must equal the calculated movement of the split.

SPRING-LOADED METHOD (2). An alternative spring-loaded design which is suitable for the larger type of split is shown in Figure 8.31. A bolt (or bolts) is fitted to the locking angle of the split. The bolt extends through a plate securely attached to the mould plate. A spring is mounted between the plate and the head of the bolt. Here again, the strength of the spring must be such that it nominally holds the splits open, and it does not put undue strain on the cams during the closing stroke.

A stop-pin is used to control the maximum movement of the split. The stop-pin is essential with this design, so that in the open position the hole in the split is in alignment with the operating cam, in preparation for the closing stroke.

For shallow type splits (Figure 8.31 for example) it is necessary to provide grooves or cut-away portions in the locking heels to provide a suitable clearance space within which the bolts can be accommodated. For deep splits there is usually sufficient space between the top of the locking heels and the moving mould plate to incorporate the bolts without modifications to the locking heels being necessary.

STOP-PINS AND STOP-PLATES. It is good design practice to incorporate a positive stop for the outward movement of the split. As will have been noticed in the previous section, a positive stop is essential for spring-loaded designs.

The positive stop may take the form of a stepped pin, termed a stop-pin (Figure 8.32a) or small plate, termed a 'stop-plate' (Figures 8.32b and c). The stop-pin is a light drive fit in a suitable hole machined in the mould plate as illustrated in Figure 8.31. Note that the hole is drilled completely through the mould plate for extraction purposes.

The stop-plate is drilled and counterbored to enable the plates to be attached to the mould plate by screws. As the screws are on the outside surface, the stop-plates are easily removed when required.

Note that the stop-pins or stop-plates must be in the split's path in order to function. It is essential, therefore, that these positive stops are easily removable to enable the split to be withdrawn for cleaning and maintenance purposes.

SPLITS

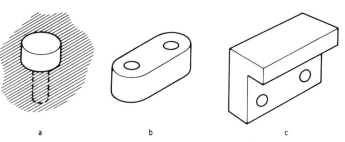

Figure 8.32—Stop-pin and stop-plates

8.2.5 Splits—Stripper Plate Design. It is sometimes necessary to incorporate stripper plate ejection for components which also necessitate splits. In this design instead of mounting the splits on the moving mould plate, the splits are mounted on the stripper plate. To ensure a positive action a latching system is necessary, either to hold the stripper plate in contact with the fixed mould plate while the core is pulled, or to hold the stripper plate in contact with the moving plate while the splits are actuated. Both designs will now be discussed.

CORE PULLED FIRST. It is often desirable to pull the core through the stripper plate before opening the splits. This system is useful where the ejection area is particularly small. By keeping the splits closed the projections in the splits form an extra ejection surface.

The latching arrangement required for this system is shown in Figure 8.33. A spring plunger box holds the latch against a stop. A stud fitted to the stripper plate engages in the latch. When the mould opens, the cam plate lifts the latch at a predetermined point, thereby releasing the stripper plate. The opening sequence of the mould is as shown:

(a) The mould is in the closed position.

(b) Actuation of the moving plate with respect to the closed split allows the core to be withdrawn from the moulding.

(c) Subsequent to the latch being lifted the stripper plate is pulled back by length bolts which are attached to the moving plate thereby causing the splits to be actuated. (The splits actuating method is not shown.)

SPLITS OPEN FIRST. The basic mould opening sequence and the general latching arrangement for this system are shown in Figure 8.34. The purpose of the latch is to hold the stripper and moving plates together during the initial period of the mould opening stroke to allow the splits to be actuated. Subsequent movement of the moving mould plate causes ejection of the mouldings.

The latching arrangement is shown at (d). The latch is pivoted on a shouldered pin which is fitted in the back plate. The latch is held against a stop by a spring loaded plunger. A stud fitted to the stripper plate engages in the latch. When the mould is opened the stripper plate is caused to move back with the moving mould plate thereby causing the splits to be opened (b). After the cam plate lifts the latch, the stripper plate's movement is arrested by length bolts.

SLIDING SPLITS

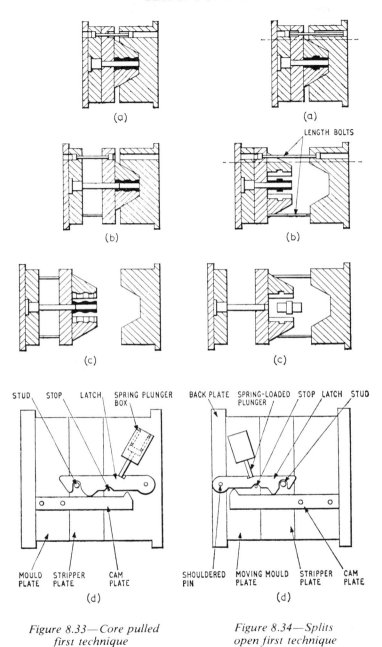

Figure 8.33—Core pulled first technique

Figure 8.34—Splits open first technique

SPLITS MOUNTED ON STRIPPER PLATE

SPLITS

Continued movement of the moving half causes the core to be withdrawn through the stripper plate (c), and the moulding is extracted from the mould.

8.3 ANGLED-LIFT SPLITS

In this design the splits are mounted in a chase-bolster which forms part of the moving half of the mould. The splits are caused to move out with an angular motion, the outward component of which relieves the undercut portion of the moulding. The splits are normally actuated by the ejector system. A typical design is shown in Figure 8.35. This shows the moving half of an angled-lift split mould for producing a spool. It will be noted from this illustration that the guiding of the angled-lift is not as critical as for guiding the sliding splits. The alignment of the splits, when closed, is accomplished by their being seated in the chase-bolster. The main requirement of the guiding system is that the split must be restrained to move smoothly in the required plane.

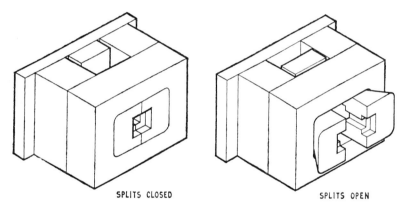

Figure 8.35—Angled-lift splits

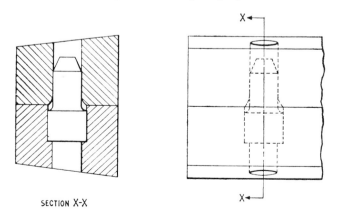

Figure 8.36—Alignment between angled-lift split halves

ANGLED-LIFT SPLITS

A substantial chase-bolster, which may be of the enclosed or open channel designs (see Section 8.2.3), locks the splits against the applied injection force. When the open channel design is used it is necessary to provide some means of alignment between the two split halves as the ends of the splits are not located. Alignment is normally accomplished by incorporating dowels in one split which fit into holes in the other (Figure 8.36).

To simplify the discussion of sliding splits we separated the guiding, operating and locking functions and discussed these separately. However, with the angled-lift design we cannot easily separate these functions as they are closely interrelated. We will therefore discuss the various alternative systems with respect to particular actuating methods as follows:
  (i) Angled guide dowel actuating system.
  (ii) Cam track actuating system.
  (iii) Spring actuation.

8.3.1 Angled Guide Dowel Actuating System. A typical design of a mould of this type is illustrated in Figure 8.37. The mould is shown closed (a) and fully open (b). In this design the ejector system is utilised to open the splits as illustrated.

Guide dowels are fitted at an angle to the underside of each split. These guide dowels pass through holes machined at an angle, in the enclosed chase-bolster. These holes are often bushed as shown. When the ejector system is actuated, the relative movement between the ejector plate and the enclosed chase-bolster causes the guide dowels to be moved forward, and, as these are fitted at an angle, the splits are caused to open (b).

A convenient angle for the guide dowel is 10° but this may be increased if a large opening movement is required. The actual opening movement of each split may be computed from:

$$M = E \tan \phi \tag{8.12}$$

where E = effective ejector plate movement,
  ϕ = guide dowel angle.

An important design point to note is that there is a transverse movement of the guide-dowel head with respect to the ejector plate as this latter member is actuated. For this reason the head of each guide dowel is domed and the working surface of the ejector plate is hardened.

For the same reason, in an alternative design (Figure 8.38) the face of the ejector plate is angled as shown, the operating face being at right angles to the centre line of the guide-dowel. This permits flat-headed guide dowels to be used, which gives a larger contact surface than for the previous design. The operating movement of each split with this design, using the same symbols as before, can be computed from

$$M = E \sin 2\phi / 2 \tag{8.13}$$

Further examples of design features which allow for the relative movement between the guide dowels and the ejector plate are given in Section 10.2.2.

SPLITS

The angle of the locking chase must exceed the operating angle ϕ by at least 5° to prevent the splits fouling the chase as they are opened. Wear strips (Figure 8.37) are normally incorporated on the working faces of the chase-bolster as shown.

Immediately the mould closing stroke commences the spring around the ejector rod causes the ejector plate to return to its rear position. The splits are returned progressively to their nest by the action of the mould closing. However, to minimise the possibility of damage occurring either to the face of the

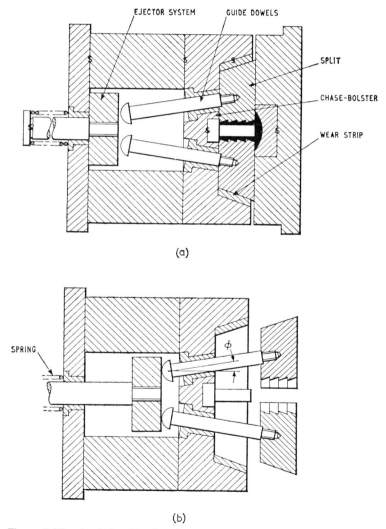

Figure 8.37—Angled guide-dowel actuating method: (a) mould closed; (b) mould open, splits actuated by ejector system

ANGLED-LIFT SPLITS

split or to the face of the fixed mould plate, springs may be fitted around each guide dowel (Figure 8.38) to nominally return the splits before the mould is finally closed.

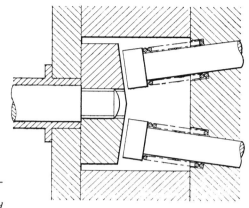

Figure 8.38—Alternative design of ejector plate for use in conjunction with angled guide-dowel actuating method

8.3.2 Cam Track Actuating System. The opening movement of the splits in this design is controlled by a cam track. When the splits are actuated, studs fitted to each end of the split slide along this cam track. Actuation of the splits is by means of a conventional pin ejector system.

Let us consider a typical design. Figure 8.39 is a composite drawing the top and bottom halves of which show the splits before and after actuation, respectively. The splits are fitted into an open channel type chase-bolster which incorporates wear plates. Studs, screwed into the splits, protrude into the cam track machined in the cam track plate (see also lower cross-sectional drawing through A-A). This cam track plate is securely attached to the chase-bolster. A conventional pin ejector system, including push-back pins (not shown), is fitted below the chase-bolster. When the ejector system is actuated the ejector pins press against the splits which then move out at an angle.

Note that a gap is left between the top of the ejector pin and the bottom of the split when the ejector system is fully returned. This ensures that there is no possibility of the split being held off its seating in the chase-bolster by the pins.

The number of ejector pins used depends upon the size of the split. For small and medium-sized moulds, two to four ejector pins may be used. The pins should be at least 13 mm (½ in) diameter and hardened. This is essential as when the splits are actuated they slide across the top surface of the pins.

The opening movement of each split can be calculated from the following relationship:

$$M = E \tan \phi \tag{8.14}$$

where E = effective ejector plate movement,
ϕ = cam track angle.

SPLITS

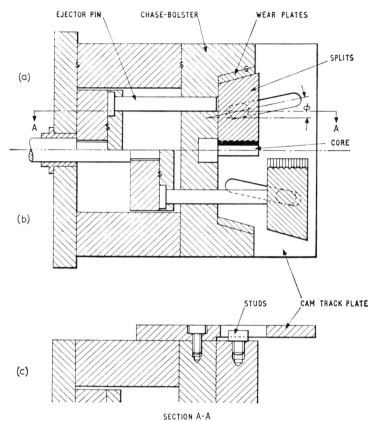

Figure 8.39—Cam track actuating method: (a) split shown seated in channel bolster; (b) split shown in forward, ejected, position; (c) part-section A-A to show stud, cam plate arrangements

Fifteen degrees is a convenient angle for the cam track, although this figure may be exceeded quite safely in order to keep the ejection and splits forward movement to a minimum.

The open-channel bolster (Section 8.2.3) is particularly convenient for use with this actuation method as the cam truck plate can be fitted across the open ends of the channel (Figure 8.40). When the stronger enclosed chase-bolster design is preferred, either the cam track is machined into the end walls of the bolster or false ends are fitted.

One advantage of the cam track design is that a bank of splits in one or more lines can be incorporated. Figure 8.41 shows a cam plate split mould with three pairs of splits in line. The operation of this mould is identical to that previously described using a standard ejector pin system. The advantages of using banks of small splits containing a number of multi-impressions are as follows:

(i) Small splits are easier to make.

ANGLED-LIFT SPLITS

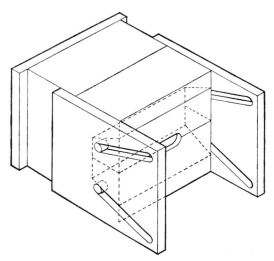

Figure 8.40—Cam plates fitted to open-channel chase-bolster

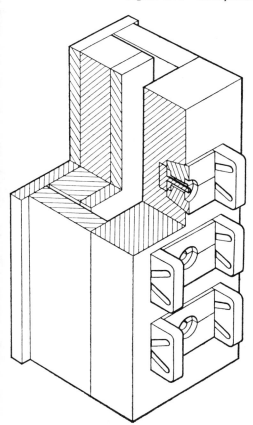

Figure 8.41—Multi-split, cam plate operated mould

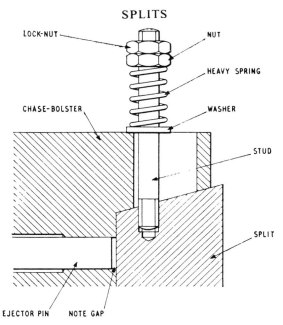

Figure 8.42—Spring actuation method

(ii) Hardening difficulties related to distortion are reduced.
(iii) Width of the split is reduced.
(iv) Damage to one pair of splits does not necessarily mean that production from the mould must stop.

8.3.3 Spring Actuation. A simple method of actuating angled-lift splits is by the use of springs. The forward movement of the split is limited, and because of this feature the design is used for components which have comparatively shallow undercuts.

Figure 8.42 shows a part-section through the relevant parts of the mould. A

TABLE 7

SUMMARY OF ACTUATING METHODS

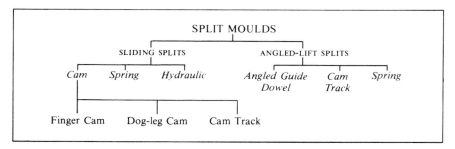

SUMMARY

stud screwed into the side of the split passes through a slot in the chase-bolster. A washer, heavy spring, nut and lock-nut complete the assembly as shown. When the split is actuated by the ejector pin the spring exerts a force which maintains contact between the split and the angled wall of the chase-bolster. Thus the splits progressively open. The splits are returned to their seating by the action of the mould closing. The studs may be fitted at 50 mm (2 in) intervals along the side of each split: the number of assemblies used therefore depends on the length of the split. Note that individual slots are machined into the chase-bolster wall for each stud. The force applied by the spring can be varied by adjustment of the nuts.

8.4 SUMMARY

Table 7 gives a general summary of the more important operating methods for split moulds. The decision as to which operating method to adopt depends on the following:
- (i) amount of splits movement required;
- (ii) whether a delay period is required;
- (iii) length of the delay period;
- (iv) ease with which the moulding can be removed;
- (v) whether a short or long production run is anticipated;
- (vi) whether the available machines are programmed for ancillary cylinder control;
- (vii) whether moulding inserts are to be incorporated.

9

Side Cores and Side Cavities

9.1 GENERAL

A *side core* is a local core which is normally mounted at right angles to the mould axis for forming a hole or recess in the side face of a moulding. This side core prevents the in-line removal of the moulding and some means must be provided for withdrawing the side core prior to ejection. The basic principle is illustrated in Figure 9.1, which shows the side core in the forward and the withdrawn positions, respectively.

The *side cavity* performs a similar function to the side core, in that it permits the moulding of components which are not in-line-of-draw. This element caters for components with a projection or projections on one or more

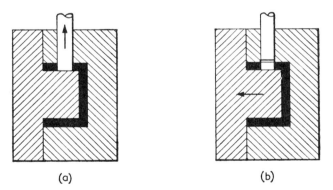

Figure 9.1—Principle of side cores: (a) moulding position; (b) withdrawn position

of their side faces. The side cavity is a segment of an otherwise solid cavity insert or plate which can be withdrawn to permit the moulding to be ejected in line. Figure 9.2 illustrates the side cavity in the forward and withdrawn position.

Typical examples of components which necessitate the use of either a side core or side cavity are shown in Figure 9.3. The arrow in each case indicates where a side element (side cavity or side core) will be required.

9.1.1 Mouldings Embodying Side Holes, Recesses or Slots. In general, any component which has a local recess, hole or slot which is not in-line-of-draw will necessitate the incorporation of a side core in the mould design.

GENERAL

There are, however, a few exceptions to this generalisation, and we will discuss these first. For example, a hole in the side face of a component could be moulded in-line-of-draw by astute component design. In example (a) (Figure 9.4) the hole is formed by a part of the core abutting on to the sloping face of the cavity. To achieve this condition the component must be designed with a definite step as shown. This step is normally equal to the general wall section. The resulting aperture in the moulding is illustrated in the lower drawing.

When a step is not permissible an alternative design may be adopted (Figure 9.4b). In this the side wall of the component is caused to slope at an obtuse angle with respect to the base. This permits the hole to be formed by a projection from the core which abuts on to the cavity as shown. Note that the top face of the projection (at X) must be such that it does not create an undercut. The resulting aperture formed in the moulding is illustrated in the lower drawing.

Moulded holes of this type, while they make the mould design more simple, are often unacceptable to the component designer as they inhibit his style. It

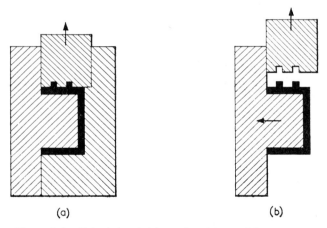

Figure 9.2—Principle of side cavity: (a) moulding position; (b) withdrawn position

is, however, used quite extensively for the cheaper lines of manufacture where mould costs have to be kept to a minimum.

A hole can also be incorporated into the side face of the moulding by a subsequent machining operation. This method should always be considered by the mould designer before proceeding with a side core design as it has the following advantages:
  (i) The mould is simple and therefore relatively cheap.
  (ii) Ease of operation of the mould creates less likelihood of production difficulties.
  (iii) Whereas a side core breaks up the normal flow of material entering the impression and there is a probability of flow lines developing, with this method it is impossible for this to happen.

SIDE CORES AND SIDE CAVITIES

Regardless of flow lines and mould cost, however, economics will usually dictate that the hole must be moulded in by means of a side core, rather than incorporating it as a separate operation, at extra cost.

The general form of the component may make the use of the split design more appropriate than the side core design. Note, however, that if the component necessitates a slender core (discussed later in this chapter) the side core design can be used with advantage.

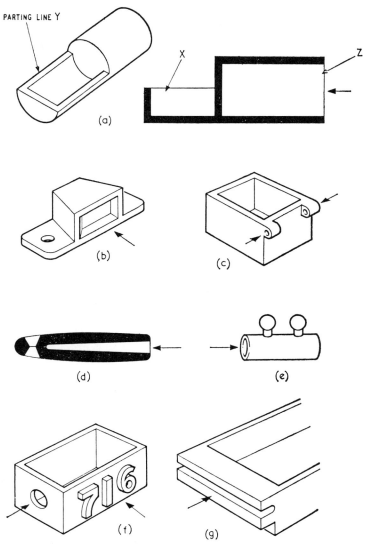

Figure 9.3—Examples of components which necessitate either side core or side cavity

GENERAL

SIDE CORE MOULDINGS. Now that we have discussed other possible methods of producing holes in the side walls of mouldings, we will consider the types of component normally produced by the side core method.

Figure 9.3 (a) illustrates a component which incorporates a recess in two planes. While the recess at X can be moulded in line of draw by making the parting line at Y, the circular recess at Z must be formed by the side core. Note that, while it is possible to make the circular recess Z in line of draw and to use

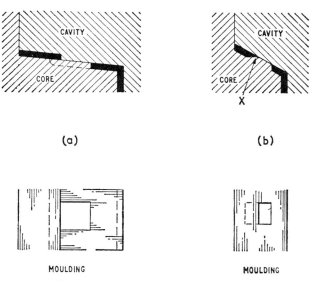

Figure 9.4—Holes formed in-line-of-draw

a side core for recess X, the resulting mould design will be far more complex.

The viewfinder moulding (Figure 9.3b) incorporates a rectangular hole formed by a complementary shaped side core that abuts on to the normal core which forms the inside surface of the moulding.

The box (c) incorporates a hole in each of two lugs which form parts of hinges. These holes may be formed by side cores which are operated in opposite directions to each other.

The previous examples have illustrated components which incorporate holes, the centre lines of which are parallel to the main centre lines of the component. This permits the side core to be actuated at right angles to the main core and in a plane adjacent to the parting surface. However, when the centre-line of the hole is not parallel to the component's main centre-line, the side core must be actuated in a suitable plane which permits its straight withdrawal from the moulding. Consider the box-shaped component (Figure 9.5a) which incorporates a slot positioned at an angle ϕ to the centre line. If one attempted to withdraw the side core at right angles to the main core (direction Y) the undercut created at X would cause cracking. By arranging to withdraw the core in direction Z, no difficulties will be encountered.

SIDE CORES AND SIDE CAVITIES

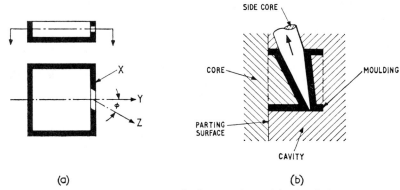

Figure 9.5—Components which necessitate side core being actuated at an angle

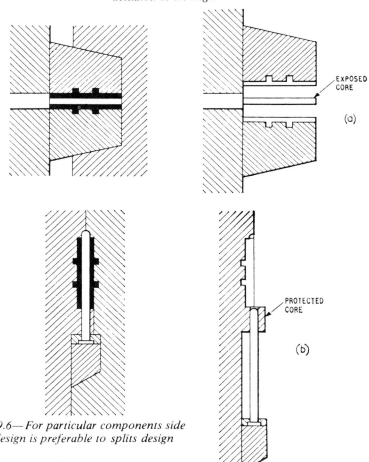

Figure 9.6—For particular components side core design is preferable to splits design

GENERAL

If the component has a hole that is not parallel to the *parting line* as shown in Figure 9.5b (this illustrates a section through a mould for a pencil sharpener body) the side core must be withdrawn at a suitable angle to the parting surface of the mould. Curved holes can be moulded, providing the component design allows the curved core to be withdrawn on a radiused path. An example which illustrates this point is given in Figure 9.33. This shows a mould for a pipe fitting.

Now consider the case where it is preferable to use the side core technique for what would normally be considered a split mould component. The component, a section of which is shown in Figure 9.6, is circular in cross-section, and has a long slender hole through its centre. It incorporates a peripheral undercut on the exterior as shown. Now, if this component is moulded in splits (Figure 9.6a), when the splits are open the long slender core which forms the hole is exposed and is liable to be bent, with serious consequences when the splits are subsequently closed. By adopting the side-core design the outside shape of the moulding is formed in the two-mould plates, the component being moulded at right angles to the mould's axis.

The advantages of this latter design are as follows:
(i) The long slender core is guided at all times in one of the mould plates. This protects the core against accidental damage.
(ii) As the component is adjacent to the parting surface, a great saving in mould height can be achieved as compared with the splits method. Reduction in mould height often allows the component to be moulded in a smaller injection machine than would otherwise be possible.

9.1.2 Mouldings which Incorporate Projections. When a moulding is required which has a projection below the general parting line, the mould designer has three possible designs to consider.

These are: (i) a stepped parting surface, (ii) splits, and (iii) a side cavity. Which of these he chooses will depend very much on the number of projections and their position and shape.

The component shown in Figure 9.7a can be moulded quite simply by the first of these methods. The mould's parting surface is stepped locally so that it passes through the centre of the projection. (Note the thick line on the end of the component.) The projection does not, therefore, create an undercut in the mould and can be ejected in-line-of-draw. As there are no moving parts within the mould, the mould design, manufacture and subsequent operation are more simple than for the other methods. There is a disadvantage, however, in that the mark left by the mould's parting surface is apparent on the finished moulding, and the deeper the projection is located from the top surface the more pronounced will be this mark (see also Figure 5.10).

The parting surface cannot be stepped to accommodate two projections, one above the other, as for example in Figure 9.7b. This component would be formed in splits which, as we have already shown in Chapter 8, may be used for components that incorporate projections below the general parting line. Generally speaking, splits are used where the projections are continuous on the periphery (Figure 9.7b), or where the projections occur on diametrically opposite sides.

SIDE CORES AND SIDE CAVITIES

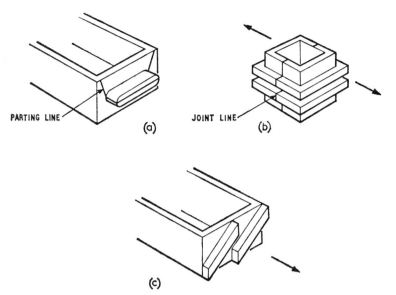

Figure 9.7—Mouldings which incorporate projections: (a) moulded with stepped parting line; (b) moulded with splits; (c) moulded with side cavity

The side cavity design is used for components which incorporate projections on one or more sides and which cannot be more simply moulded by either of the previous methods. Consider, for example, Figure 9.7c, where two projections occur, one overlapping the other. The stepped parting line design cannot be used because of the overlap, and as the projections occur on one face only it would not be desirable to use the splits design either.

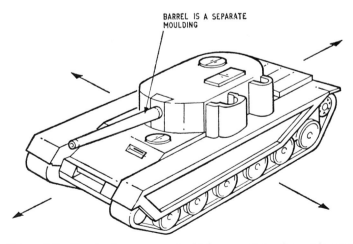

Figure 9.8—Toy tank component which necessitates four side cavities

GENERAL

A further example, a box, is shown in Figure 9.3f. This incorporates figures which project from one of the side faces. For similar reasons to those above, the side cavity design would be chosen. Figure 9.3g shows a part of a sliding tray. A runner groove is incorporated at both ends. Two side cavities would be used for this component, each being operated independently. Similarly, if a component has projections on two adjacent sides, the side cavities would be arranged on a plane at right angles to one another.

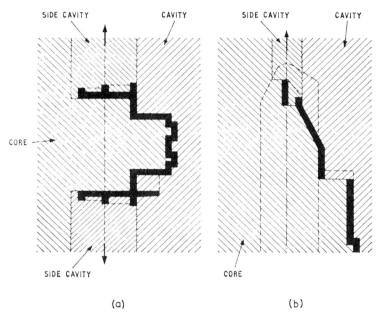

Figure 9.9—Transverse and longitudinal cross-sections through part of mould for toy tank

Finally let us consider the type of component which has projections and undercuts on all four sides, as for example the toy tank (Figure 9.8). A side cavity would be required for each of the four sides of this component. Note that those parts of the component which are in-line-of-draw are formed by the cavity and core (Figure 9.9). This shows a transverse and longitudinal part-section through the centre-line of the mould. (The barrel is a separate piece fitted to the body after moulding.)

9.1.3 Side Core and Side Cavity. The terms *side core* and *side cavity* strictly relate to the core or cavity portion that form part of the impression. In practice the former term, in particular, has come to mean the whole moving assembly. However, for the purpose of this discussion the terms will be kept in their original sense, and the whole moving assembly will be termed *side core assembly* and *side cavity assembly* as applicable.

SIDE CORES AND SIDE CAVITIES

The design requirements are similar to those specified for splits, i.e.:
(i) Arrangements must be made for guiding the assembly.
(ii) Means must be incorporated for actuating the assembly.
(iii) The assembly must be securely locked when in the moulding position.

Mould designs incorporating side cores and side cavities for components having holes or projections vary considerably according to the position, number and form of the restriction. The design can be placed in one of the following classifications.

(i) Designs for mouldings with peripheral undercuts, having a slender centre hole.
(ii) Designs for moulding with holes or slots in one or more sides.
(iii) Designs for mouldings having recesses or projections in one or more sides.
(iv) Designs for mouldings with curved holes.

9.2 DESIGN FEATURES

Having once determined that a side core or a side cavity is necessary for a particular component the next step is to decide how it can be incorporated in the mould. This is most simply achieved by considering the side core or side cavity as an assembly consisting of the movable part of the impression (hereafter called the *side core element* or *side cavity element*) and a carriage to which the element is attached. The carriage is the part of the assembly which provides for the guiding and operating functions of the design. The carriage may be mounted in guides on or below the mould's parting surface, or may be mounted externally on an outrigger system. To differentiate between these, the former is termed an *internal side core* or *side cavity assembly*, and the latter an *external side core* or *side cavity assembly*.

9.2.1 Internal Side Core (or Side Cavity) Assembly. In this design the side core (or side cavity) assembly is similar to splits (Chapter 8) and like methods are adopted for guiding, locking and operating it. The essential design features are illustrated in Figure 9.10. The drawings show sections through a typical assembly, in two planes at right angles to each other. The carriage is T-shaped in transverse cross section and is mounted in guides which are securely attached to the moving mould plate. The side core element is secured by the retaining plate to the carriage. (If the design had necessitated a side cavity element this would have been secured directly to the carriage.) The actuation of the carriage is by means of a finger cam although, as for splits, other forms of actuation are often used. The carriage is locked in the forward position by a locking heel.

The operation of the mould is as follows:

When the mould is opened, the finger cam causes the carriage assembly to move away from the impression and the side core element to be withdrawn from the moulding. As the cam is not in permanent contact with the carriage, provision should be made to ensure that the carriage remains in the withdrawn position when the mould is open. The spring detent method (discussed

DESIGN FEATURES

in Chapter 8) may be used for this purpose (Figure 8.29). When the mould closes, the cam re-enters the angled hole in the carriage and the assembly is progressively returned to the moulding position. The final movement and lock is attained by the locking heel. A mould of this type is shown in Plate 9.

INTERNAL SIDE CORE (OR SIDE CAVITY) ASSEMBLY DETAILS. As previously stated, the assembly consists of a carriage and either a side core element or a side cavity element. These elements may be secured to the carriage either directly or by means of a retaining plate. The carriage has the same T-shape form as the sliding split and the various methods of build-up shown in Figure 8.5 apply here also. However, in practice, the carriage is normally relatively shallow and the one-piece design (a) is usually preferred. The carriage, unlike the split, rarely carries the moulding form so it may be made from a low-carbon steel, suitably heat-treated, to give a wear-resisting surface.

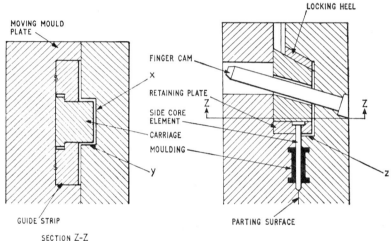

Figure 9.10—Internal side core assembly. Longitudinal and transverse cross-section through part of mould

CARRIAGE FITTING. The nature of the side core assembly (or side cavity assembly) does not normally permit the carriage to be mounted on top of the moving mould plate, and room for the carriage is provided by a pocket machined in both mould plates. To illustrate the construction a sketch of the side core assembly, mounted in the moving mould plate, is shown in Figure 9.11. The major part of the carriage is accommodated below the parting surface in a pocket (Y). The width of this pocket is sufficient to accommodate the guide strips, which are secured by screws (not shown). The portion of the carriage which projects above the parting surface must be accommodated in a pocket in the fixed mould plate (Figure 9.10, right view). One face of this pocket is angled to form the locking heel.

The remainder of the pocket in the fixed mould plate is dimensioned so

SIDE CORES AND SIDE CAVITIES

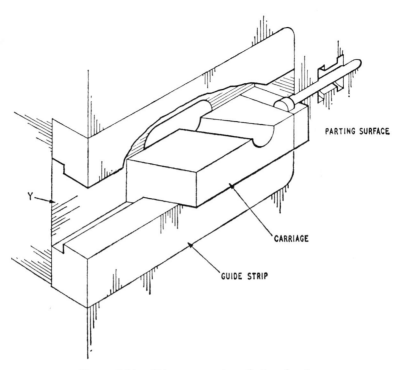

Figure 9.11—Side core carriage fitting details

that there is a definite clearance all around the top section of the carriage—x, y, z.

GUIDING ARRANGEMENT. The guiding arrangement for the internal side core (or side cavity) assembly is similar to that adopted for guiding splits (Section 8.2.1). The assembly is a relatively small unit, however, which allows a local guiding arrangement to be used. This contrasts with the splits design where the guides normally extend completely across the mould (Figure 8.3). The female T-form can be machined from the solid mould plate using a T-type milling cutter, but the built-up assembly (Figure 9.11), utilising the guide strips, is generally preferred as it allows for a simple machining and fitting operation.

METHODS OF ACTUATION. In our discussion, so far, we have shown that the internal side core (or side cavity) assembly can be actuated by means of a finger cam. Amongst other methods, dog-leg cam actuation and spring actuation are commonly used. The formulas given in Chapter 8 relating to the above actuating methods also apply here. The finger cam method of actuation is used when a short delay period only is required. The amount of clearance between the cam and the cam hole determines the actual delay period (see Figure 9.10). The finger cam must be of sufficient length to withdraw the

DESIGN FEATURES

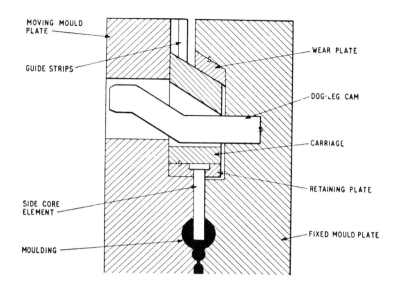

Figure 9.12—Dog-leg cam method of actuating side core assembly

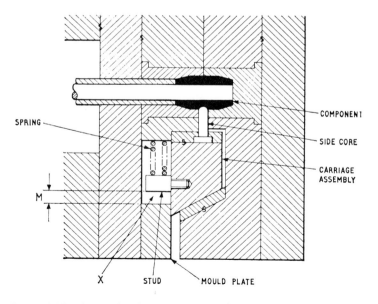

Figure 9.13—Spring-loaded system for actuating side core assembly

255

SIDE CORES AND SIDE CAVITIES

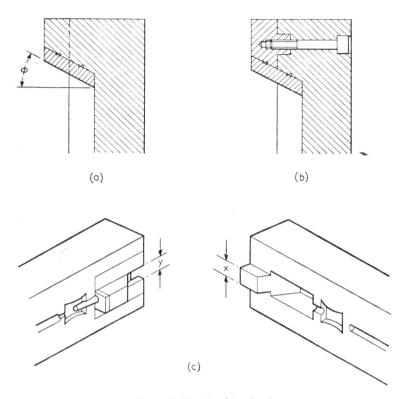

Figure 9.14—Locking heel

side core (or side cavity) from the moulding completely. It will be apparent, therefore, that a very deep moulding undercut necessitates long finger cams. These cams must be of substantial cross-section to withstand the forces applied during the operating period.

The dog-leg cam method of actuation, the general design of which is shown in Figure 9.12, may be specified if an extensive delay period is required. It is often desired, for example, to withdraw the moulding completely from the fixed half before retracting the side core. This ensures that the moulding remains in the moving half in readiness for ejection.

The spring-loaded system is an operating method confined to those mouldings that have very shallow undercuts or projections. Figure 9.13 shows a section through a side core assembly of this design. The component has a small indentation in one side formed by a side core mounted in a standard carriage assembly. For spring actuation a stud is attached to the underside of the carriage (as shown) and this is accommodated in a slot (X) machined in the mould plate. A spring or springs are fitted in the slot and cause the side core assembly to withdraw immediately the mould opens. Note that the locking heel is used to progressively return the assembly when the mould is being closed.

256

DESIGN FEATURES

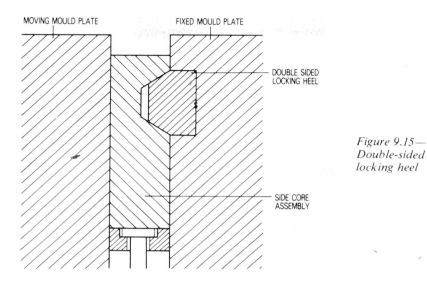

Figure 9.15—
Double-sided
locking heel

The maximum withdrawing movement of the assembly must be controlled, and this may be achieved by restricting the movement of the stud as shown (M).

LOCKING THE CARRIAGE ASSEMBLY. The final closing movement and lock of the carriage assembly in all of the actuating methods discussed above are effected by means of a locking heel (Figure 9.14). The undesirable method of using the operating cams to perform this locking function should be avoided.

The angle ϕ specified for the locking heel, and the corresponding angle of the carriage, are normally between 15° and 40°. When finger cams are used to actuate the carriage the angle must always be greater than the operating angle (Section 8.2.3). To provide wear-resisting surface for the locking heel, and to provide some means of adjustment for wear, etc., wear plates are fitted on the angled face. These plates are normally made from low-carbon steel and subsequently carburised. The locking heel normally protrudes above the parting surface and two alternative designs are shown (Figure 9.14). The preferred design, with the heel an integral part of the mould plate (a), is costly because a large amount of material is removed in manufacture. The alternative method is to fabricate the heel by securing a heel block on to the mould plate as shown (b). A projection is provided on the underside of this heel to withstand the applied forces.

Accommodation must be provided in the moving mould plate for that part of the locking heel which projects above the parting surface. In practice, the heel is accommodated in the space provided for the carriage movement in the moving mould plate (c). The width of the heel (x) must therefore be less than the distance between the guide strips (y).

An alternative design for the locking heel is shown in Figure 9.15. The functional member in this case is basically trapezoidal in cross-section and is

SIDE CORES AND SIDE CAVITIES

secured to the fixed mould plate as indicated. This *'double sided locking heel'* engages with a complementary shaped recess machined across the side core assembly.

The object of this design is to ensure that the side core assembly is locked in position, in two directions, when the mould is shut.

A typical example of the use of this design is illustrated in Plate 9.

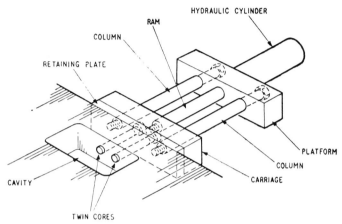

Figure 9.16—External side core assembly. (Note—In this drawing columns are shown longer than required, for reasons of clarity. In practice, the length of these columns depends on side core movement required)

9.2.2 External Side Core (or Side Cavity) Assembly. In this design the side core (or side cavity) is coupled to an externally mounted carriage. This carriage, often supported by an outrigger arrangement, is generally actuated by hydraulic or pneumatic means, although cam and spring actuation are occasionally used.

Figure 9.16 shows a typical mould which illustrates the main features of this design. The mould is for a box-shaped component having two holes in one side. To form these holes, twin cores project through the side walls of the cavity. The side cores are attached to the carriage by a retaining plate. The carriage is supported on two columns which, together with the platform, are termed the *outrigger*. The ram of a hydraulic cylinder (suitably mounted on the platform) is attached to the carriage which by a suitable control system can be actuated when required. We will now consider the assembly in detail.

SIDE CORE (OR SIDE CAVITY) ASSEMBLY DETAILS. A side core or side cavity assembly consists basically of a side core or side cavity element, retaining plate and a carriage (Figure 9.16). Positive alignment of the side element is essential and in the design illustrated the side cores are fitted in close tolerance holes, machined in the side wall of the cavity. It is essential to allow these side cores to float in the retaining plate in a similar manner to the ejector pin float discussed in Chapter 3 (see Figure 3.24).

DESIGN FEATURES

In most designs, the carriage is supported on the columns of the outrigger as shown. The columns pass through suitably bushed holes in the carriage. The carriage must be locked in its forward position to withstand the force applied to the side element by the pressurised melt. This force is the product of the pressure (within the cavity) and the effective projected area of the side element. To clarify this point we will consider some examples (Figure 9.17).

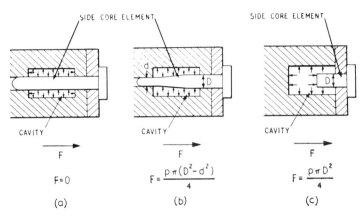

Figure 9.17—Force applied to side core (side cavity) by melt

(a) In this design a parallel side core passes completely through, and slightly beyond, the end of the cavity. The effective projected area being zero (i.e. in the direction of side core movement) the force applied to the side core will also be zero.

(b) A tapered side core which also passes completely through the cavity is shown in this design. The effective projected area in this case is the difference between the cross-sectional areas of the two sections of the core. The force applied can thus be computed from:

$$F = \frac{p\pi(D^2 - d^2)}{4}$$

where p is the intensity of pressure of the melt within the cavity and D and d are the large and small diameters of the side core, respectively.

(c) In this design the side core does not pass completely through the cavity and thus the effective projected area is the total cross-sectional area of the side core (i.e. $\pi D^2/4$ for a circular side core).

Most side cavity designs come within this latter category, as the pressurised melt is normally applied to the entire 'impression' face of the element. Thus if A is the impression area in question the force applied equals pA.

When hydraulic or pneumatic actuating systems are adopted, the locking force is provided by the actuating medium. The size of the actuator must therefore be calculated so as to withstand the force applied by the pressurised

SIDE CORES AND SIDE CAVITIES

melt. A safety margin should be included in the calculations to allow for:
  (i) instantaneous cavity pressures in excess of the theoretical maximum.
  (ii) slight compression of the actuating medium.

When side core elements are not directly in line with the actuating ram the design of the carriage must be such that bending tendencies are resisted. For example, when injection pressure is applied to the side cores shown in Figure 9.18 a bending moment will be applied to the carriage, tending to bow the member as shown. It is essential, therefore, that the depth (x) of the carriage is sufficient to withstand this bending force without undue deflection.

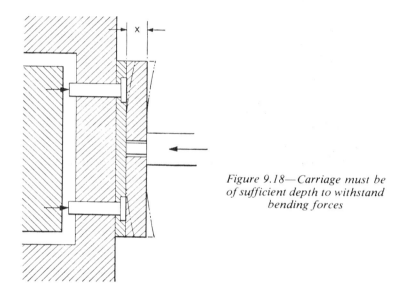

Figure 9.18—Carriage must be of sufficient depth to withstand bending forces

THE OUTRIGGER. The primary purpose of the outrigger is to provide a platform upon which to mount a hydraulic or pneumatic actuator. The secondary purpose of the outrigger is to provide means of guiding and supporting the carriage. This is of particular importance when cam and spring methods of actuation are used.

Several designs for this unit are shown in Figure 9.19. The system shown at (a) is used in conjunction with a light actuator. Two columns project from the mould wall and a platform is mounted on them as shown. As the ram of a light actuator is relatively slender it is usually desirable to support the weight of the carriage, which is indicated by the chain-dotted line, on the columns. (An isometric sketch of a complete assembly including the actuating cylinder is shown in Figure 9.16.) For heavier cylinders, a four-column design may be adopted (b). With this system the proportions of the actuating ram are usually sufficient to support the carriage. However, if extra support is thought to be desirable, half-shoes may be fitted which slide on the columns. The relative

DESIGN FEATURES

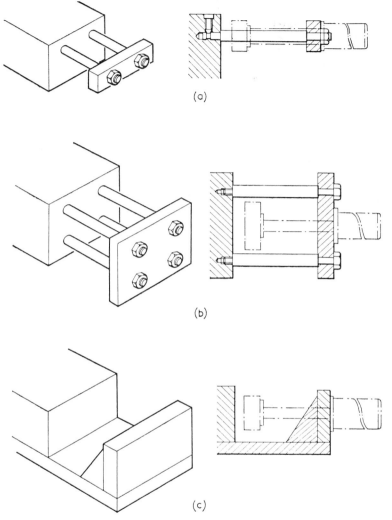

Figure 9.19—Outriggers for side core and side cavity designs

position of the carriage and the actuator in this example is shown by the chain-dotted line in the sectional drawing.

Plate 10 shows a two-impression mould which incorporates an outrigger of the four-column design. In this example side cores are incorporated on two opposite sides of the mould.

An alternative design is shown at Figure 9.19c. This design does not incorporate columns but is of a fabricated, welded steel construction with the two triangular members giving the required rigidity to the platform.

The platform in all designs may be subjected to a considerable force and

SIDE CORES AND SIDE CAVITIES

the designer should ensure that the thickness of this platform is sufficient to withstand this force without undue deflection.

A mild steel is suitable for both the platform and the columns, although the latter items should be given a case-hardened surface if they are to act as guides for the carriage.

METHODS OF ACTUATION.

The actuator. The most common method of operating the external carriage is by means of an actuator. The ram of the actuator may be either directly coupled to the carriage or indirectly coupled via a toggle linkage system. In the first case, the actuator is mounted on an outrigger platform and the ram coupled to the carriage either by a screw or by a flange connection. A typical arrangement (Figure 9.20) illustrates a side core mounted on the parting surface of a mould. The side core is shown in the forward position (a) and in the withdrawn position (b). To prevent undesirable bending forces the actuator should be mounted in line with the core so that it can apply a direct pull (or push) to the side element. The actuator may be operated either by hydraulic or pneumatic means.

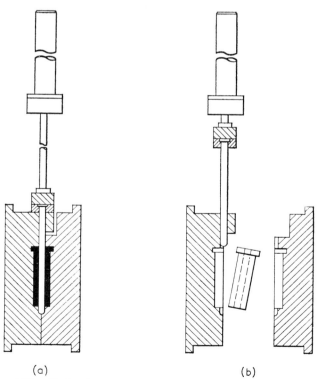

Figure 9.20—*Hydraulic (or pneumatic) actuation of side core assembly*

DESIGN FEATURES

The hydraulic supply is obtained either by coupling the side core actuator to the injection machine's hydraulic circuit or by coupling to an independent portable pumping unit. In either case, pressures of between 7×10^6 and 14×10^6 N/m^2 (1×10^3 and 2×10^3 lbf/in^2) are usually available.

The pneumatic supply, on the other hand, is normally tapped from the factory air line system, the pressure of which is seldom more than 7×10^5 N/m^2 (100 lbf/in^2).

To apply a corresponding locking force, the size of a pneumatic actuator must be many times larger than that of a hydraulic one. It follows, therefore, that the pneumatic method is usually limited to cases where a relatively small locking force is required, for example where the core passes completely through the moulding (Figure 9.17a).

The advantages of using a hydraulic or pneumatic actuator as against other methods of actuation include:

(i) The carriage can be operated at any suitable time (e.g. side core or side cavities can be withdrawn before the mould is opened).
(ii) Long side core withdrawal strokes can be achieved very easily.
(iii) Design and mould making are simplified.

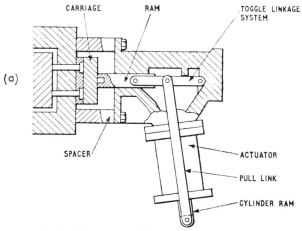

Figure 9.21—Toggle actuator unit for side core withdrawal

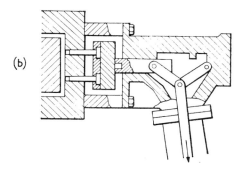

263

SIDE CORES AND SIDE CAVITIES

There is a disadvantage, however, in that the setting-up time of the mould is extended due to the necessity of coupling the actuator to the power supply.

A mould which incorporates actuators for operating side cores is shown in Plate 11. Note that a finger cam is also incorporated to actuate the third side core assembly, transversely.

In the second case a toggle actuator unit (manufactured by Engineering Tool and Product Design Co., Ltd) is attached to the mould. The basic assembly is shown in Figure 9.21a. The ram is actuated by a hydraulic or a pneumatic actuator via a toggle linkage system. When the cylinder ram is

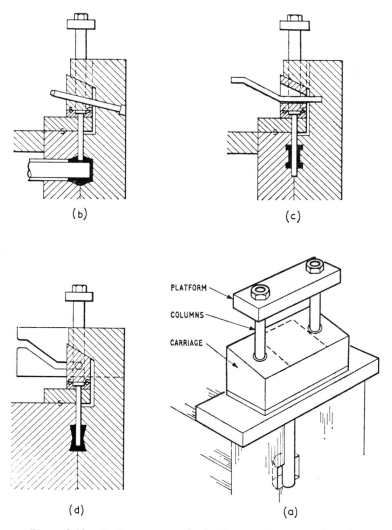

Figure 9.22—Various cam methods of actuating external carriage

DESIGN FEATURES

actuated, the pull link causes the toggle links to withdraw the ram, which slides in phosphor bronze bearings within the frame.

To allow for the movement of the carriage the toggle actuator unit is mounted on spacers. The lower drawing (Figure 9.21b) shows the ram in the withdrawn position. The additional advantages of this system over the direct thrust actuator are:

(i) Maximum pull at the moment of initial movement of approximately 10 times the cylinder force.
(ii) The side element is mechanically locked by the toggles within the frame.
(iii) The standard unit permits simple replacement should damage occur.

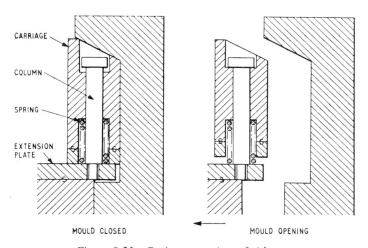

Figure 9.23—Spring actuation of side cores

Cam actuation. Finger, dog-leg and plate cams can be incorporated in certain designs to operate the external carriage in a similar manner to that described for operating the internal carriage (Section 9.2.1).

These methods are normally confined to those designs where the side core is mounted on the parting surface; examples of these are shown in Figure 9.22.

The basic design (Figure 9.22a) and the operating technique are common to all three methods. The carriage is guided on the columns; the platform in this design acts solely as a cross-beam to maintain the alignment of the columns. Assuming the carriage is mounted on the moving mould half (as shown), when the mould is opened the cam (fitted to the fixed mould half) causes the carriage to be withdrawn. The reverse action takes place as the mould is being closed. The carriage is locked by the locking heel coming into contact with its angled face. The dotted lines indicates the position of the locking heel.

With systems using finger cams (Figure 9.22b) or dog-leg cams (c) a suitable hole is provided in the carriage to accommodate the cam, whereas for the

SIDE CORES AND SIDE CAVITIES

cam plate system (d) a stud is provided on either side of the carriage to fit into the cam plate. Details of the cams, materials, and formula for movement are given in Section 8.2.2.

Spring actuation. If it is required to move the side core only a short distance then spring actuation may be considered. Figure 9.23 shows a part-section through the relevant details of this system. The carriage is mounted on columns screwed to the extension plate. Suitable recesses in both the carriage and the retaining plate accommodate a spring mounted on each column. When the mould is being opened the springs exert a force on the carriage which progressively withdraws the side element to which it is attached (not shown). The shouldered heads on the columns limit the outward movement of the carriage. As the mould closes the locking heel contacts the sloping face of the carriage which progressively returns the side element to the moulding position.

9.3 TYPES OF SIDE CORE AND SIDE CAVITY

The component form determines the type of side element which may be used. For example, a component which incorporates a hole on the parting line requires the side core to be mounted on the parting surface of the mould. If the hole is below the parting line or the component incorporates a projection, the side coring becomes more difficult as there are various factors to take into consideration, as will be discussed later.

When the designer has to deal with a curved hole in a component an even more complex system is required as the curved core, to form the hole, must be withdrawn on a radius.

When flash is liable to develop it is important that the mould design is such that wherever possible the flash is in line of draw. Unless close attention is paid to this point the ejection of the moulding may be impeded and build-up of the flash may prevent the side element from operating efficiently. Figure 9.24 illustrates the undesirable and preferred designs, respectively.

9.3.1 Side Cores on the Parting Surface. When a hole or recess occurs on the parting line of a moulding a side core can be mounted on the mould's parting surface. This is the ideal condition as it permits the carriage assembly to be mounted on either mould half, whichever is the more convenient.

The two basic designs to be considered are (i) where the side core element is semi-located, and (ii) where the side core element is positively located.

SEMI-LOCATION. This method is where the side core element is permanently located in a slot in one mould half. An example where this method has been used is shown Figure 9.25.

The component, a knife handle, incorporates a centrally placed hole for the knife tang. The female form of the handle is formed by a pair of cavity inserts mounted in suitable bolsters. The side core element which forms the hole is a rectangular section blade fitted into a slot in the carriage and secured by a pin.

TYPES OF SIDE CORE AND SIDE CAVITY

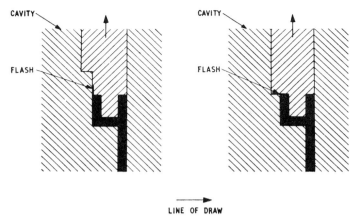

Figure 9.24—*Fitting of side elements should be such that if flash occurs, it is in line of draw: (a) undesirable design; (b) preferred design*

That part of the blade which is outside the moulding form (X) is housed an equal depth in both mould inserts. The cross-sectional shape of the blade and of the corresponding slots in the cavity inserts are shown enlarged at (c). A 1° angle should be incorporated on each side of the top half of the blade (Y) to avoid excessive wear occurring. A similar angle is not required on the bottom of the blade (Z) as this portion is in permanent contact with the cavity insert.

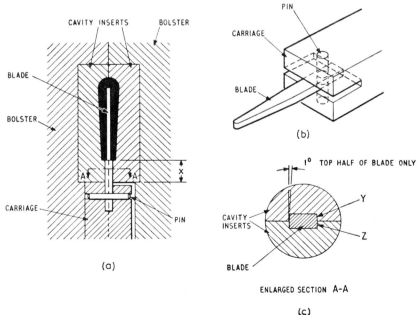

Figure 9.25—*Semi-location of side core element*

267

SIDE CORES AND SIDE CAVITIES

POSITIVELY LOCATED SIDE CORE. It is often desirable to give a more positive location to the side core than that given in the previous example. This is particularly so when the design necessitates very slender round cores. With the previous design there is the possibility that a very slender core may be accidentally moved out of its locating recess, with the likelihood of the mould being damaged. To obviate this possibility the side core can be positively located in a block adjacent to the insert.

An example of a positively located side core design is shown in Figure 9.26a. The side core is positively located by the location block secured to the cavity insert; both are mounted in the mould plate. To allow the side core to

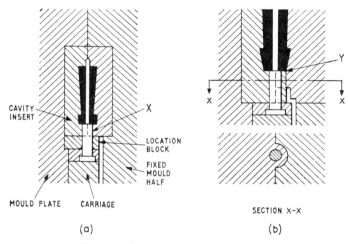

Figure 9.26—Positively located side cores

pass through the location block, the block must protrude beyond the parting surface of the mould and therefore a suitable pocket is provided in the other mould half as shown. Note, however, that that part of the side core which is immediately adjacent to the impression is located in a groove in both mould inserts in an identical manner to the previous example (i.e. at X).

An alternative design for this component is given in Figure 9.26b. This design is a little more complex but it does allow for greater side core support. The location block is integral with the cavity insert and the face of the block (Y) now forms part of the impression. A section taken through the side core and the locating cavity insert is shown in the lower drawing and from this cross section it can be seen that the form of the projection follows that of the component. To accommodate the projection a complementary shaped pocket is provided in the opposite cavity insert. This means, in effect, that the parting surface is locally stepped. As the hole in this component runs the entire length it is good practice to continue the core beyond the end of the component and to provide a nest for this in both cavity inserts. The end of the side core should be conical to provide for self-alignment.

TYPES OF SIDE CORE AND SIDE CAVITY

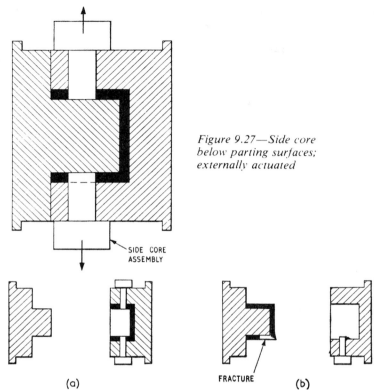

Figure 9.27—Side core below parting surfaces; externally actuated

Figure 9.28—(a) Side cores hold moulding in cavity from which it is ejected. (b) With an unbalanced side core system moulding may be damaged as mould is opened

9.3.2 Side Cores Below the Parting Surface. When a hole or any other form of undercut occurs below the parting line of a component, the side core incorporated in the mould to form this restriction must inevitably be positioned below the cavity parting surface. The side core assembly can still be positioned on either mould half, but a considerable difference in design is required, depending upon which mould half is chosen.

EXTERNALLY ACTUATED SIDE CORES. Consider the case of a simple component, such as a box, which incorporates a hole in one side wall. A schematic drawing of a mould for this component is shown in Figure 9.27. The side core in this case operates through the cavity wall and is normally withdrawn prior to the mould being opened.

In certain instances it may be preferable not to withdraw the side core immediately the mould is opened. This will cause the moulding to be left in the cavity, from which it can quite simply be ejected (Figure 9.28a). There is a tendency, however, for the moulding to remain on the core, and if this design is contemplated for a component which necessitates only a single side core the

269

SIDE CORES AND SIDE CAVITIES

possibility that the component may be fractured during the opening stage should not be overlooked (Figure 9.28b).

The side core is attached to an externally mounted carriage assembly and is

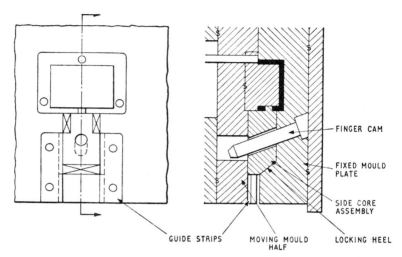

Figure 9.29—Internally mounted side core

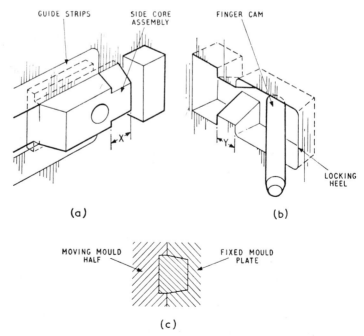

Figure 9.30—Sketch of important details of Figure 9.29

TYPES OF SIDE CORE AND SIDE CAVITY

normally hydraulically actuated. A typical assembly of a mould with side cores mounted in this manner is shown in Figure 9.16. The side cores pass through the cavity wall and are retained by the retaining plate, which is attached to the carriage. In this example the carriage is actuated and locked by a hydraulic cylinder.

INTERNALLY MOUNTED SIDE CORES. It is often convenient to adopt the internal carriage design so that, for example, a cam method of actuation can be used. A typical finger cam actuated design is shown in Figures 9.29 and 9.30. The side core assembly, which is shown constructed as a solid member for simplicity, is mounted on the moving mould half in guide strips. When the mould is closed the assembly nestles in a pocket which is machined in the fixed mould plate. The assembly is withdrawn by the finger cam as the mould is opened, thereby permitting the moulding to be ejected. During the mould's closing stroke the assembly is returned approximately to its moulding position by the cam, the final movement and lock being accomplished by the locking heel.

In this internal design, where we are mounting the side core assembly on the core plate, it is essential that the side element forms part of the cavity in addition to forming the hole.

To ensure a leak free joint between the side element and the cavity, the front end of the side element (at X) is designed to nestle accurately in a complementary slot, machined into the cavity plate (at Y). To reduce manufacturing difficulties the length of this matching region is usually limited to 10 mm (⅜ in). The cross-sectional shape of this region is normally trapezoidal, as shown in inset Figure 9.30c. Outside this matching region, and apart from the rear face which forms the locking heel, the remainder of the pocket, machined in the cavity plate to accommodate the projecting portion of the side element assembly, can be given a generous clearance.

9.3.3 Angled Withdrawal. A component which has a hole in its side at an angle (other than a right angle) generally requires the side core assembly to be mounted at a corresponding angle in the mould. Apart from this feature the remainder of the design is identical to that previously described.

Now consider when it is essential to use an angled side core. The components shown in Figure 9.31 incorporate conical holes, the centre lines of which (A-A) are not at right angles to the component's centre line (ZZ). Whereas in one case (b) the line Y-X forms an obtuse angle with the centre line (ZZ) permitting vertical withdrawal, the acute angle formed in the second case (at a) necessitates angled withdrawal. An example of an angled side core design is shown in Figure 9.32. This is a part section through a mould for a model boat. The angled hole (to accommodate the propeller shaft) is formed by the core attached to the carriage. This carriage is suitably mounted at an angle on the core plate, and is actuated by the finger cam mounted in the fixed mould plate. The side core assembly is locked in the moulding position by means of the locking heel.

SIDE CORES AND SIDE CAVITIES

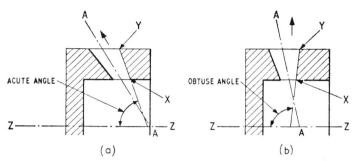

Figure 9.31—Illustrating where side core must (a), and need not (b), be withdrawn at an angle

9.3.4 Curved Side Core. The curved side core is a particular type of side core where provision must be made for withdrawing the core along the arc of a circle. The curved side core is necessary for components such as pipe fittings and telephone hand sets, etc., where a smooth curved internal hole is required. These components are such that the side core can be mounted on the parting surface of the mould.

Figure 9.33 shows a plan view of a mould for a pipe fitting. The small inset shows a cross section through the carriage and actuating arm. The curved

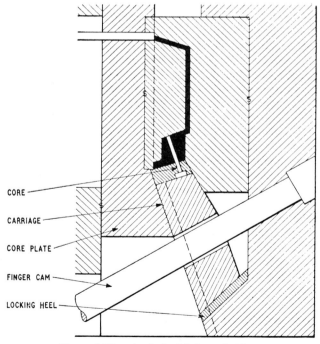

Figure 9.32—Angled withdrawal side core

272

TYPES OF SIDE CORE AND SIDE CAVITY

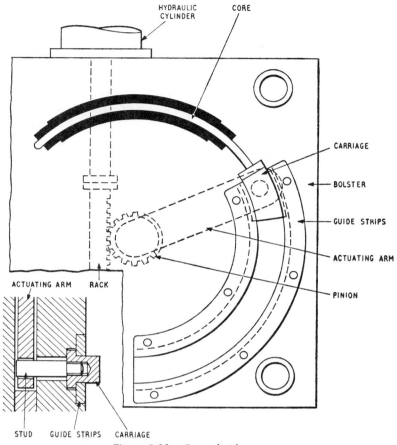

Figure 9.33—Curved side core

side core is attached to the carriage which is mounted in a curved pocket machined into the bolster. The carriage is retained and guided by curved guide strips. The centre line chosen for the system must be the same as the centre-line radius of the component, to allow the side core to be withdrawn. Apart from the radius, the basic assembly is the same as that for a side core mounted on the mould's parting surface. None of the previously discussed direct actuation methods are, however, suitable for this design, due to the curved operating path.

One method of actuation is the hydraulically actuated rack and pinion shown in the figure. The ram of the hydraulic cylinder is attached to a rack, the latter being a free-running fit in a bored hole in the bolster below the parting surface. Movement of the rack actuates a pinion attached to the actuating arm. A slot in the actuating arm accommodates a stud secured to the carriage. Movement of the actuator ram thereby operates the side core on a curved path.

273

SIDE CORES AND SIDE CAVITIES

9.3.5 Side Cavities. When a projection or recess occurs on the side face of a component, preventing it from being moulded in-line-of-draw, it is necessary to design the cavity so that a part of it is movable, to allow the restriction to be relieved. That part of the cavity which is movable is termed the *side cavity*. The basic principle of this design is shown in Figure 9.2 which shows the side cavity in the moulding position and the withdrawn position, respectively.

The side cavity design can be treated in a similar manner to that of the side core, and in fact with certain designs the two terms are synonymous.

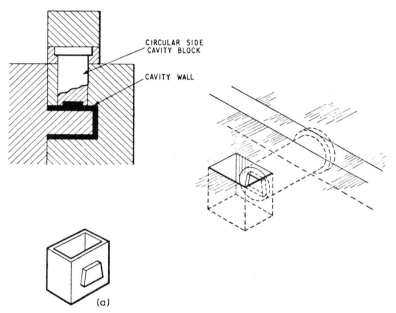

Figure 9.34—Side cavity mounted on cavity side, design 1

SIDE CAVITIES MOUNTED ON THE CAVITY SIDE. When it is required to mount a side cavity on the cavity side, the identical design to that described for a side core, mounted on the cavity side, is used. Figure 9.34 shows an example to illustrate the similarity. The component in this case is a box moulding which incorporates a projection on one of the side walls (a). The projection is formed by a recess machined into a circular side cavity block which extends through the cavity wall. A circular cavity block is preferred as it is simple to make and fit. Circumstances, however, often dictate that square or rectangular side cavity blocks must be incorporated.

The side cavity block is attached to an external carriage (Section 9.2.2) and the actuation can be by any of the methods discussed in this section.

When the length, shape or position of the restriction is such that it cannot be incorporated in a cavity block, which passes through the cavity wall, then the alternative designs shown in Figure 9.35 may be considered. In this design the side cavity block is mounted in a pocket which is machined adjacent to the

TYPES OF SIDE CORE AND SIDE CAVITY

cavity insert. The width of the pocket (W) is sufficient to permit the side cavity block to be actuated so as to relieve the restriction. By actuating the side cavity immediately prior to opening the mould, the moulding will remain on the core and can be ejected in the conventional way.

As the side cavity must be operated while the locking force is applied to the two mould plates, care must be taken in the dimensioning of the side cavity and the associated slot to ensure that the unit can, in fact, function.

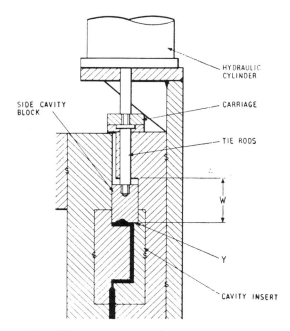

Figure 9.35—Side cavity mounted on cavity side, design 2

The side cavity block is generally returned prior to the mould being closed to ensure that it abuts the cavity insert at Y. Tie-rods connect the side cavity block to the carriage which is normally actuated by a hydraulic cylinder as shown. A considerable force will be applied within the impression by the plastic melt, which will tend to move the side cavity block back. This force must be resisted and, therefore, a relatively large hydraulic cylinder will be required. For the reason mentioned in the next section this design is only occasionally used.

SIDE CAVITIES MOUNTED ON THE CORE SIDE. In this design (Figure 9.36) the side cavity block is attached to a carriage mounted in guides (not shown) on the moving half. A suitable pocket is machined in the fixed mould half, adjacent to the cavity, to accommodate the side cavity assembly when the mould is closed. A finger cam is shown as the actuating method. If a comparison is made between this design and the side core design (illustrated in Figure 9.29)

SIDE CORES AND SIDE CAVITIES

it will be seen that the two designs are similar, and the same general comments are therefore applicable to both.

This design is generally preferred to the previous design in that a positive lock is applied by the locking heel to the side cavity assembly when the mould is closed. (In the previous design the lock is achieved hydraulically.)

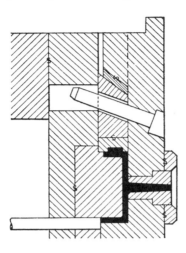

Figure 9.36—Side cavity mounted on core side

MULTIPLE-SIDE CAVITIES. Certain components, because of their form, will necessitate the use of more than one side cavity. Consider for example the toy tank shown in Figure 9.8. This component has numerous projections and undercuts on all four sides. (Note that the barrel is a separate component.) The turret and top of the tank, being in-line-of-draw, are incorporated in a solid cavity as shown (Figure 9.9).

A section taken transversely through the mould (Figure 9.9a) illustrates how the sides of the tank are formed, while the longitudinal section (Figure 9.9b) shows the side cavity which forms the front of the component. (A similar side cavity is required for the rear.) Note that in all cases the forward movement of the side cavity is positively stopped by the side cavity block coming up against the core insert.

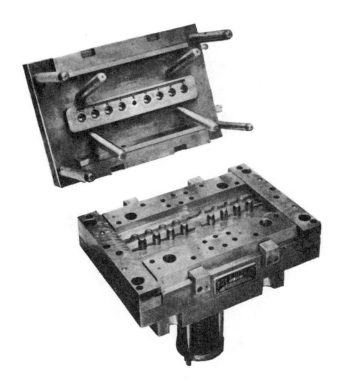

Plate 7   Eight-impression split mould for lens holder (Courtesy Fox and Offord Limited)

Plate 8 Single-impression split mould for a jug top (Courtesy Bakelite Xylonite Limited)

Plate 9 Six-impression side core mould for composite toothbrush head (Courtesy Fox and Offord Limited)

Plate 10 Hydraulically actuated side core mould (Courtesy E. Elliot Limited. Mould manufactured by M.C.M. Tools Limited)

Plate 11 Three-impression side core mould for sink trap parts (Courtesy Fox and Offord Limited)

Plate 12   Two-impression mould for cursor (Courtesy Fox and Offord Limited)

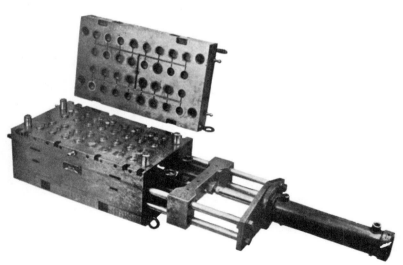

Plate 13   Hydraulically actuated 32-impression unscrewing mould for threaded caps (Courtesy Fox and Offord Limited)

Plate 14 Cylindrical hot-runner plate manifold mould (Courtesy Tooling Products (Langrish) Limited: photo by Mann Brothers)

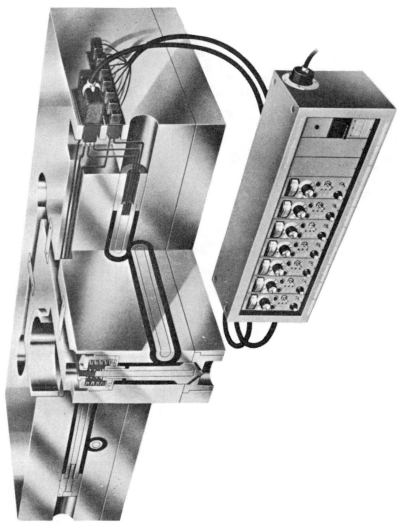

Plate 15   Hot-runner plate mould (Courtesy DME (USA and Europe))

10

Moulding Internal Undercuts

10.1 GENERAL

An internal undercut is any restriction which prevents a moulding from being extracted from the core in line of draw. Various methods are used for relieving internal undercuts; the specific design adopted depends upon the shape and position of the restriction. Figure 10.1 shows examples of components with internal undercuts; where necessary part of the component only is shown to reveal the undercut portion.

Provided the undercut is local (Figure 10.1a), the restriction may be incorporated on a *form pin* which is moved forward during the ejection phase. In the simplest version of this design, the form pin has a straight action (similar to an ejector pin) and the moulding is subsequently removed from the form pin at right angles to the mould's axis. The alternative method is to mount the form pin at an inclined angle so that during ejection the top face of the pin moves inwards, thereby relieving the undercut.

When an undercut extends completely along one internal wall of a component, Figure 10.1b, it may be moulded by incorporating a core which is constructed of two parts. The part which incorporates the restriction is subsequently moved forward during ejection, thereby permitting the moulding to be extracted. The movement of the moving part, termed the *split core*, may be either straight or inwardly inclined. This latter design may also be used for components which incorporate undercuts on opposing faces as illustrated by the components (c) and (d).

The conventional side-core design (Chapter 9) may be used for certain internal undercut components but this method is restricted to those components which incorporate an open side, as for example (e) and (f). Certain types of undercut components may be stripped from the core and as this method allows for relatively cheap mould construction it should not be overlooked. Two typical components which would be moulded in this manner are shown at (g) and (h).

Each of the above designs for dealing with the internal undercut type component will now be dealt with in more detail.

10.2 FORM PIN

A component which has a local undercut portion can be successfully moulded in the conventional mould by incorporating the undercut form on a form pin. Two examples are shown in Figure 10.2. The top moulding (a) has a

MOULDING INTERNAL UNDERCUTS

small projection on one of the inner walls, while the lower moulding (b) has a recess in a similar position. Enlarged views of the form pins which form these restrictions are shown at (c) and (d), respectively. Note that the face of the pin (X) forms part of the impression and must, therefore, have the same surface finish as the remainder of the core. The form pin will leave a *witness mark* at Y on the mouldings and careful mould fitting is essential to prevent flash occurring at this point. With this design there are two basic alternatives; the

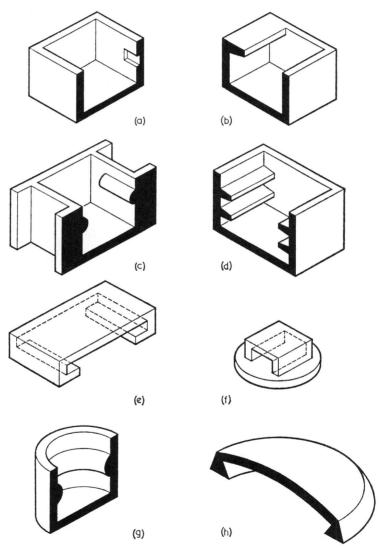

Figure 10.1—Internal undercut components; to illustrate undercut only half component is shown in some examples

FORM PIN

form pin can either have a straight action, i.e. as for an ejector pin, or it can have an angled action.

10.2.1 Form Pin: Straight Action. This design is normally used for components which incorporate an undercut on one internal wall only. It is impracticable to use this design for a component which has an undercut on two facing walls.

The mould assembly for this design is identical to the basic pin ejection design (Figure 3.22), except that one (or more) of the ejector pins is replaced by a form pin which contains the undercut form. An enlarged view of the relevant parts is shown in Figure 10.3, which also illustrates the principle of operation. The closed mould is shown at (a) and the form pin, in its rear position, forms part of the impression. The moulding shrinks on to the core and is withdrawn from the cavity when the mould is opened. The form pin is moved forward during ejection allowing the moulding to be removed in the direction indicated (b). The form pin is attached to the ejector assembly in a similar manner to an ejector pin, but a locking dowel pin must be incorporated to prevent rotation.

10.2.2 Form Pin: Angled Action. The basic feature of this design is that the working face of the form pin is caused to move inwards relative to the core during ejection, thereby relieving the undercut. It can be used for components which incorporate internal undercuts on one or more walls.

The basic design and method of operation is illustrated in Figure 10.4. The form pin which incorporates the undercut form is fitted at an inclined angle ϕ in the mould plate and it is maintained in contact with the ejector assembly by means of a key plate suitably attached to the mould plate. When the ejector assembly moves forward (relative to the core) the undercut is relieved by the

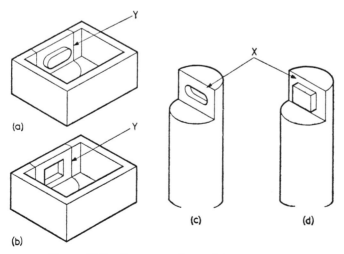

Figure 10.2—Form pins for undercut components

MOULDING INTERNAL UNDERCUTS

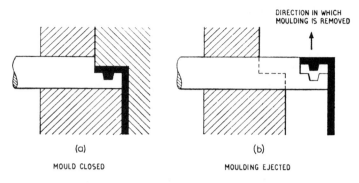

Figure 10.3—Form pin—straight action

lateral movement of the form pin and the moulding can be lifted or blown clear (Figure 10.4b). The amount of withdrawal obtained may be computed from the following relationship:

$$M = E \tan \phi$$

where (see Figure 10.4) M = the withdrawing movement,
E = ejection movement,
ϕ = fitting angle of the form pin.

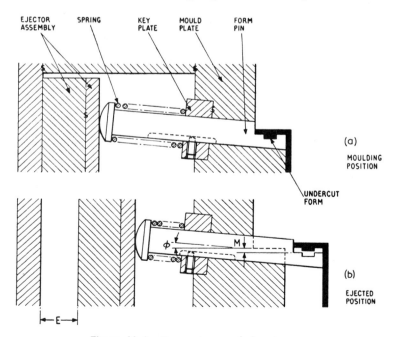

Figure 10.4—Form pin—angled action

FORM PIN

The form pin rests on the ejector assembly in this design and, because of the angle, a relative movement takes place between them during ejection. Because of this movement, either the retaining plate should be hardened or a local insert incorporated below the form pin to prevent undue wear occurring. The actual movement across the retaining plate is equal to the withdrawal movement at the top of the form pin.

This is illustrated in Figure 10.6, which shows the form pin in the forward position (shown dotted) superimposed on the form pin in the rear position.

The basic spherical headed form pin design (Figure 10.4) is not favoured by many designers because there is only point contact with the retaining plate and the following alternative designs are commonly used.

Figure 10.5 shows a spherical headed form pin mounted in a *nest block* which is free to slide on the retaining plate. All the other features are the same as for the basic design. The advantage of using the nest block is that a greater contact surface is obtained while the self-aligning feature of the spherical end is retained.

Figure 10.6 shows another variation on the basic approach but in this design the form pin head is angled to present a large contact surface to the retaining plate. Note that very careful fitting is necessary with this design to ensure that the improved surface contact is in fact achieved. Though it is not shown (for reasons of clarity), the form pin is spring-loaded as in the previous designs.

In the design in Figure 10.7 the form pin, which has a standard ejector pin head, is attached to a slide block by a small retaining plate. The slide block is of a general T-form and is mounted in guides which are attached to the ejector plate. Rotation of the form pin is prevented by the dowel pin. When the ejector plate is actuated, because the form pin is constrained to move at an angle, the slide block will move in direction V across the face of the ejector plate. Similarly, as the ejector plate is returned, the slide block will be progressively returned to its original position. It does not rely on a spring to return the form pin to the moulding position and, because of the guides, permanent flat sliding surfaces in contact are assured.

Let us now consider a design (Figure 10.8) where the face of the ejector

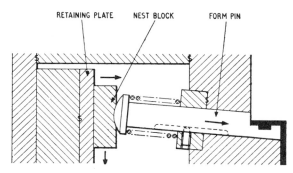

Figure 10.5—Form pin—angled action; alternative head, design 1

MOULDING INTERNAL UNDERCUTS

assembly is angled to correspond to the actuating angle of the form pin. This face may be obtained either by machining at an angle and then inserting a hardened steel wear plate (a), or a hardened steel angle block can be mounted on a plain ejector plate (b). Contact between the form pin, which has a standard ejector pin head, and the angled face is maintained by the spring. As the ejector plate is actuated there is relative movement between the two parts in direction W.

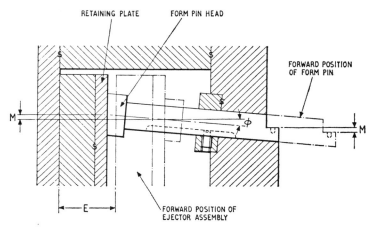

Figure 10.6—Form pin—angled action; alternative head, design 2

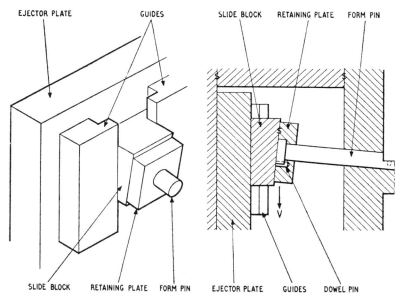

Figure 10.7—Form pin—angled action; alternative head, design 3

SPLIT CORES

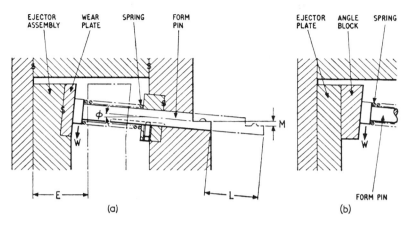

Figure 10.8—Form pin—angled action; alternative ejector plate design

The withdrawal movement obtained with this design is different from that with the previous design, because the ejector plate is now angled. The following relationship applies

$$M = (E \sin 2\phi/2) \tag{10.2}$$

where (see Figure 10.8) M = the withdrawal movement,
E = the movement of the ejector plate,
ϕ = the operating angle of the form pin.

Note that with this particular design it is undesirable to incorporate ejector pins in addition to form pins as the two sets of pins will have a different movement. The forward movement of the ejector pin will be equal to E, whereas the forward movement of the form pin = $E \cos^2 \phi$. This difference in forward movement, if permitted, would result in damage to mouldings.

10.3 SPLIT CORES

The split core design is used for components which have extensive internal undercuts that cannot be incorporated on a form pin. Typical components in this category are shown in Figure 10.1b, c, d. To permit this type of component to be moulded successfully it is necessary to manufacture the core in two (or more) parts and that part in which the undercut is formed is subsequently moved forward during ejection to permit the moulding to be extracted. This split core, as it is termed, may be moved forward either in a straight plane or an angled plane. In both designs the core is constructed as shown in Figure 10.9. The movable part, the split core, nestles in a pocket machined in the moving mould plate. The sides of the split core and of the complementary pocket walls are angled as shown. This minimises the wearing action which would occur if the sides were straight. When in position, the fit between the split core and the main core must be precise to prevent the very undesirable

MOULDING INTERNAL UNDERCUTS

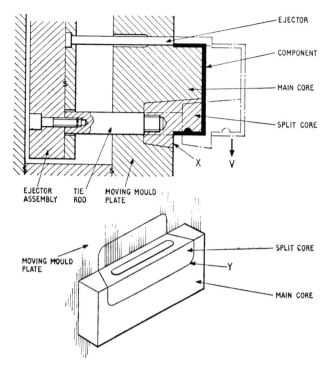

Figure 10.9—Split core design—straight action

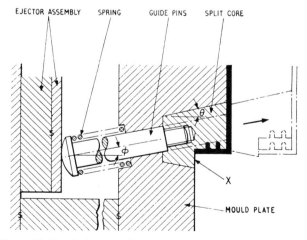

Figure 10.10—Split core design—angled action, design 1

SPLIT CORES

feature of flash developing between the two parts (i.e. along line Y). The main core may be machined from the mould plate, as with the integer method, or it may be incorporated as an insert. This latter method is normally preferred as it permits more simple machining of the split core pocket.

10.3.1 Split Cores—Straight Action. This design is used for components which incorporate an external undercut on one wall only. It is not a practicable design for components which have undercuts on opposite walls. The design and operating principle are shown in Figure 10.9. The component chosen to illustrate this design is box-shaped and incorporates a projecting bead on one of the inside walls. This bead is formed by a recess which is machined into the split core as shown in the lower drawing. The split core is directly attached to the ejector assembly by means of a tie rod, so that when the ejector assembly is actuated the split core moves forward in a straight line. The moulding can then be removed at right angles to the mould axis (i.e. in direction V). Face X is used as an ejector face but additional ejection will be required on the other moulding walls as indicated.

10.3.2 Split Cores—Angled Action. In this design the split core is caused to move inwards during the ejection stroke, thereby withdrawing the restriction and allowing the moulding to be extracted in-line-of-draw. The advantage of this is that the withdrawing action is automatic and the moulding does not have to be removed at right angles to the mould's axis as in the previous case (Section 10.3.1). This design can therefore be used for components with undercuts on opposing faces in addition to being used for components with an undercut on one wall only.

There are two principal ways by which the split core can be actuated and there are many variations of these two methods. The first design is shown in Figure 10.10. The split core is constrained to move at an angle with respect to the line of draw by means of guide pins attached to it as shown. The operating angle ϕ must be less than that chosen for the fitting angle θ to prevent fouling when the core is actuated. The guide pins are not attached to the ejector assembly as, due to the angle, there is relative movement between these two members. When the mould opens and the ejector plate is actuated, the dome-headed guide pins and the split core to which they are attached move forward inwardly inclined, thereby relieving the moulding's undercuts (indicated by the chain-dotted line). When the mould is closed the split core is returned nominally by the spring but it is finally held in position by the fixed mould half acting on face X.

The alternative design is shown in Figure 10.11; here again the split core is constrained to move in a path at an angle to the line of draw by means of guide pins. But in this design the pins are used only for guiding and they are independent of the ejector system. The split cores are actuated by ejector pins fitted to the ejector assembly in the normal way.

When the mould opens and the ejector plate is actuated, the ejector pins act on the base of the split core, and because of the angle pins the split core moves forward obliquely, thereby relieving the undercuts. There is relative move-

MOULDING INTERNAL UNDERCUTS

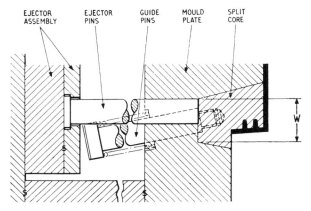

Figure 10.11—Split core design—angled action, design 2

ment between the top face of the ejector pin and the split core, therefore the base of the latter must be sufficiently wide (W) to allow for this. When the mould is closed the ejector assembly is returned to the rear position by pushback pins (not shown).

In either design two guide pins per split are normally used, and if the mould plate through which the pins are to pass is to be left unhardened, then bushes should be fitted. Where springs are used to nominally return the split core to its rear position it is important to check that with the maximum possible ejector stroke the springs do not become coil-bound.

When undercuts occur on faces of a component directly opposite each other care must be exercised in the design to ensure that the two split cores do not foul each other as they are moved obliquely forward. The limit of this movement is determined by the point at which the split cores just touch. From the illustration (Figure 10.12) it will be observed that the maximum forward movement of the split cores is dependent on both the operating angle ϕ and the width (G) of the main core.

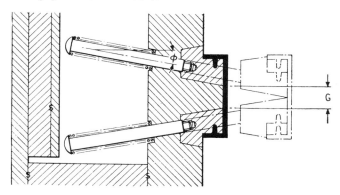

Figure 10.12—Split core design—angled action; twin split cores for relieving undercuts on directly opposite faces

10.4 SIDE CORES

Some components with internal undercuts can be moulded using a conventional side core design as discussed in Chapter 9. However, this method is limited to components which have at least one open side. This permits the side core to be withdrawn through this side to relieve the undercut and allow the moulding to be ejected in-line-of-draw. Typical examples of components of this type are shown at (e) and (f), Figure 10.1.

A part-sectional drawing of a mould for producing a typical component is shown in Figure 10.13. The cavity for the component (a) is machined into the moving mould plate. Note that this allows the moulding surface of the fixed mould half to be completely plain. The internal form of the component is machined on the side core, which is mounted in guides on the moving mould

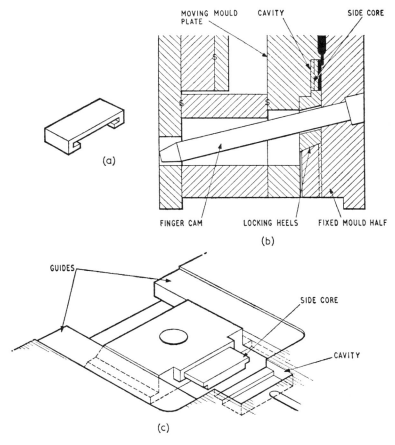

Figure 10.13—Side core design; (a) typical component; (b) section through relevant parts of mould; (c) pictorial view of moving mould half with side core withdrawn

MOULDING INTERNAL UNDERCUTS

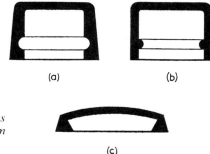

Figure 10.14—Components which can be stripped from the core

half. The side core is withdrawn as the mould opens by the action of the finger cam (see Chapter 9). Continued mould movement causes the ejector system (not shown) to operate and the moulding to be ejected. When the mould closes the side core is progressively returned to its forward position by the finger cam but the final movement and locking action is by the locking heels. A mould of this type is shown in Plate 12.

10.5 STRIPPING INTERNAL UNDERCUTS

A simple and effective way of dealing with a particular type of internal undercut is to strip the moulding off the core. Whether or not a moulding can be moulded in this way depends on several factors which include: (i) the shape of the undercut, (ii) the elasticity of the material, and (iii) whether the external form permits expansion during ejection. Ideally the undercut should be of the form shown in Figure 10.14 (a) or (b) but relatively sharp-cornered undercuts are stripped in practice, an example being shown at (c). (Note that while the components illustrated in Figure 10.14 are circular and the undercuts continuous, non-circular and non-continuous undercuts can also be moulded in this way.)

The stripper plate design is the most common method used for stripping this type of component but valve ejection and sleeve ejection are also used. An example of a stripper plate design is shown in Figure 11.5. This illustrates a mould for stripping mouldings with internal threads, which are of course a

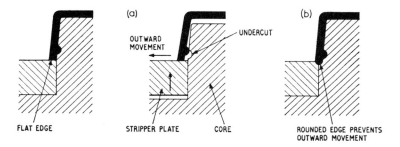

Figure 10.15—Edge shape limitations for stripping components

STRIPPING INTERNAL UNDERCUTS

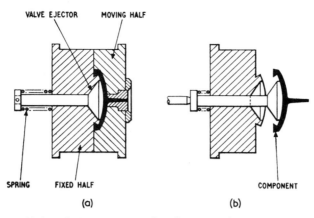

Figure 10.16—Stripping internal undercuts; valve ejector method

type of internal undercut. The mould opens at P1 and the moulding which incorporates the undercut is withdrawn from the cavity. At a later stage the ejection is actuated and the stripper plate strips the moulding from the core.

As the component must be free to expand during ejection, it is important that its bottom edge is flat. The reason for this is illustrated in Figure 10.15. At (a) a cap component, which has an internal bead, is shown being extracted by a stripper plate. As the stripper plate is moved forward the bead is forced from the recess in the core and the moulding is expanded slightly in the process as shown. However if the component has a rounded edge (b), which must be partly formed in the stripper plate, expansion is prevented and the moulding may be damaged during its extraction.

An alternative stripping design is illustrated in Figure 10.16. A large valve ejector is shown extracting the moulding from the core. For efficient ejection the diameter of the head of the valve ejector should be as large as possible consistent with maintaining sufficient strength in the remainder of the core. With some materials, too small a valve would tend to push a hole in the moulding instead of stripping it.

Before attempting a stripping design for a multi-impression mould it is desirable to design a prototype single-impression mould to ensure that the component can, in fact, be stripped.

11

Moulds for Threaded Components

11.1 GENERAL

It is frequently necessary to design a mould for a component which incorporates a thread. Now a thread is a form of undercut and, as we have seen in previous chapters, an undercut normally increases the intricacy of the design. The extent of complication varies and it is dependent on a number of factors, namely:

(i) *Type of thread:* internal, external, continuous or discontinuous.
(ii) *Method:* thread form moulded or incorporated by the use of a metal insert.
(iii) *Type of production:* manual, semi-automatic or fully automatic.
(iv) *Other considerations:* does the required thread form allow for stripping, etc?

As we move on to consider the various types of mould design which cater for these factors it will become apparent that, apart from the unscrewing moulds, the designs discussed are similar to, or extensions of, those already covered in earlier chapters.

Before proceeding we first deal with those aspects of component design which have a direct bearing upon the type of mould required.

11.1.1 Component Design. The moulded component which incorporates a thread may be primarily classified as *external* (male) or *internal* (female). An example of each type is shown in Figure 11.1a, b. The small-diameter internal thread (under, say, 7 mm (¼ in) diameter) is normally incorporated in a moulded part to permit the attachment of other parts, usually by means of metal screws. For example, the threaded internal holes in the instrument case (c) permit a cover to be secured. Another method of achieving this is to mould plain holes using a self-tapping screw to fasten the relevant parts together.

Small diameter internal threads may be incorporated by moulding in a metal insert which embodies the required thread (Figure 11.1d). This design obviates the necessity for long slender threaded cores which are very susceptible to damage. The larger internal thread would normally be moulded without this metal insert and be used primarily in conjunction with moulded male threaded components. An example of this type is illustrated at (e).

When a relatively large component incorporates one or more small diameter projecting male threads (see, for example, the junction box (Figure 11.1f)) then again the use of a metal insert, this time with a threaded male form, should be considered (Figure 11.1g).

GENERAL

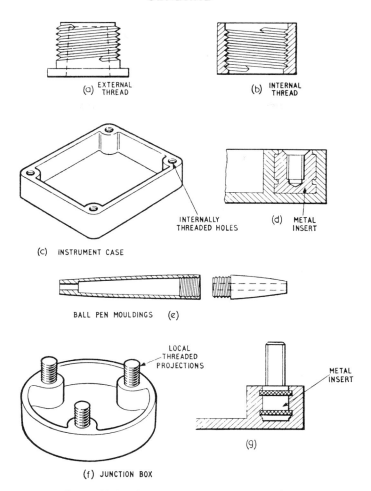

Figure 11.1—Components incorporating threads

When the moulded thread (male or female) is to be used in conjunction with a threaded metal component then pitch shrinkage compensation must be considered. We have noted previously that the size of any impression must be increased to allow for the shrinkage of the plastic material as it cools. Thus, if we require a normal fit over the complete length of engagement between a metal and a plastic component, the pitch of the mould thread must be suitably increased to allow for this shrinkage.

Let us now consider the actual threaded portion of the component. Unless this section is correctly designed, problems will be created for both the mould-maker and the production engineer. An example of a poor design is illustrated in Figure 11.2a. The thread is shown running out to a sharp edge at the mouth of the cap. This thread would quickly become ragged in use, and it is

MOULDS FOR THREADED COMPONENTS

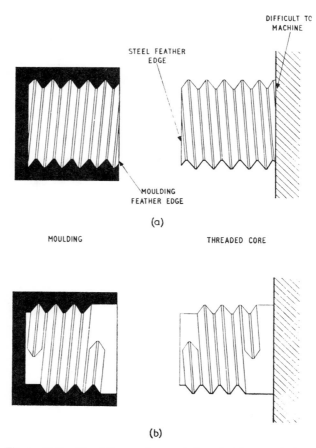

*Figure 11.2—Moulding an internal thread: (a) poor design
—note feather edges; (b) preferred design*

difficult to machine on the core. (See the complementary core design.) The thread is also shown running out at the bottom of the cap and to produce this would give the edge of the core a fragile feather edge which would break down in production, with the result that subsequent mouldings would be produced with a damaged thread. The preferred design is shown at (b). Here the thread has a definite start and finish. This is simpler for the mouldmaker to produce and it obviates feather edges in core and component alike.

These general comments apply also to male threaded components and are illustrated by Figure 11.3.

11.2 MOULDS FOR INTERNALLY THREADED COMPONENTS

The internal thread comes within the broad definition of an internal undercut in that the thread forms a restriction which prevents the straight draw

MOULDS FOR INTERNALLY THREADED COMPONENTS

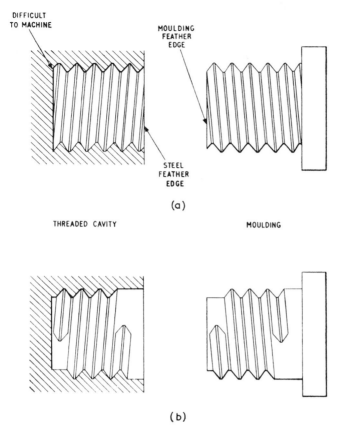

Figure 11.3—Moulding an external thread: (a) poor design —note feather edges; (b) preferred design

removal of the moulding from the core. While certain of the methods given in Chapter 10 can be used for particular thread forms (for example the stripping method for roll threads) the majority of internally threaded components are formed in moulds which incorporate threaded cores which are rotated to eject the component.

Where the total production requirements are relatively small, less complicated designs may alternatively be used. One such method is the fixed core design. In this, the moulding is unscrewed manually from the core after being extracted from the cavity. Another method is to use a loose-threaded core to form the thread, subsequently ejecting this with the moulding.

11.2.1 Fixed Threaded Core Design. This mould design is the male counterpart of the threaded cavity design (Section 11.3.1) and in general the same comments apply.

293

MOULDS FOR THREADED COMPONENTS

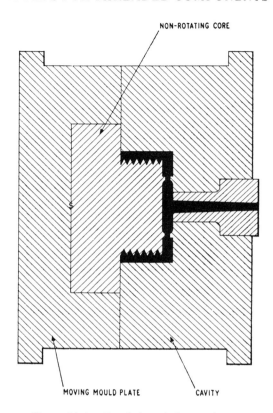

Figure 11.4—Fixed threaded core design

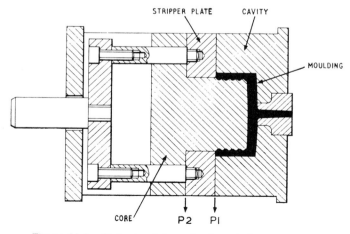

Figure 11.5—Stripping internally threaded components

MOULDS FOR INTERNALLY THREADED COMPONENTS

Figure 11.4 shows a section through a simple mould of this type. The thread form is incorporated on a non-rotating core attached to the moving mould plate. An integer type cavity forms the external shape of the moulding.

In operation, when the mould is opened, the moulding remains on the core and is subsequently unscrewed by the operator. The external shape of the component must be such that it provides facility for this, commensurate with the devices available to the production department concerned.

The advantages of this design, compared with the unscrewing type mould design, are as follows:
(i) Mould cost: Considerably cheaper (no unscrewing or ejector mechanism is required).
(ii) Servicing: No moving parts within the mould (servicing costs are kept to a minimum).

The major disadvantage of this design, particularly for multi-impression moulds, is that the individual mouldings must be unscrewed manually, thereby increasing considerably the moulding cycle time.

11.2.2 Stripping Internal Threads. The internally threaded component may be stripped from the core using the basic stripper plate design providing a roll thread is required and the plastic material has sufficient elasticity during the ejection phase. The same principle applies here as to stripping any internal undercut, i.e. the moulding must be free to expand during ejection to permit the moulded undercut to ride over the restriction on the core. This means that the outside form of the component must be such that it can be formed in a cavity which is fully contained in one half of the mould. (If part of the moulding form is incorporated in the stripper plate it would tend to restrict the required expansion.)

A mould of this type is illustrated in Figure 11.5. This diagram shows a single impression mould for an internally threaded cap. The moulding is formed by the cavity and the core. Ejection is normally by means of a stripper plate as shown. Let us briefly consider the mould operation. The mould opens initially at P_1 and, because of moulding shrinkage and the undercut form, the moulding is withdrawn from the cavity. During the latter stages of the opening movement the stripper plate is checked, thereby causing the second opening to occur at P_2. The moulding is, therefore, progressively stripped as the core is withdrawn through the stripper plate.

The aperture in the stripper plate must be slightly larger in diameter than the major diameter of the thread in order to prevent scoring of the thread. While the illustration shows a single-impression mould, the same principle can be applied to multi-impression moulds. (See also Section 11.3.3.)

11.2.3 Loose Threaded Cores. In cases where a large component incorporates a local internally threaded hole (see, for example, Figure 11.1c) or has several internally threaded holes in close proximity to each other, the loose threaded core technique should be considered. This technique obviates automatic unscrewing, thereby considerably reducing the cost of the mould.

MOULDS FOR THREADED COMPONENTS

Where a number of holes are closely spaced, automatic unscrewing often becomes impracticable anyway.

The basic principle of the loose threaded core design is illustrated in Figure 11.6. At (a) the mould is shown closed. The threaded hole in the moulding is being formed by the loose threaded core. This loose core, which has a valve head type seating, is accommodated in a pocket machined into the main core. When the mould is opened (b) the moulding is ejected by an ejector pin system. The loose threaded core is ejected with the moulding, and is subsequently unscrewed.

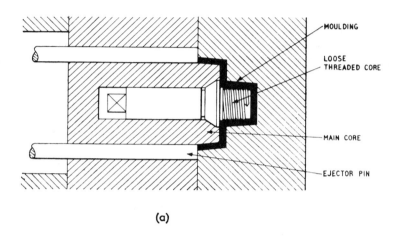

(a)

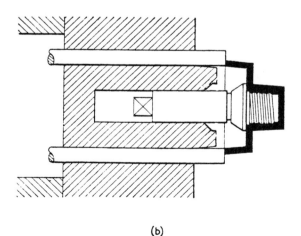

(b)

Figure 11.6—Loose threaded cores: (a) in moulding position; (b) in partly ejected position

MOULDS FOR INTERNALLY THREADED COMPONENTS

Two sets of loose cores should be used during production. At the end of the first moulding cycle, the second set of cores can be inserted into the mould and the next cycle commenced. During this cycle the first set of loose cores can be removed from the first moulding and made ready for insertion into the mould immediately the next component is ejected and so on.

Consider the case of a component where a threaded hole is incorporated with a boss (Figure 11.7). If the design shown previously (Figure 11.6) is used, the boss will tend to restrict the ejection of the moulding from the core. This restriction could cause distortion of the component. To overcome this, the boss must be positively ejected from the core. A standard ejector pin fitted behind the loose threaded core would not be suitable in this instance because, after ejection, the pin will be in the forward position preventing reinsertion of the loose core. Figure 11.7 shows one method by which this problem may be solved. The loose threaded core is contained in a pocket machined into the ejector pin. Positive ejection is achieved and reinsertion of the next loose core is possible with the pin in the forward position; again, using two sets of cores, a fast moulding cycle can be achieved. On many modern injection machines

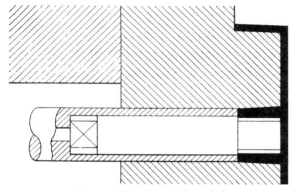

Figure 11.7—Loose threaded core mounted in ejector pin

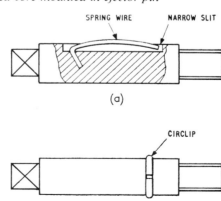

Figure 11.8—Methods of retaining threaded core in mould

MOULDS FOR THREADED COMPONENTS

the ejection is operated hydraulically, in which case the above problem does not arise as the ejector system can be returned while the mould is open.

When loose threaded core moulds are used on a conventional horizontal injection moulding machine, a simple catch device must be provided to prevent the loose cores from falling out during closure of the mould. Two such devices are shown in Figure 11.8. Device (a) uses spring wire, and device (b) a circlip. The principle is the same in both methods, i.e. a frictional retaining force is exerted on the loose core, keeping it in the required position. To facilitate unscrewing the loose core from the moulding, a square portion is provided at the lower end.

11.2.4 Unscrewing Moulds. High labour costs and other modern production requirements demand the maximum use of automatic operation. Manual unscrewing of individual components from the mould is thereby precluded from all but experimental runs or small batch production.

The more complex unscrewing mould must therefore be considered where the component design does not permit stripping of the threads.

In an unscrewing type mould either the cores or the cavities are rotated to automatically unscrew the mouldings from the mould. To provide the required rotary motion an unscrewing unit is fitted behind the moving mould plate in place of the conventional ejector unit. In certain designs however, where positive ejection is required, an ejector system may also be incorporated.

From the impression construction standpoint we have three designs to consider. In one, the threaded core is merely rotated to remove the moulding, in the second in addition to being rotated it is also withdrawn and in the third, the cavity is rotated. Which particular design to adopt depends largely upon the design of the component.

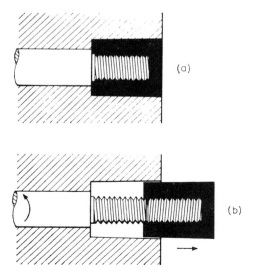

Figure 11.9—Principle of axially fixed rotating core design

MOULDS FOR INTERNALLY THREADED COMPONENTS

There are two basic impression layouts applicable to this section; they are (i) the pitch circle diameter (PCD) layout, and (ii) the in-line layout.

Various power systems are available to actuate the unscrewing mould and there are also alternative methods of linking the power source to the mould. These may be broadly classified as follows:

Power source
  (i) Manual
  (ii) Machine
  (iii) Hydraulic or pneumatic
  (iv) Electric

Transmission system
  (i) Gear train
  (ii) Chain and sprocket
  (iii) Rack and pinion
  (iv) Worm and worm-wheel

From the foregoing it will be evident that extensive variation of design can be achieved by alternative combinations of the various layout, power and transmission systems applied to the basic types of design which we will now proceed to consider in detail.

AXIALLY FIXED ROTATING CORE. This design is particularly suitable for a component whose external form permits the cavity to be located in the same mould half as the threaded core (Figure 11.9). When the threaded core is rotated, in an axially fixed position, with respect to the cavity, the moulding is progressively ejected provided the cavity is maintained in a stationary position. The external shape of the moulding must be such that it cannot rotate with the core during the ejection phase. (Let us relate this to the function of unscrewing a nut from a bolt. If we rotate the bolt with the left hand, and prevent the nut from rotating with the right hand then the nut will progressively ride up the bolt until the threads disengage.) Smooth cylindrical components would obviously be unsuitable subjects for this technique because the moulding will tend to rotate with the core once the initial cavity-moulding adhesion is broken.

The length of the threaded section of the core should, ideally, be slightly less than the depth of the cavity so that the cavity exercises a restraining effect

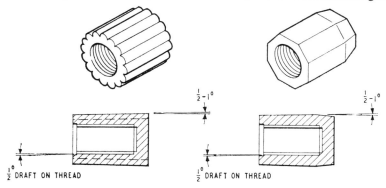

Figure 11.10—Typical components which could be moulded utilising axially fixed rotating core design

MOULDS FOR THREADED COMPONENTS

during the complete ejection phase. If the length of the threaded section is relatively small compared with the cavity depth, it will be necessary to incorporate an ejector pin through the centre of the threaded core to positively eject the moulding after the unscrewing operation is complete. Alternatively, if the length of threaded section of the core exceeds the depth of the cavity, the moulding will not have been completely unscrewed when the restraining effect of the cavity is lost. One solution here is to increase the size of the gate

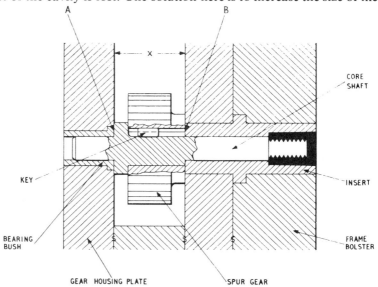

Figure 11.11—Illustrating how axially fixed rotating core is mounted in mould

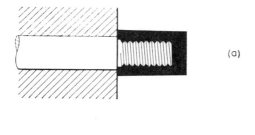

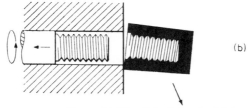

Figure 11.12—Principle of withdrawing rotating core design

MOULDS FOR INTERNALLY THREADED COMPONENTS

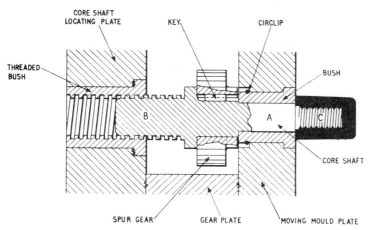

Figure 11.13—Illustrating how withdrawing type of rotating core is mounted in mould

so that the feed system prevents the moulding rotating during the final unscrewing stage.

Two typical components which would be moulded by the axially fixed rotating core technique are shown in Figure 11.10. Note particularly that draft is incorporated to facilitate the ejection of the mouldings.

Now turn to features of the mould design, the relevant portion being shown in Figure 11.11. The cavity form is machined into an insert which is extended to form one bearing for the core shaft. The cavity insert is mounted in a standard type of frame bolster. A spur gear is locked to the core shaft by a key. The rear end of the core shaft is located in a bearing bush mounted in the gear housing plate. Axial movement of the core shaft must be prevented. It is, therefore, essential that the axial length between the steps (A, B) on the core shaft, and the depth of the pocket (x) in the gear housing assembly, are accurately dimensioned to give minimum working clearance.

WITHDRAWING ROTATING CORE. The principle of this design is illustrated in Figure 11.12. The threaded core is unscrewed from the moulding by progressively withdrawing the core through the mould plate as it is rotated. (If we relate this action to the simple nut and bolt analogy discussed in the previous section, this time we hold the nut firmly in the right hand and progressively unscrew the bolt with the left hand. Note that when we rotate the bolt one complete revolution it moves out of the nut a distance equal to the lead of the thread. Thus in our mould design we must arrange for the threaded core to be withdrawn through the mould plate a distance equal to the lead of the moulded thread for each revolution of the threaded core.)

Now we turn to the mould design (Figure 11.13) to see how to achieve this withdrawing action. The front end (A) of the core shaft is located in the moving mould plate bush. The rear end (B) of the core shaft is threaded and is accommodated in an internally threaded bush. This latter bush is mounted in

MOULDS FOR THREADED COMPONENTS

the core shaft locating plate. A spur gear is fitted on to the core shaft and secured to it by a key. Thus, when the spur gear is rotated, the rear threaded portion of the core shaft progressively screws into the threaded bush, thereby withdrawing the impression thread (C) from the moulding. The depth of the pocket formed by the gear plate must be sufficient to accommodate the axial movement of the spur gear. This movement will be slightly in excess of the length of the impression thread. Also, the length of the bush must be sufficient to ensure that the core shaft is still supported in the withdrawn position.

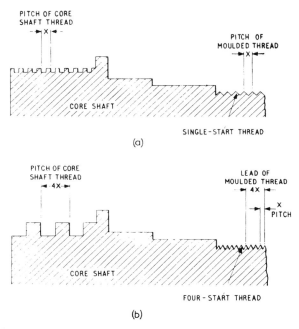

Figure 11.14—Part-section through a withdrawing type of rotating core shaft: (a) single-start thread; (b) multi-start thread

The axial distance the impression thread is withdrawn per revolution is controlled by the pitch of the operating thread at the rear end of the core shaft (B). For a single-start moulded thread, the pitch of the thread at either end of the core shaft is made the same, i.e. at B and C. For multi-start moulded threads, however, the pitch of the operating thread, i.e. at B, must equal the lead (pitch times number of starts) of the impression thread (C). Figure 11.14 illustrates the two types of core shaft. At (a) the pitches of both threads are the same whereas at (b) the pitch of the operating thread is four times the pitch of the impression (four-start) thread.

A considerable force may be applied to the operating threads of the core shaft by the melt and, therefore, apart from fairly small-diameter components where the V-thread may be used, the square thread (as shown) or buttress thread should be adopted to withstand the applied force.

MOULDS FOR INTERNALLY THREADED COMPONENTS

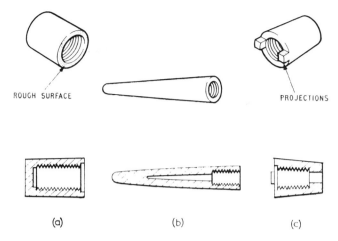

Figure 11.15—Typical components which could be moulded utilising withdrawing rotating core design

Once the moulding has been ejected the core shaft must be returned to its forward position. This means reversing the direction of rotation of the spur gear.

Typical components which would be moulded using the withdrawing rotating core design are shown in Figure 11.15. The adhesion between the moulding and the mould plate is usually sufficient to prevent the moulding rotating while the core is withdrawn. However, it is an advantage to specify that the surface of the mould plate which forms the end of the moulding is slightly roughened (a, b). Some designers prefer a more positive approach and specify small projections (or recesses) to be included on the component as illustrated by component (c).

In the pen cap type component (b) it will be noted that the thread forms only a small part of the core proper. For such components the core shaft is often constructed in two parts. A separate core is attached to the threaded core in Figure 11.16.

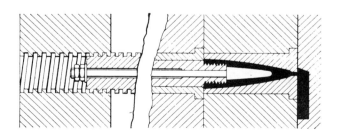

Figure 11.16—Illustrating core construction where thread forms only small part of complete core

MOULDS FOR THREADED COMPONENTS

ROTATING CAVITY.   When gate marks are not permissible on the external surface of the moulding then the rotating cavity design may be considered. In this design the feed is arranged to pass through the centre of the core thereby permitting the pin gate to be adopted (Figure 11.17). For convenience the core is usually mounted on the injection half. The type of feed system adopted can be based on either the underfeed design (as shown) or on one of the runnerless designs.

Let us now consider the sequence of operation for this type of mould. In Figure 11.17 the impression is filled via the feed system as shown at (a). The rotation of the cavity insert starts immediately the mould opening commences, and, providing the moulding cannot rotate within the cavity, the moulding is progressively unscrewed (b). The final drawing (c) shows the mould fully open. The moulding, having been unscrewed from the core, is ejected from the cavity (in this example air ejection is adopted). The feed system is removed from between the floating cavity plate and the feed plate. The mould is closed and the cycle continues.

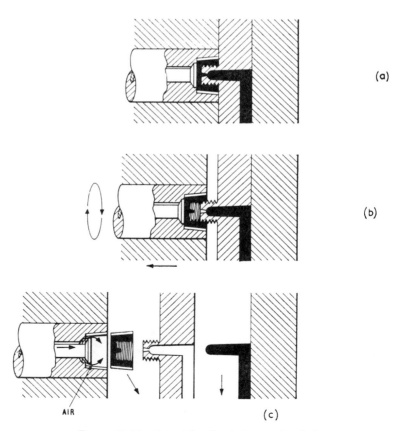

Figure 11.17—Principle of rotating cavity design

MOULDS FOR INTERNALLY THREADED COMPONENTS

The rate at which the moulding is unscrewed must be synchronised with the opening movement of the mould. For if the mould opens too fast the moulding (because of the thread) will be pulled from the cavity and rotation of the cavity will be pointless. Alternatively, if the mould is opened too slowly the rotating cavities will tend to jack open the mould, the mouldings acting as jacks. If this occurs there is the probability that the moulded threads will be damaged. The desired synchronisation can in fact be achieved quite simply by incorporating specially designed screw jacks within the mould.

The more important features of the rotating cavity designs are illustrated in Figure 11.18. The external shape of the component is formed in the cavity insert which is suitably mounted in a bearing bush. The bearing bush is fitted in the moving mould plate and secured by a circlip. A spur gear is attached to the cavity insert as shown. The hardened thrust bush is provided as a wear resisting thrust surface for the rotating cavity insert. The air ejector valve passes through the centre of the thrust bush and the cavity insert. The valve is held on its seating in the cavity insert by a spring. Air is introduced to the system via the entry port in the back plate assembly. By adopting an in-line layout the movement of a rack may be utilised to rotate the spur gear cavity insert assembly.

The core insert is fitted in the floating plate and secured by a circlip. Rotation of the insert is prevented by a suitably positioned grub screw (not shown). The core insert is supported by the feed plate assembly which con-

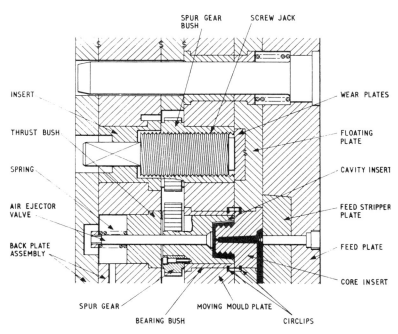

Figure 11.18—Cross-section through rotating cavity design

305

MOULDS FOR THREADED COMPONENTS

sists in this case of the feed plate and the feed stripper plate. For further details of underfeed arrangements see Chapter 12.

Let us now turn to the screw jack assembly. Normally three or four jacks are provided in the mould to ensure smooth operation. The screw jack is a hardened threaded member in permanent engagement in the spur gear bush. The parallel section of the latter is a close rotating fit within the moving mould plate. The screw jack incorporates a square projection which slides in a complementary shaped hole in an insert suitably mounted in the back plate assembly. Movement of a rack (or other actuating method) causes the spur gear to rotate at the same time as the cavity inserts. Thus the screw jacks are thrust outwards against the wear plates mounted in the floating plate. If perfect synchronisation between mould opening and the moulding unscrewing rate is required then the pitch of the screw jack threads must be the same as the moulded thread pitch. The screw jacks control the opening rate of the mould by pushing against the spring-loaded floating plate. Some designers prefer to have a slightly larger pitch for the screw jack threads than for the incorporated moulded thread. With this arrangement the moulding is partly withdrawn from the cavity as the unscrewing progresses.

11.2.5 Layout of Impressions. When more than one impression is required the question of layout must be considered at a very early stage in the design irrespective of whether the axially fixed or withdrawing rotating core design is used. There are two possibilities: (i) the pitch circle layout, (ii) the in-line layout.

In the pitch circle layout the impressions are positioned in such a manner that the respective pinions, attached to the individual core shafts, can be

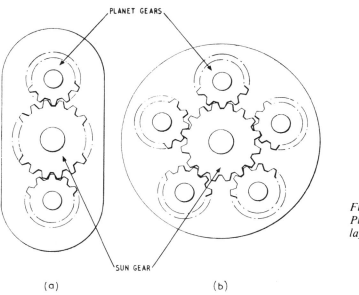

Figure 11.19— Pitch circle layout

MOULDS FOR INTERNALLY THREADED COMPONENTS

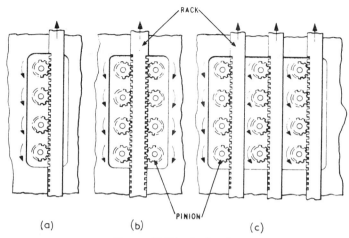

Figure 11.20—In-line layout

driven by one central driving gear (Figure 11.19). This system is usually referred to as the *sun-and-planet system*. When the central sun gear is rotated, the pinions are operated, thereby rotating the core shafts in unison. The actual pitch circle diameter chosen for the mould depends primarily on the number of impressions and the diameter ratio of the gears used. Several alternative methods can be used to drive the sun gear and these are discussed later in this section.

The alternative arrangement is to have the impressions in one or more lines. The core shaft pinions can then be actuated directly by the vertical movement of a rack which may be coupled to the ram of a hydraulic or pneumatic cylinder. Several in-line layouts are shown in Figure 11.20; (a) and (c) illustrate single- and multi-line systems, respectively. The centre drawing (b) shows two sets of pinions being rotated by one rack. This method is used to simultaneously unscrew right-hand and left-hand threaded components.

11.2.6 Power and Transmission Systems (Pitch Circle Layout). In this section we will review the alternative methods which may be adopted for driving the sun-and-planet system. Which actual system is used depends on a number of factors which must be considered before the design is started. These factors include

 (i) The number of threads to be unscrewed.
 (ii) The production requirements: is the number of mouldings required small or large?
 (iii) The component design: will an axially fixed or withdrawing rotating core design be required?
 (iv) The available machine: is it designed to control additional actuators? Has the machine been modified previously to accommodate a particular power or transmission system?
 (v) Mould cost.

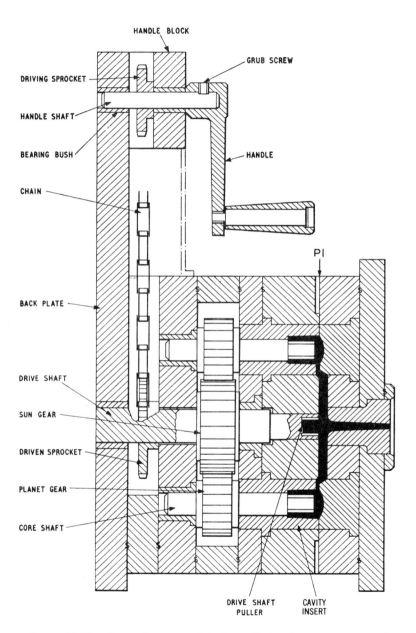

Figure 11.21—Manually powered design with fixed rotating core and chain-and-sprocket transmission

MOULDS FOR INTERNALLY THREADED COMPONENTS

The four power systems, with their associated transmission systems, which will be discussed are as follows:
  (i) Manually powered.
  (ii) Machine powered.
  (iii) Hydraulically or pneumatically powered.
  (iv) Electrically powered.

MANUALLY POWERED SYSTEMS. The majority of manually powered systems allow for the machine operator to unscrew the mouldings by rotating a handle which is normally positioned at either the top or the front of the mould, whichever is the more convenient. The handle assembly is usually mounted on an extension of the back plate (Figure 11.21) and the assembly consists essentially of a U-shaped handle block, a handle shaft suitably located in bearing bushes, and a handle which is locked to the handle shaft by a grub screw. The following designs, which illustrate manually powered systems, differ only with respect to the transmission system used to connect the handle assembly to the sun gear drive shaft.

Figure 11.21 illustrates a manually powered design with a *chain-and-sprocket* transmission. A brief description of the operation and relevant comments on the design follow. The mould opens at P_1, the mouldings remaining in the cavities. The feed system is pulled by the puller incorporated in the drive shaft. When the mould is fully open the operator rotates the driving sprocket via the handle and the handle shaft. A chain connects the driving and driven sprockets, the latter mounted on the drive shaft. The sun gear, also mounted on the drive shaft, meshes with the planet gears (or pinions) secured to the individual core shafts. Thus in this design, which incorporates axially fixed rotating cores, rotation of the handle causes the individual core shafts to rotate, which in turn causes the mouldings to ride up their respective threads and thereby be ejected. All the shafts which are rotated are mounted in bearings, as shown.

Figure 11.22 shows the same basic mould but incorporating an alternative gear train transmission. In this design a driving gear is mounted on the handle shaft and a driven gear on the drive shaft. Suitable idler gears complete the train. Thus, as in the previous design, manual rotation of the handle causes the sun gear, pinions and core shafts to rotate and hence eject the mouldings.

When either of the above drive systems is used in conjunction with the axially fixed rotating core design (as illustrated) the operator simply rotates the handle until all the mouldings are ejected. However, if these drive systems are used in conjunction with the withdrawing rotating core design, once the mouldings are ejected the operator must return the cores to their forward positions by turning the handle in the reverse direction.

Another transmission system which may be used is the *rack-and-pinion* (Figure 11.23). Because of the component's form, the withdrawing rotating core is adopted in this case, but, as in the designs reviewed above, the technique applies to both rotating core systems.

A description of the transmission system follows. A pinion is secured to the

MOULDS FOR THREADED COMPONENTS

main drive shaft on which is also mounted the sun gear. A rack, suitably guided within the transmission plate, meshes with the pinion, and also with the driving pinion located within the handle block assembly. Manual rotation of the driving pinion by the handle causes the rack to move upwards, which in turn causes the driven pinion and hence the sun gear, to rotate. The planet gears are thereby rotated, which causes the core shafts to be progressively withdrawn through their respective bushes and the mouldings ejected.

As the conventional ejector system is unnecessary with the designs shown,

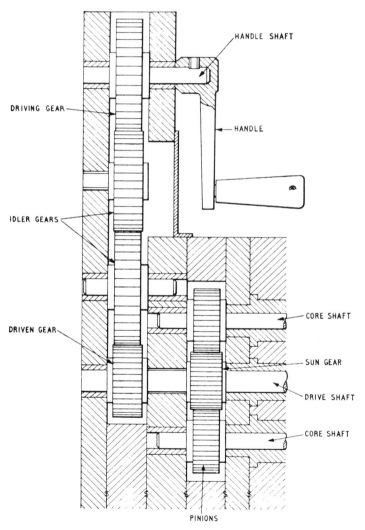

Figure 11.22—*Manually powered, fixed rotating core design with gear transmission*

MOULDS FOR INTERNALLY THREADED COMPONENTS

some thought must be given to the ejection of the sprue puller. In all the above designs the cold slug well, machined into the drive shaft, is threaded. The thread provides a positive pull on the feed system when the mould opens and, when the drive shaft rotates, the puller is extracted at the same rate as the mouldings. Note that the thread chosen must be of the opposite hand to the moulding thread, and to ensure that the puller is extracted at the same rate as the moulding some adjustment of pitch is necessary if the sun gear and pinions are not of the same diameter.

A completely different approach to manual operation is to utilise a ratchet spanner within the mould as shown in Figure 11.24. Any suitable commercially available ratchet spanner may be used. The spanner is mounted on the

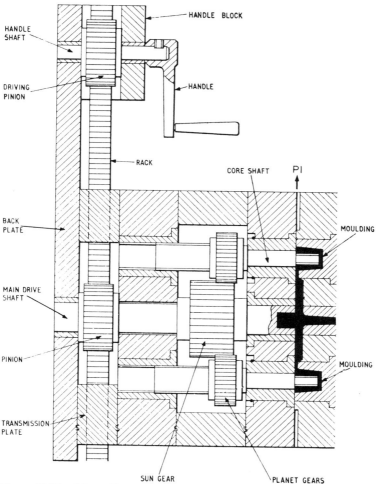

Figure 11.23—Manually powered, withdrawing core design, with rack-and-pinion transmission

MOULDS FOR THREADED COMPONENTS

main drive shaft and the handle projects beyond the top surface of the mould. The operator simply unscrews the moulding by a rapid short reciprocating movement. The length of the movement is controlled by the shape of the aperture cut into the transmission plate.

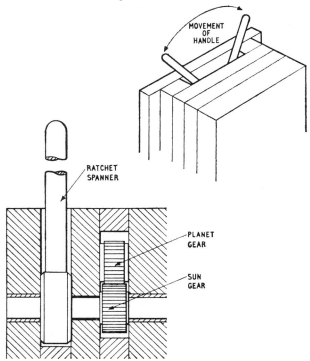

Figure 11.24—Manually operated ratchet spanner design

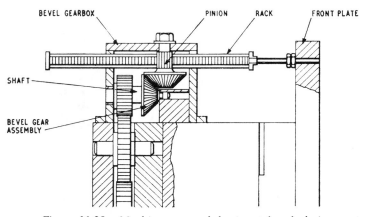

Figure 11.25—Machine powered, horizontal rack design

MOULDS FOR INTERNALLY THREADED COMPONENTS

MACHINE-POWERED SYSTEMS. These systems utilise the opening movement of the injection machine to actuate the gearing system and so unscrew the mouldings. This has certain advantages in that the unscrewing is automatically accomplished while the platens are opening and therefore the overall cycle time is not increased as with the various manual methods. There are limitations, however, particularly with respect to the number of turns of thread which can be unscrewed. There are two basic designs to consider: one utilises a horizontal rack, while the other exploits the Archimedean screw.

The basic rack-and-pinion design is shown in Figure 11.25. The rack is fitted parallel to the axis of the mould and attached to the front plate. The rack is located in the bevel gearbox attached to the moving mould half as shown.

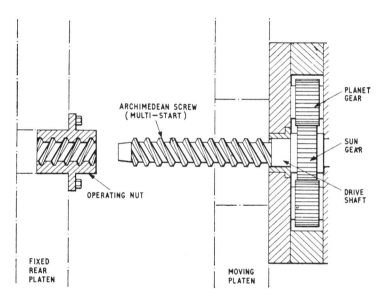

Figure 11.26—Machine powered, Archimedean screw design

When the mould opens, the rack via the pinion and bevel gear assembly causes the shaft to rotate. Now we can incorporate any of the previously described transmission systems to couple this rotary motion to the sun gear. (In the illustration the gear train method is adopted.) A fine axial method of adjustment of the rack is necessary if the withdrawing rotating core design is required to set the return position accurately.

The principle of the automatic unscrewing mould which utilises an Archimedean screw is shown in Figure 11.26. The Archimedean screw is an extension of the drive shaft to which the sun gear is attached. Note that the Archimedean screw projects beyond the rear face of the mould. A complementary nut is attached to a rear platen, or auxiliary platen, in alignment with the screw. While shown disconnected in the illustration for reasons of clarity, in practice the Archimedean screw remains in contact with the operating nut.

313

MOULDS FOR THREADED COMPONENTS

Rearward movement of the mould actuates the Archimedean screw thereby causing rotation of the sun-and-plant systems. Closing the mould causes the screw to be partially withdrawn from the nut, rotating of course, in the opposite direction, thereby returning the core shafts to their former positions.

HYDRAULICALLY OR PNEUMATICALLY POWERED SYSTEMS. These designs are extensions of the manual rack-and-pinion design which allow the unscrewing operation to be made fully automatic. The operating handle assembly (Figure 11.23) is replaced by an actuator, the ram of which is directly coupled to the rack (Figure 11.27). The internal design of the mould is identical to the basic rack-and-pinion design. The actuator is mounted on a platform which is supported on columns suitably secured to the moving half of the mould. The rack, which extends through the mould, is coupled to the ram of the actuator. Independent control of the actuator is possible and therefore the unscrewing operation can take place at any convenient time. This can often save valuable seconds of the cycle time.

The hydraulic medium is normally chosen in preference to the pneumatic medium for operating the rack-and-pinion for the following reasons:

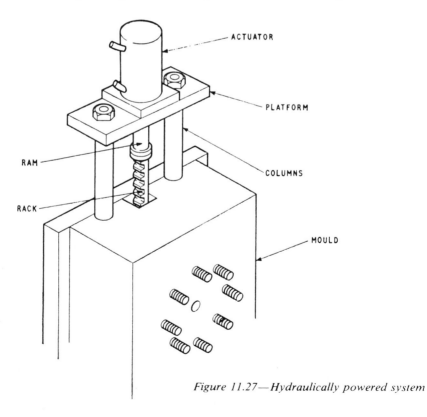

Figure 11.27—Hydraulically powered system

MOULDS FOR INTERNALLY THREADED COMPONENTS

(i) A greater force is available for a given size of actuator.
(ii) A hydraulic oil supply is available on most injection machines, whereas a separate air supply must be arranged unless the moulding shop is already piped for high pressure air.
(iii) On modern injection machines provision is made to programme for extra actuator operations in the correct sequence as part of the machine cycle.

ELECTRICALLY POWERED. A geared electric servomotor may be used to drive the rotating core system. Generally the motor is coupled to the cores either via a chain and sprocket transmission or via a gear train transmission. A sprocket or gear is fitted to the motor shaft as applicable. In principle the electric motor replaces the operating handle assembly of the manual systems illustrated in Figures 11.21 and 11.22, respectively.

The requirements of the servomotor for this purpose are:
(i) At the commencement of the unscrewing operation a high torque should be available.
(ii) A torque-limiting device should be included to ensure the motor stops immediately the core shaft reaches its rear or forward position (for withdrawing core systems). This is also a safeguard should a seize-up occur with either core system.
(iii) The servomotor must be reversible if it is to be used for the withdrawing core system.
(iv) It should be possible to rotate the cores manually in case of electrical failure; a declutching arrangement is therefore desirable.
(v) Operating speed should be adjustable to suit requirements.
(vi) It should be reliable in operation and have a long life.
(vii) It should be possible to mount the motor in any position on the mould (top or sides).

The above requirements are equally applicable to the motorisation of fluid valves, and therefore certain servomotors available for this purpose are also suitable for driving rotating cores in unscrewing type moulds.

An example of an electric motor drive system is shown in Figure 11.28. The servomotor is mounted on the top surface of the moving half. In this design a gear transmission is used to transmit the power to the sun gear and, hence, to the planet gears mounted on the rotating core shafts.

11.2.7 Power and Transmission Systems (In-Line Layout). In this section we will review the alternative methods which may be adopted when the mould impressions are arranged in line. In general these are very similar to the methods described in the previous section and some cross referencing is used to prevent undue repetition.

The power systems adopted are:
(i) Manually powered.
(ii) Hydraulically or pneumatically powered.
(iii) Machine powered.
(iv) Electrically powered.

MOULDS FOR THREADED COMPONENTS

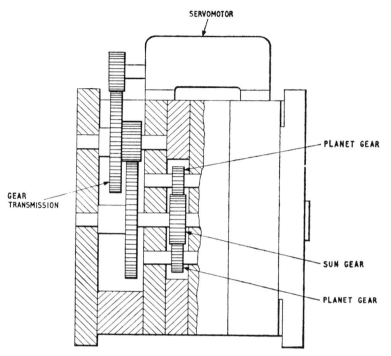

Figure 11.28—Electrically powered system with gear transmission

In general, because the rotating core shafts are positioned in-line, the most suitable transmission system is normally the rack-and-pinion design. However, other systems are used and one example is given in the electrically powered section where worm gears suitably mounted on a vertical shaft are used to operate the core shafts via worm wheels.

MANUALLY POWERED SYSTEM (IN-LINE LAYOUT). In operation this design is identical to the PCD layout design in that the machine operator is required to unscrew the mouldings by rotating a handle. However, as the impressions are in line the core shafts can be directly operated by a rack and pinion arrangement, which allows for a somewhat simplified design, as shown in Figure 11.29 (compare this drawing with Figure 11.23).

The operation of the mould is as follows: when the mould is fully open the operator rotates the handle (1M) which is suitably mounted in a U-shaped handle block assembly (4M) attached to the back plate. The rack is thereby actuated by the rotation of the driving pinion (2M). Movement of the rack causes rotation of the core shaft pinions and of the core shafts to which they are securely attached. Thus, in this axially fixed rotating core design, rotation of the core shaft causes the mouldings to ride up their individual impression threads and so be ejected. The rack is then returned to the lower position in preparation for the commencement of the next cycle of operations. This

MOULDS FOR INTERNALLY THREADED COMPONENTS

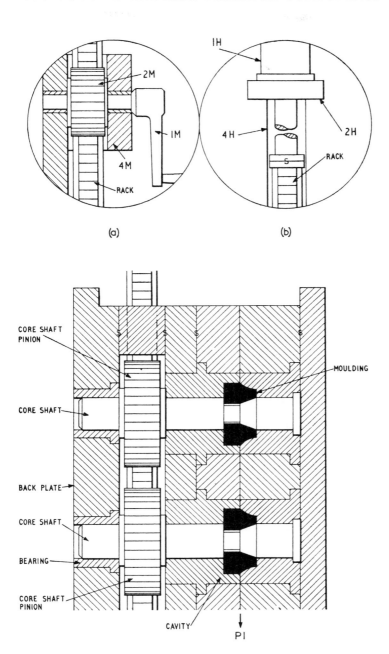

Figure 11.29—*Rack-and-pinion transmission for the in-line layout:
(a) manually powered; (b) hydraulically powered*

MOULDS FOR THREADED COMPONENTS

manual operation is very time-consuming and, except for moulds used for short production orders, is seldom used. It does, however, form the basis of the hydraulically powered system discussed in the next section.

HYDRAULICALLY AND PNEUMATICALLY POWERED SYSTEMS (IN-LINE LAYOUT). In this design, the operating handle assembly (of the preceding section) is replaced by a hydraulic (or pneumatic) actuator (Figure 11.29b). The actuator (1H) is fitted on a platform (2H) mounted on columns (4H). The columns are suitably secured to the moving mould half. (A pictorial view of a hydraulic actuator mounted on a mould is shown in Figure 11.27.) The distance between the platform and the mould body depends on the length of stroke required for the rack. The length of stroke can be determined given the number of revolutions that the core shaft must make and the number of teeth on the core shaft pinion.

The rack may be directly attached to the actuator ram via a coupling (Figure 11.29b) or, when several racks must be actuated simultaneously, the racks may be fixed to a crosshead which, in turn, is attached to the ram. A mould of this latter type is shown in Plate 13. When the rack is actuated the mouldings are unscrewed in a manner identical to that described in the previous section.

Providing the actuator can be coupled to the machine's control circuit, the unscrewing operation can be achieved automatically. The alternative method is to arrange for the moving platen to actuate limit switches which, via solenoid-operated valves, cause the ram to be moved in one direction or the other.

The hydraulic medium is normally chosen in preference to the pneumatic medium, and the reasons for this choice are given in Section 11.2.6.

MACHINE POWERED SYSTEM (IN-LINE LAYOUT). An example of a machine powered arrangement suitable for an in-line layout is illustrated in Figure 11.30. This system utilises two racks mounted at right angles to each other as shown. The horizontal rack and pinion transforms a linear motion (of the platen) to a rotary motion, and the second vertical rack serves to transmit this rotary motion to the core shaft pinions.

Consider the opening sequence of this mould. When the mould is initially opened, the mouldings are withdrawn from the cavities because a slight delay is provided for on the front part of the rack. Immediately this delay movement has been taken up the horizontal rack is stopped by the stop-bar mounted on the fixed mould plate assembly. The horizontal rack passes through a slot machined in the back plate. Continued movement of the moving mould half causes the operating pinion to rotate. A plate box construction suitably mounted on the moving mould half forms a housing for the operating pinion and also acts as a guide for the horizontal rack. Rotation of the operating pinion causes the vertical rack to be moved upwards. Note that gear teeth are cut into the rear of the vertical rack at the top end whereas at the lower end the gear teeth are cut in one side. Thus the vertical movement of

MOULDS FOR INTERNALLY THREADED COMPONENTS

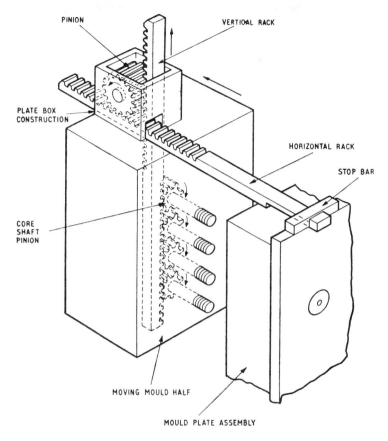

Figure 11.30—Machine powered in-line unscrewing system

this rack causes the core shaft pinions to be rotated and the mouldings are thereby unscrewed.

To obtain a sufficient number of revolutions of the core shaft a relatively long horizontal rack may be required. Either the closed height of the mould must be sufficient to accommodate this or a machine must be chosen such that the rack can overhang the platen, as shown.

ELECTRICALLY POWERED SYSTEM (IN-LINE LAYOUT). Electrically powered operation of an in-line layout arrangement may be achieved in one of several ways, two of which are discussed below. The first system is an extension of the manually powered system (described above) and the basic designs are identical (Figure 11.29). However, in the electrically powered system the operating handle assembly (Figure 11.29a) is replaced by a servomotor assembly. A schematic diagram (Figure 11.31) is given to illustrate the basic principle. The drive pinion is mounted on the drive-shaft of the servomotor, which is

MOULDS FOR THREADED COMPONENTS

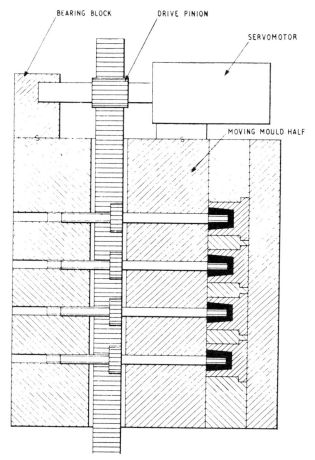

Figure 11.31—Electrically powered in-line unscrewing system with a rack-and-pinion transmission (schematic drawing)

securely bolted to the moving mould half. To provide rigidity, the outer end of the drive shaft is located in a bearing block. When the drive pinion is rotated, the rack (which is suitably guided within the mould) is actuated and the individual core shaft pinions are rotated.

In the alternative system (Figure 11.32) the servomotor is mounted on the mould so that a direct vertical drive (as shown) can be achieved. The vertical operating worm shaft is attached to the drive shaft via a coupling. The worm is in mesh with worm wheels mounted on the core shafts. Thus the rotary motion of the main drive shaft, in either direction, is transmitted to the individual core shafts. The vertical operating worm shaft is suitably mounted in bearings (not shown) within the mould.

The main advantage of electrical operation over hydraulic operation is that

MOULDS FOR EXTERNALLY THREADED COMPONENTS

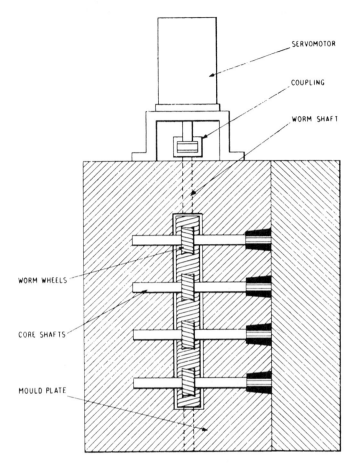

Figure 11.32—Electrically powered in-line unscrewing system with worm and worm-wheel transmission (schematic drawing)

mould setting is somewhat simplified, and a more rapid change-over of mould is possible. One disadvantage to be weighed against this is that servomotors are less robust than hydraulic actuators and can more easily be damaged in transit to and from the machine.

11.3 MOULDS FOR EXTERNALLY THREADED COMPONENTS

An external thread comes within the broad definition of an external undercut, namely 'a restriction situated in the outside form of a moulding which prevents straight draw removal of the moulding from the cavity.' However, unlike any other form of undercut component, the threaded type of moulding can be unscrewed from the cavity and in certain cases this allows a simplified design. But if automatic operation is required some form of rotary

MOULDS FOR THREADED COMPONENTS

motion within the mould is necessary to perform the unscrewing operation automatically.

Another method often used is the stripping design. This method, while allowing for fast production cycles, is limited to those components which incorporate roll threads.

Finally, as the thread forms an external undercut, a threaded component may be treated in the same way as any other external undercut type component, namely, by the use of the splits design. In this the threaded portion is machined in the split halves which are opened prior to the moulding being extracted.

Let us consider these methods in more detail.

11.3.1 Fixed Threaded Cavity Design. The simplest form of mould for producing an externally threaded component is one in which the threaded portion is machined directly into the cavity insert (Figure 11.33a). When the mould is opened the moulding can be unscrewed from the mould plate as shown (b). The unscrewing operation may be manual or power assisted; in the latter case the operator is provided with a suitable power tool. In either case the compo-

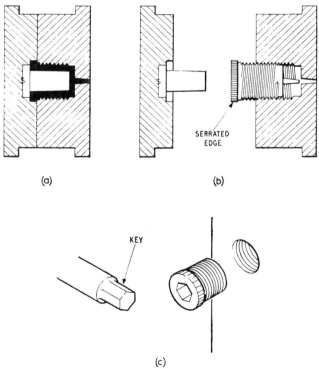

Figure 11.33—Fixed threaded cavity design: (a) mould closed; (b) mould open; (c) if suitable, a key may be used to unscrew moulding

MOULDS FOR EXTERNALLY THREADED COMPONENTS

nent's form must be such that a grip can be obtained for unscrewing. For example, in the illustration the parting line has been chosen to allow the head of the component to project above the mould's parting surface. In addition, the edge of the head is serrated to provide a grip. An alternative method is to design the component with either a square or a hexagonal aperture (c). This permits the use of a key (as shown) for unscrewing the moulding.

Let us consider what advantages this design has over the more automatic splits or rotating cavity designs.
 (i) *Mould Cost:* this is far less than for the automatic designs.
 (ii) *Joint Line:* the mouldings produced do not suffer the disadvantage associated with the split design in that a joint line (formed by the splits) does not occur.
 (iii) *Undercuts:* Undercuts created by certain thread forms when using the split design are avoided.
 (iv) *Servicing:* as there are no moving parts within the mould servicing costs are kept to a minimum.

The major disadvantage of this design is that the moulding cycle is relatively long because it involves a manual unscrewing operation.

11.3.2 Automatic Unscrewing. One method of automatically unscrewing a moulding from a fixed threaded cavity is illustrated in Figure 11.34. This design is basically the same as the rotating cavity design (Section 11.2.4), for automatically unscrewing internally threaded components. However, for externally threaded components the cavity is fixed, and the core is rotated to provide the unscrewing action.

The operating sequence of the mould illustrated is as follows:

After the impression is filled (a), and the moulding cooled sufficiently, the mould is opened (at P1) and the core shaft rotated. Thus the moulding is progressively unscrewed from the cavity. The gate is broken in the process (b). While the moulding is being unscrewed the second opening occurs between the floating cavity plate and the feed plate. This allows for the removal of the feed system (see underfeed moulds). Finally, the moulding is ejected from the core (c).

The rate at which the moulding is unscrewed must be synchronised with the opening movement of the mould and this can be achieved by the use of the screw jack (Figure 11.18).

It is essential that the shape of the core is such that it acts as a key when the core shaft is rotated, to unscrew the moulding. For example, small projecting ribs on the internal bore of a component provide a suitable grip for unscrewing purposes.

The general build-up of the moving half of the mould is basically the same as for the fixed rotating core arrangement and similar layout, power unit and transmission systems may be adopted (Section 11.2.4).

11.3.3 Stripping External Threads. External threads may be stripped from a cavity providing (i) the component has a rolled thread form; (ii) the thread

MOULDS FOR THREADED COMPONENTS

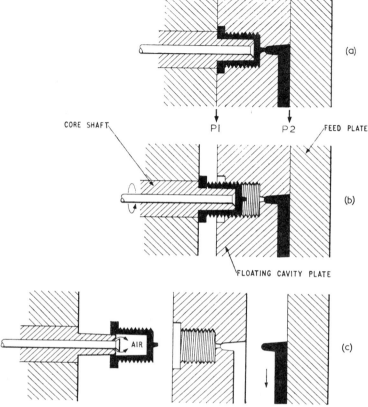

Figure 11.34—Automatic unscrewing method for externally threaded mouldings

depth is relatively shallow in relation to the diameter; (iii) the moulding material sufficiently elastic to return to its original shape after being deformed.

The stripping design eliminates the necessity of either unscrewing the moulding, or using splits, and fully automatic operation is practicable.

The principle of the design is illustrated in Figure 11.35. The external shape of the moulding is die-sunk into the cavity insert which is mounted in the moving mould plate. The core is mounted in the fixed mould plate. When the mould is opened the component is retained in the cavity by the threads and is, therefore, pulled from the core. A relatively large ejector pin is positioned below the lower face of the moulding and when the ejector assembly is actuated the moulding is ejected from the cavity. The pocket in the component, formed by the core, permits slight contraction which allows the moulded threads to ride over the complementary cavity threads.

Figure 11.36 illustrates the roll thread form for which this technique is suitable. Production, by this method, of conventional V- or square threads should not be attempted.

324

MOULDS FOR EXTERNALLY THREADED COMPONENTS

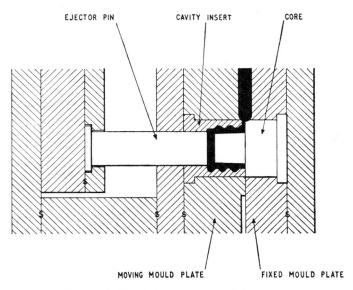

Figure 11.35—Stripping external threads

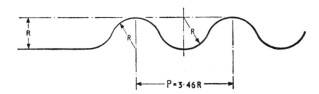

Figure 11.36—Typical roll thread

11.3.4 Threaded Splits. When automatic production is required for an externally threaded component and where the thread form is such that it cannot be stripped the splits design is used.

The design and operation of the mould is identical to that described earlier on in this volume (Chapter 8, Splits) except that in this case the undercut is a thread form and the joint line occurs along the thread (Figure 11.37a). Extreme accuracy is essential, therefore, in fitting the splits. Otherwise flashing may occur along the joint line and this, of course, will seriously impair the efficiency and quality of the produced component. In the case of a component which has an interrupted thread form (Figure 11.37b) the joint line can be positioned on the plain section, thereby avoiding the necessity of requiring such extreme accuracy in the fitting of the splits.

A split mould for an externally threaded component is shown in Figure 11.38. In this design the dog-leg cam operating method was chosen to permit the use of a simple pin ejector system. The dog-leg cams provide a sufficient delay, before the splits are opened, to allow the moulding to be pulled clear of

MOULDS FOR THREADED COMPONENTS

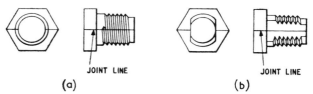

(a)   (b)

Figure 11.37—Joint line on externally threaded components moulded in splits

the core. For a complete description of the operation of a split mould consult Chapter 8.

Care must be taken when designing splits for threaded components because certain thread forms create an undercut in the split halves. Provided the hazard is appreciated, the thread profile can be adjusted to obviate the undercut. If the undercut is small, the adjustment can be made local to the area adjacent to the splits face and this is often left to the mouldmaker's discretion. Severe undercuts, however, particularly with brittle materials, should be avoided by suitably modifying the thread form.

The square thread profile will always create an undercut of varying severity

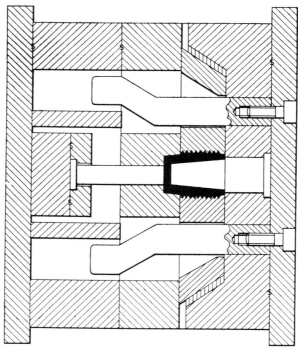

Figure 11.38—Split mould for externally threaded component

MOULD CONSTRUCTION

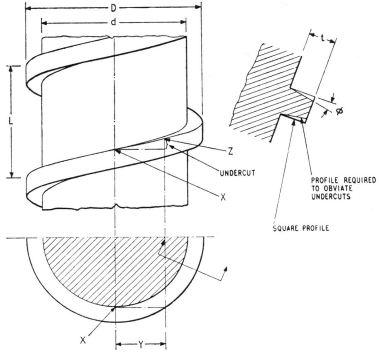

Figure 11.39—Certain thread forms create an undercut when moulded in splits

according to the pitch and depth of thread (Figure 11.39). Consider point X situated on the minor thread diameter at the joint line. The complementary point on the splits must travel distance Y (see plan) before it is clear of the thread. The point Z at which the split loses contact with the moulding is higher than point X, therefore an undercut must result. To avoid this, the thread profile should be changed to a trapezoidal form having a flank thread angle ϕ. From the formula a value for ϕ can be calculated

$$\tan \phi = \frac{L}{\pi D t}\sqrt{\frac{(D^2-d^2)}{2}} \qquad (11.1)$$

where (Figure 11.39) L = lead of thread (= pitch for single-start thread),
D = major thread diameter,
d = minor thread diameter,
$t = 0.5\,(D - d)$.

11.4 MOULD CONSTRUCTION

The construction of the unscrewing mould is somewhat different from the basic mould construction so, for those readers who find cross-sectional drawings difficult to follow, exploded views of two typical build-ups are given to

MOULDS FOR THREADED COMPONENTS

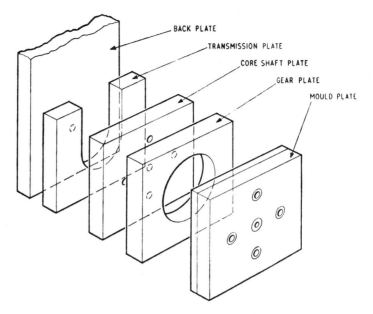

Figure 11.40—Typical mould construction for pitch circle layout

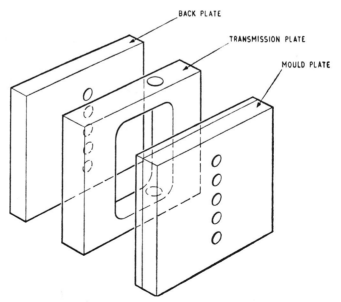

Figure 11.41—Typical mould construction for in-line layout

MOULD CONSTRUCTION

assist in the general visualisation of the mould. One is based on the pitch circle layout, and the other on the in-line layout. It will be appreciated that the build-up of plates used in an actual mould design will also depend on such other factors as the type of transmission chosen, the type of core system used (axially fixed or withdrawing), etc.

11.4.1 Mould Construction for the Pitch Circle Layout. An exploded view of the build-up of the various plates in a typical unscrewing mould utilising a chain and sprocket transmission is shown in Figure 11.40. The gear plate, which accommodates the sun-and-planet gear system, is mounted immediately to the rear of the mould plate. The core shaft plate backs directly on to the gear plate. It will be recalled that for the axially fixed rotating core and the rotating cavity designs the core shaft plate incorporates bearing bushes, whereas for the withdrawing rotating core design internally threaded bushes are incorporated. The transmission plate must be designed to suit the individual transmission system used. The back plate must be designed to suit the machine or machines to which the mould will be attached. Because of the large number of plates which are used in this assembly, particular attention must be paid to the position and number of screws and dowels for fixing these plates together.

11.4.2 Mould Construction for the In-Line Layout. An exploded view of the moving mould half of a mould which incorporates a rack and pinion transmission is shown in Figure 11.41. The impressions are incorporated in-line in the mould plate. The transmission plate positioned immediately to the rear of the mould plate has a central aperture which accommodates the core shaft pinions. It also has a vertical hole or slot through which the rack operates. This hole or slot acts as a location for the rack. The back plate serves as a coreshaft plate in this design. The plate incorporates bushes which simply support the core shafts or have internal threads to allow for core withdrawal, according to the type of rotating core design used.

12

Multi-daylight Moulds

12.1 GENERAL

We saw in Chapter 2, that the basic mould consists of two parts or halves, namely a fixed half and a moving half, respectively (Figure 12.1a). When these two parts are opened (b) the mouldings can be extracted. Such an assembly is sometimes referred to as a *single-daylight mould* because, when the mould is open, there is only one space or, as it is normally termed, *daylight*, between the two mould halves.

On more complex moulds more than one daylight occurs when the mould is opened. Consider, for example, a stripper plate mould. This type of mould consists of three parts, namely (Figure 12.1c) a fixed mould plate, a moving mould plate and a stripper plate. When the mould is fully open (d) there are two daylights. The moulding and feed system are removed from the first daylight, that is from between the stripper plate and the fixed mould plate as shown. Another type of mould where more than one daylight occurs is shown at (e). This mould also consists of three main parts: a feed plate, a floating cavity plate and a moving mould plate. When the mould is opened (f) there are again two daylights. This design permits a particular feed technique known as *underfeeding*, and the double daylight is necessary in this case to permit the feed system to be removed from the mould. The mouldings are removed from between the floating cavity plate and the moving half while the feed system is removed separately from between the feed plate and the floating cavity plate.

If the stripper plate and underfeed designs are combined, as is often necessary for multi-impression circular cap components, then an even more intricate design results. Figure 12.1 shows schematically such a mould in its closed (g) and open (h) positions respectively. This mould is called a triple-daylight mould.

In this section we will confine our discussion to the underfeed and to triple daylight designs. The stripper plate design is dealt with in Chapter 3 as it is primarily an ejection method.

The terminology for moulds is somewhat complicated by the word *plate*, which is used freely when discussing mould types. Normally this term means a block of steel used in the mould assembly; for example, a cavity plate. However, it is quite common to use the terms *two-plate mould* and *three-plate mould* for moulds which obviously contain more than two or three plates. To prevent confusion we suggest that the reader uses the term *part* instead of plate. Thus a two-plate mould becomes a *two-part mould* (single daylight), a three-plate mould becomes a *three-part mould* (double-daylight), and so on.

UNDERFEED MOULDS

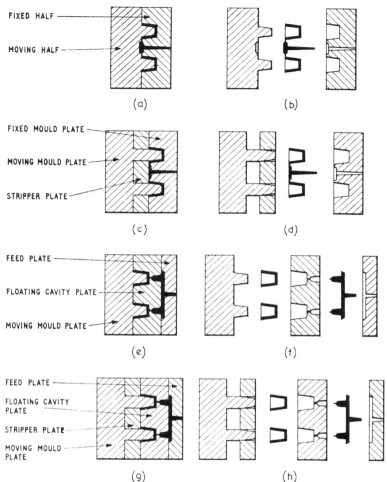

Figure 12.1—Basic mould types; each mould is shown in closed and open positions: (a, b) single-daylight mould; (c, d) double-daylight, stripper plate mould; (e, f) double-daylight, underfeed mould; (g, h) triple-daylight, underfeed-stripper plate mould

12.2 UNDERFEED MOULDS

An underfeed mould is one in which the feed system is arranged to feed into the underside (or occasionally inside) of the component. The advantages of underfeeding were discussed in Chapter 4, where we saw, for example, that a centre feed position is desirable for a cup type of component. For single impressions, this centre feed can be achieved quite simply by using a sprue gate. For multi-impressions, however, a centre feed necessitates a special mould design, one of which is the underfeed system (see also sections on nozzle types and hot runner systems).

MULTI-DAYLIGHT MOULDS

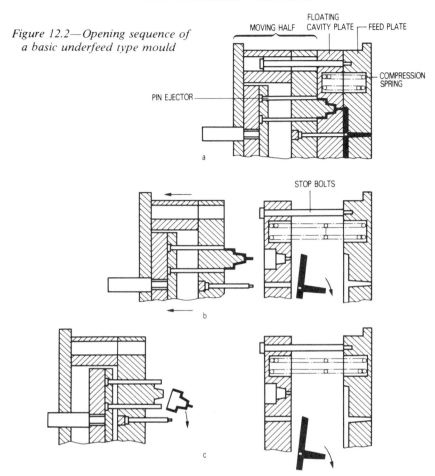

Figure 12.2—Opening sequence of a basic underfeed type mould

The underfeed mould is basically a two-part mould with an additional plate mounted on the front. This plate, termed the *feed plate,* normally incorporates the runner which, by suitable configuration, can be directed to any point on the cavity plate.

The main uses of the underfeed type of mould are as follows:
  (i) Mouldings produced on multi-impression moulds can be centre-fed.
  (ii) Off-centre feeding can be achieved for both single- and multi-impression moulds.
  (iii) Multi-point feeding can be accomplished on single-impression moulds and on multi-impression moulds.

12.2.1 Basic Underfeed Mould. A simple mould of this type is shown in Figure 12.2, and a description of the design and of the opening sequence follows. The mould consists of three basic parts, namely: the moving half, the floating cavity plate and the feed plate, respectively.

UNDERFEED MOULDS

The moving half consists of the moving mould plate assembly, support blocks, back plate, ejector assembly and the pin ejection system. Thus the moving half in this design is identical with the moving half of basic moulds discussed in Chapters 2 and 3.

The floating cavity plate, which may be of the integer or insert-bolster design (the latter design is illustrated), is located on substantial guide pillars (not shown) fitted in the feed plate. These guide pillars must be of sufficient length to support the floating cavity plate over its full movement and still project to perform the function of alignment between the cavity and core when the mould is being closed. Guide bushes are fitted into the moving mould plate and the floating cavity plate respectively.

The maximum movement of the floating cavity plate is controlled by stop bolts. The moving mould plate is suitably bored to provide a clearance for the stop bolt assembly. The stop bolts must be long enough to provide sufficient space between the feed plate and the floating cavity plate for easy removal of the feed system. The minimum space provided for should be 65 mm (2½ in) just sufficient for an operator to remove the feed system by hand if necessary.

The desired operating sequence is for the first daylight to occur between the floating cavity plate and the feed plate. This ensures the sprue is pulled from the sprue bush immediately the mould is opened. To achieve this sequence, springs are incorporated between the feed plate and the floating cavity plate. The springs should be strong enough to give an initial impetus to the floating cavity plate to ensure it moves away with the moving half. It is normal practice to mount the springs on the guide pillars (see Figure 12.3) and accommodate them in suitable pockets in the floating cavity plate. Extra length-bolts are also normally included in the design (Figure 12.3) to pull the plate the remaining distance if the floating cavity plate stops before completing the required movement. These length-bolts are accommodated in suitable bored holes.

The major part of the feed system (runners and sprue) is accommodated in the feed plate and to facilitate automatic operation the runner should be of a trapezoidal form so that once it is pulled from the feed plate it can easily be extracted. Note that if a round runner is used, half the runner is formed in the floating cavity plate, where it would remain, and be prevented from falling or being wiped clear when the mould is opened.

Now that we have considered the mould assembly in some detail, we look at the cycle of operation for this type of mould.

The impressions are filled via the feed system (Figure 12a) and, after a suitable dwell period, the machine platens commence to open. A force is immediately exerted by the compression springs, which causes the floating cavity plate to move away with the moving half as previously discussed. The sprue is pulled from the sprue bush by the sprue puller. After the floating cavity plate has moved a predetermined distance it is arrested by the stop bolts. The moving half continues to move back and the mouldings, having shrunk on to the cores, are withdrawn from the cavities. The pin gate breaks at its junction with the runner (Figure 12.2b).

MULTI-DAYLIGHT MOULDS

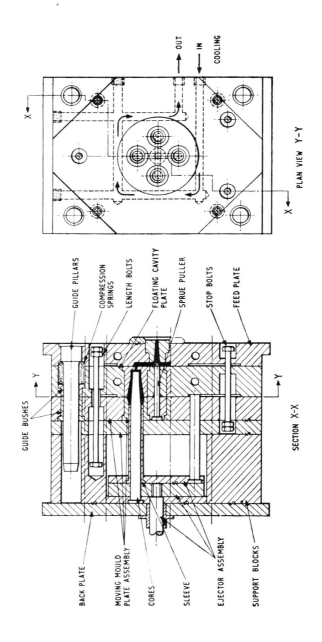

Figure 12.3 — Drawing example of underfeed mould

UNDERFEED MOULDS

The sprue puller, being attached to the moving half, is pulled through the floating cavity plate and thereby releases the feed system which is then free to fall between the floating cavity plate and the feed plate. The moving half continues to move back until the ejector system is operated and the mouldings are ejected (Figure 12.2c). When the mould is closed the respective plates are returned to their moulding position and the cycle is repeated.

An example of a simple underfeed design is illustrated in Figure 12.3. This is a four impression mould for a pen-cap type component.

12.2.2 Types of Feed. In Figure 12.3 we used a direct pin gate to connect the runner to the impression and while this is the simplest form of feed system it is not practicable for all types of components. This is because the shape of the component affects the design of the floating cavity plate, which in turn determines which form of feed system can be adopted. The criterion of a good floating cavity plate design is that the depth of steel situated below the cavity impression is sufficient to withstand the applied injection force without undue deflection. With reference to Figure 12.3, we see that theoretically, when the mould is closed, the floating cavity plate is supported by the feed plate. However, in practice flash or other foreign matter may be trapped between these two plates, in which case this support is lost.

A graph of the depth of steel required below the cavity impression plotted against the cavity diameter is given in Figure 12.4. The cavities superimposed on this graph are included to give a visual indication of the depth of steel required below cavities of particular dimensions. The graph is based on Reuleaux's formula with values inserted for allowable deflection, load and modulus of elasticity.

Reuleaux's formula for a circular flat plate secured all around the edge with a uniformly distributed load over the unsupported area of the plate is given by

$$t = \sqrt[3]{(PR^4/6Ed)} \tag{12.1}$$

where t = the depth of the plate (m or in)
  P = the load (N/m² or lbf/in²)
  R = the radius (m or in)
  d = the deflection (m or in)
  E = the modulus of elasticity (N/m² or lbf/in²)

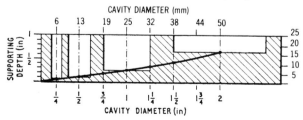

Figure 12.4—Amount of steel required below unsupported cavities of various diameters (based on Reuleaux's formula)

MULTI-DAYLIGHT MOULDS

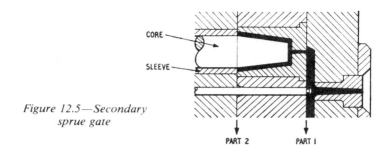

Figure 12.5—Secondary sprue gate

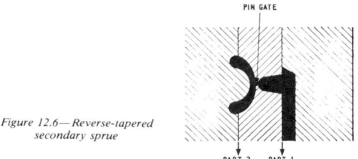

Figure 12.6—Reverse-tapered secondary sprue

12.2.3 Secondary Sprue Gate. This gate follows the standard design (Chapter 4) but in this particular case it connects the runner to the impression instead of the nozzle exit to the impression. Figure 12.5 shows a part-section through a mould which incorporates this type of gate. The general design and operation of the mould is identical to the basic underfeed design discussed in a previous section (12.2.1). This design has, however, the undesirable feature common to all sprue gates in that the moulding, when extracted, must be degated in an extra operation.

12.2.4 Reverse-Tapered Secondary Sprue. This design was developed to allow a pin gate to be used irrespective of the depth of steel below the cavity impression. The basic design is shown in Figure 12.6. The secondary sprue has a reverse-tapered form, as shown, which terminates a short distance from the impression. A standard type of pin gate can, therefore, be used.

There is, however, a drawback to this method in that the feed system is not free to fall when the mould is opened because the secondary sprues are retained within the floating cavity plate. The feed system must therefore be removed by hand if the basic underfeed mould design (Figure 12.3) is used. This drawback can be overcome by the adoption of one of the following methods.

UNDERCUT RUNNER SYSTEM, DESIGN 1. This design incorporates two features that are additional to the basic underfeed design: (i) The runner is undercut at

UNDERFEED MOULDS

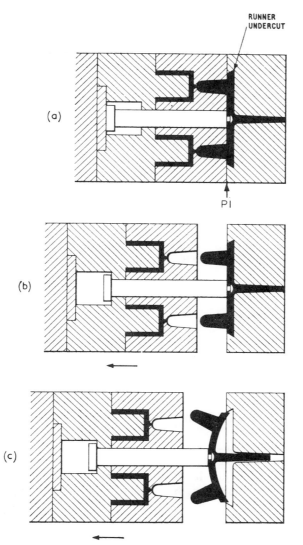

Figure 12.7—Design and operating sequence of undercut runner system, design 1: (a) mould closed; (b) mould partly opened, runner retained in undercuts; (c) further movement and the sprue puller pulls sprue from mould plate

either end in the feed plate (Figure 12.7), and (ii) a delayed acting sprue puller is incorporated.

The sequence of operations is as follows:

The mould opens at P1 (by spring pressure) and, because the runner is retained in the feed plate by the undercuts, the pin gates break immediately.

MULTI-DAYLIGHT MOULDS

The sprue is not pulled at this stage because of the float permitted by the puller design. Thus the moving mould half and the floating cavity plate move back together leaving the feed system in the feed plate (Figure 12.7b). When the puller's floating distance is exceeded, the sprue is pulled, and because of the undercuts the runner is distorted (c) as it is pulled from the feed plate. The operation now proceeds as for the basic underfeed design. The floating cavity plate is arrested by stop bolts, and the mouldings are withdrawn from the cavities. The sprue puller is pulled through the floating cavity plate, thereby releasing the feed system. Continued movement of the moving half causes the ejector system to be actuated and the mouldings to be ejected.

There are several points to bear in mind when designing this type of mould:
  (i) The float period must be sufficient to permit the floating cavity plate to clear the reverse-tapered sprues before the puller pulls the sprue.
  (ii) Considerable runner deflection is necessary to ensure that the secondary sprues are not pulled back into their respective holes when the puller is actuated.
  (iii) To ensure that the puller is pushed back to its moulding position when the mould is closed, the diameter of the puller must exceed the width of the runner.

UNDERCUT RUNNER SYSTEM, DESIGN 2. This is a similar system to the previous design in that the secondary sprues are pulled from the floating cavity plate by means of undercuts incorporated at the runner extremities. The sequence of operation, however, is somewhat different because in this design the float period is achieved by secondary pullers (Figure 12.8a).

The operation of this design is as follows:

The mould is opened at P1 (with the aid of compression springs) and the sprue is pulled immediately by the sprue puller. The entire feed system thereby moves back with the floating cavity plate, the runner withdrawing the secondary pullers from the feed plate. When the floating distance of the secondary pullers has been taken up (b) the feed system is arrested and continued movement of the floating cavity plate causes the secondary sprues to be withdrawn (c). The remainder of the opening sequence is the same as that described in the previous section.

There are several points to bear in mind with this design:
  (i) The secondary pullers must be prevented from rotating and in the design illustrated a round key is incorporated; the head of the puller is suitably recessed to accommodate this key.
  (ii) The front plate is necessary to support the heads of the secondary pullers. Note that a force is applied to the pullers by the plastic melt. The plate also prevents the secondary pullers being displaced when the mould is off the machine.
  (iii) There is a tendency for the feed system to be retained by the secondary pullers when the mould is open. To ensure fully automatic operation a pneumatically operated wiper may be incorporated to eject the feed system.

UNDERFEED MOULDS

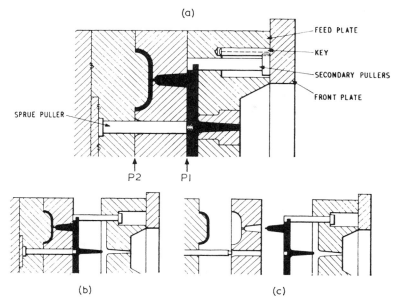

Figure 12.8—Design and operation of the undercut runner system, design 2: (a) part-section through closed mould; (b) mould partly opened; (c) secondary sprues pulled from floating cavity plate by undercut pullers

RUNNER STRIPPER PLATE DESIGN. This design provides for positive pulling of the secondary sprues from the floating cavity plate and for positive ejection of the feed system. In this design (Figure 12.9), a fixed secondary puller is fitted directly below each secondary sprue. To release the feed system from these fixed secondary pullers a runner stripper plate is incorporated between the feed plate and the floating cavity plate.

The opening sequence of the mould operation is as follows:

The mould opens at P1 assisted by compression springs (not shown). The feed system is retained by secondary pullers thereby causing the gates to be broken and the secondary sprues to be withdrawn (Figure 12.9c). The feed system is subsequently stripped from the pullers by the stripper plate connected to the floating cavity plate by length bolts. A suitable delay period is achieved by having the mould plates bored and counterbored as shown. The delay period is necessary to provide a gap of sufficient width to permit the feed system to be removed from between the stripper plate and the floating cavity plate. The movement of the stripper plate and the floating cavity plate is arrested finally when the delay distance provided by the shoulder bolt has been taken up (d). The final opening now occurs and the mouldings are withdrawn from the cavities and subsequently ejected (e). When the mould is closed, the plates are progressively returned to their original positions (b).

Let us now consider a few design features:

(i) The runner stripper plate may be made the same size as the other

MULTI-DAYLIGHT MOULDS

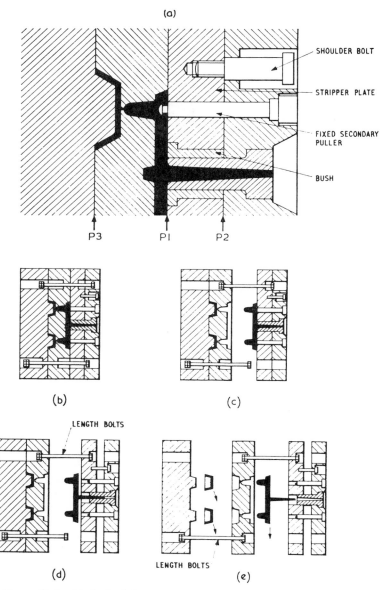

Figure 12.9—Underfeed design incorporating a runner stripper plate: (a) part-section through closed mould; (b) mould closed; (c) Part 1 occurs; (d) Part 2 occurs — runner is stripped from pullers; (e) Part 3 occurs — mouldings ejected

UNDERFEED MOULDS

mould plates and be mounted on the main guide pillars or, alternatively, it may be a narrow plate accommodated in a slot machined in the feed plate. The particular design to use will depend upon the number and layout of the impressions.

(ii) It is essential that the plastic material does not creep through the joint between the sprue bush and the accommodating hole in the runner stripper plate. It is therefore good practice to provide a bush in the plate so that should wear occur, due to the reciprocating movement of the stripper plate, the bush can, quite simply, be replaced. An alternative method is to incorporate a taper on that part of the sprue bush which protrudes through the stripper plate. The hole in the runner stripper plate bush is made of a complementary form. Thus wearing action between two parallel surfaces is avoided.

(iii) It is not generally practicable to withdraw the sprue from the sprue bush completely by the movement of the runner stripper plate with this design. The situation can be improved by keeping the sprue as short as possible, that is by sinking the head of the sprue bush well into the feed plate (Figure 12.3). In any case, to achieve positive ejection of the feed system from the mould it is desirable to arrange for an air jet to blow the feed system clear.

(iv) Extra length-bolts should always be incorporated in the design to pull the floating cavity plate positively to the correct open position should the plate stop prematurely, which would prevent the feed system being removed.

Alternative designs which overcome the problems associated with (iii) above are available and these are known as direct feed systems. Two designs are illustrated, (a) the extended nozzle design, and (b) the heated sprue bush design. Both designs dispense with the conventional sprue system.

In design (a) a heated extended nozzle (Section 13.2.1) is incorporated (Figure 12.10) and this protrudes into the front face of the mould and abuts onto the front face of the floating cavity plate. The nose of the nozzle is tapered and it fits into a complementary shaped aperture in the feed stripper plate.

The opening sequence of the mould is similar to that illustrated in Figure 12.9, the only difference being that as the 'sprue' breaks at 'A' (Figure 12.10) the complete feed system is free to fall between the floating cavity plate and the feed plate, without the probability of being held up, as would be the case with the basic design.

As a heated member (the extended nozzle) is in direct contact with internal mould plates, careful attention must be given to the design of the coolant flowway system in each plate to minimise local heating effects. Note, however, that direct contact only occurs while the mould is closed. A circumferential clearance of at least 7 mm (¼ in) should be provided in the front plate and register plate to minimise transfer of heat from the band heater to the mould.

A heated-sprue bush is incorporated in design (b) and this is similar to the previous design in that the plastics material is available as a melt at the front

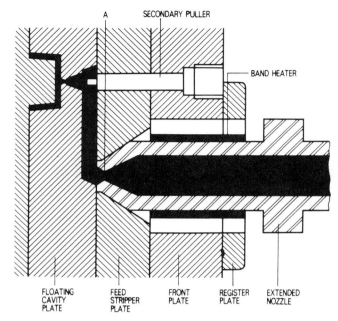

Figure 12.10—Underfeed/runner stripper plate design, incorporating an extended nozzle

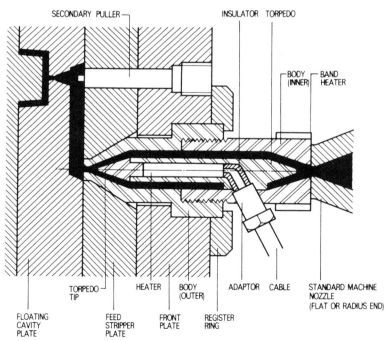

Figure 12.11—Underfeed/runner stripper plate design incorporating a heated sprue bush (Courtesy Incoe Corporation, USA)

TRIPLE-DAYLIGHT MOULDS

face of the floating cavity plate. In this current design, however, the heated-sprue bush is part of the mould structure, and the mould can therefore be fitted to any suitable injection machine fitted with a standard nozzle. A typical design using one type of heated-sprue bush is illustrated in Figure 12.11.

The design of this particular heated-sprue bush is discussed in Section 13.2.4. Note, however, that when the heated-sprue bush is used in conjunction with the underfeed/runner stripper plate design, the nose of the bush is fitted into the complementary conical shaped aperture machined in the floating cavity plate.

The operation of this mould design is, again, similar to the design illustrated in Figure 12.9, the difference being, in this case, that as the feed stripper plate moves away from the heated-sprue bush, the runner is broken from the pin gate, and is then free to fall under gravity between the two mould plates.

The addition of the runner stripper plate to the basic design results in a *triple-daylight mould* (Figure 12.9e).

12.3 TRIPLE-DAYLIGHT MOULDS

In the preceding section we saw the evolution of one type of triple-daylight mould. Another, and more common, mould of this type results when the underfeed and stripper plate designs are combined. This latter design is often adopted where a centre feed position is required on a multi-impression mould, and, because of the component's shape, stripper plate ejection is desirable. Such a mould is illustrated diagrammatically in Figure 12.1g. The four basic plates in the assembly are: the feed plate, the floating cavity plate, the stripper plate and the moving mould plate. The mould is shown in the open position at (h) to illustrate the three daylights.

A more detailed drawing of a triple-daylight mould is shown in Figure 12.12, and while a casual inspection of the drawing may give the impression that it is of a complex nature, a closer inspection will reveal that the major part of the construction has already been covered in preceding sections. For example, the *feed section* (the feed plate and the floating cavity plate) is similar to that described in Section 12.2.4, and the ejector system similar to that described in Chapter 3. Therefore, to save unnecessary repetition the author will confine this discussion to particular design points which may not be obvious to the reader.

Consider the guiding arrangements for this type of mould. In Chapter 3 (Stripper Plate Ejection), we noted that guide pillars are mounted in the moving half to support the stripper plate, whereas in the underfeed design we mounted the guide pillars in the feed plate to support the floating cavity plate. When we combine the two designs to form a *triple-daylight mould*, some compromise must be reached regarding the location and number of the guide pillars. There are several possibilities and one of these is illustrated in Figure 12.12. Here the guide pillars are mounted in the feed plate (as for a conventional underfeed type mould) and these serve the dual purpose of supporting the floating cavity plate, and provide for the alignment between the cavity

MULTI-DAYLIGHT MOULDS

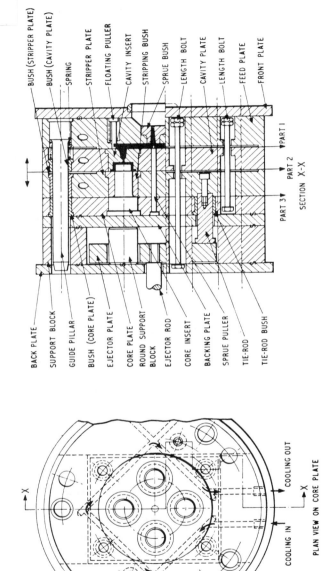

Figure 12.12 — Triple-daylight mould

and the core. The stripper plate is supported on tie-rods which connect the stripper plate to the ejector assembly. Tie-rod bushes are provided in the core plate through which the tie-rods pass. In an alternative design secondary guide pillars (two or three in number) are mounted in the core plate (suitably offset from the main guide pillars mounted in the feed plate) with the sole object of supporting the stripper plate. Suitable clearance holes are provided in the other plates for these secondary guide pillars to pass through. However, this design has certain disadvantages. For example: (i) a larger mould area is necessary to accommodate the extra guide pillars, (ii) because of the number of large-diameter holes, the temperature control system is more complex, and (iii) the guide pillars tend to box in the moulding, making its removal difficult unless special care is taken to avoid this feature. When mould height is limited this design is advantageous as direct actuation of the stripper plate is possible (Chapter 3).

Provision must be included in any design for the possibility of the mould opening in an incorrect sequence. For example, in Figure 12.12 the length-bolt ensures that the floating cavity plate is pulled if the first opening occurs at Part 2 instead of Part 1. The combined free lengths of the stop-bolts, the length-bolt and the permitted stripper plate movement (controlled by headed tie-rod) must be slightly in excess of the maximum opening stroke of the machine on which the mould is to be used, otherwise the mould will be pulled from the machine or the bolts broken.

Let us now consider the operation of the triple-daylight mould. The mould opens at Part 1 assisted by the compression springs mounted on the guide pillars. The sprue is pulled immediately by the sprue puller. The feed system moves back with the floating cavity plate. When the free distance of the floating pullers has been taken up, the feed system is arrested, and continued movement of the floating cavity plate causes the gate to be broken and the secondary sprues to be withdrawn. Mould opening continues until the free length of the length-bolt has been taken up; the movement of the floating cavity plate is then arrested and Part 2 occurs. The mouldings, having shrunk on to their respective cores, are pulled from the cavities and the sprue puller is pulled through the floating cavity plate. Continued movement of the assembly finally causes the actuation of the stripper plate (Part 3) via the ejector assembly and the mouldings are thereby ejected. The individual plates are returned to their moulding position as the mould is closed.

13

Runnerless Moulds

13.1 GENERAL

The term *runnerless mould* may be applied to any mould in which a conventional runner system is not incorporated. We have already discussed simple types of runnerless mould, for a mould which incorporates a direct feed from the nozzle comes within this classification (Chapter 2). Many variations of this basic arrangement have been developed, in order either to reduce the sprue length or to eliminate the sprue altogether (Figure 13.1 a-d).

So far we have only considered the simplest of cases, that of the single-impression mould. Now what happens with multi-impression moulds? We saw in Chapter 2 that, for a simple two-plate mould, a runner system is required to provide a flow path for the plastics material from the central

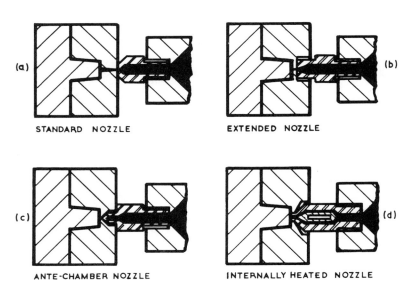

Figure 13.1—Types of nozzle

sprue to each impression, the gate being situated on the mould's parting surface. The alternative underfeed design (Chapter 12) permits the gate to be located at the base of the component, but a more complex multi-plate system

GENERAL

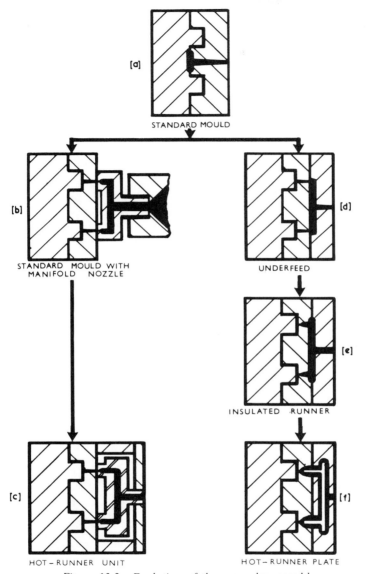

Figure 13.2—Evolution of the runnerless mould

is necessary. In both designs the runner system must be removed from the mould as part of the moulding cycle. To eliminate the necessity for this operation, various runnerless-type moulds have been developed. The following discussion follows the evolution of the runnerless mould and is intended as a primary introduction, of which the aim is to familiarise the reader with the various types of runnerless mould so that he can readily identify each type.

RUNNERLESS MOULDS

13.1.1 Evolution of the Runnerless Mould. All of the many and varied types of injection mould in current use have evolved from the standard two-part mould (Figure 13.2a) which was discussed in Part 1. This standard mould, then, is a useful starting-point for the discussion. The runnerless mould developed from the standard mould by two distinct paths, as shown in Figure 13.2.

Consider the left-hand side of the chart first. The manifold nozzle was developed in the early period of injection moulding to permit individual impressions of a standard two-part mould to be fed from the underside. This type of nozzle is shown schematically in Figure 13.2b. The manifold nozzle is external to the mould and is basically an extension of the injection machine's cylinder. It consists of a manifold block in which a series of drillings, called the 'flow-way', connect the cylinder outlet aperture to a number of secondary nozzles, the outlets of which are in turn connected to the impressions within the cavity plate by means of a sprue gate. On more recent designs, alternative types of secondary nozzle are adopted, such as the extended, ante-chamber, hot-tip and valve types, etc, examples of which are discussed later in this chapter.

The hot-runner unit mould (Figure 13.2c) evolved from the above system. The heated assembly, now termed 'hot-runner unit', is, however, mounted *within* the structure of the mould. Naturally the hot-runner unit must be suitably insulated from the rest of the mould structure to avoid undue heat loss. The internal flow-ways provide a flow path from the injection machine nozzle (not shown) to the sprue entry into the impressions. The unit is maintained at a closely controlled elevated temperature so that the polymer material remains as a melt and can be fed intermittently into the impressions.

The first stage of the alternative evolutionary sequence was via a runner type of mould known variously as an underfeed mould or three-part (three-plate) mould (Figure 13.2d). This design incorporates an additional extra plate in front of the standard two-part mould for feeding purposes. The underfeed mould permits the underfeeding of the moulding while still retaining a runner system. In operation, the mouldings and feed system are removed separately from between different daylights when the mould is open. The underfeed mould is an important design type and is discussed separately in Chapter 12.

The insulated runner mould (Figure 13.2e) developed directly from the underfeed type when it was discovered that certain materials could be moulded on the underfeed mould without removing the runner, by keeping the two plates through which the material flows latched together. This technique is possible only because polymer materials exhibit such excellent insulating properties. To achieve efficient production it was found necessary to increase the cross-sectional area of the runner and of the outlet aperture. The technique is limited to use with easily processable materials and where close control of the moulding cycle can be maintained.

The problem which often arises with the insulated runner mould is that the gate freezes if the cycle is interrupted for any reason. To minimise the effect of this limitation, local heating can be introduced; this is usually achieved by

NOZZLE TYPES

incorporating a heated probe in the outlet flow-way. In this design the insulating properties of the thermoplastic are used to minimise the transfer of heat from the internal heaters to the mould plate.

In the hot-runner plate mould (Figure 13.2f) the design features described above are developed further. In this design the flow-way is internally heated by incorporating a heating element within a tube which is mounted centrally in the flow-way. This internal heating technique is extended to include secondary and outlet flow-ways. The insulation provided by the outer layers of the polymer in the flow-way permits the outer surface of the hot-runner plate to operate at a relatively low temperature, so that the hot-runner plate can be directly attached to the cavity plate (as shown) without an intermediate insulation pad.

It should not be assumed that, because a particular design has been part of the evolutionary sequence, it is necessarily out of date and no longer used. Most of the designs illustrated are in current use, albeit in a modified form. The runnerless systems are now examined in more detail, starting with the basic nozzle types.

13.2 NOZZLE TYPES

The purpose of the nozzle is to provide a flow path for the plastic melt from the machine's cylinder to the mould. In the simplest design the nozzle butts on to the sprue bush of the mould. There are two standard designs in common use and they differ only with respect to the form of seating that is made with the sprue bush (Figure 2.32). One design incorporates a hemispherical end and the other design is flat-ended. The small length of reverse taper in the bore at the front end of the nozzle is such that the sprue is broken just inside the nozzle. This helps to keep the nozzle face clean and assists in maintaining a leakfree sealing face. The standard nozzle of either design is provided by the machine manufacturer and, so that moulds are interchangeable, it is advantageous for a company to standardise on one particular design. Special nozzles are required for certain free-flowing materials such as nylon but here, too, these can be purchased from the machinery manufacturer and are not therefore included in this work. Certain other nozzle types are not standard items, however, and as such they must be designed at the same time as the mould.

13.2.1 Extended Nozzle. We noted in Chapter 4 that it is an advantage to keep the length of the sprue gate as short as possible to minimise the pressure drop across the gate and also to minimise the blemish left on the moulding when the sprue gate is removed.

With the standard nozzle design (Figure 2.32) the length of the sprue gate is controlled by the depth of the mould plate. The length of the sprue gate can be reduced quite simply, however, by designing a special nozzle variously termed a *long-reach nozzle* or *extended nozzle* which protrudes into a pocket machined in the mould plate (Figure 13.3).

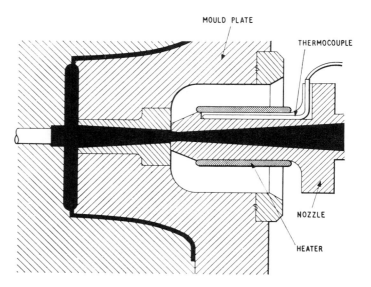

Figure 13.3—Extended nozzle

Some means of heating the nozzle must be provided to prevent undue cooling of the plastics material. One method is to fit a resistance-type band heater on to the parallel length of the nozzle and to control the temperature by means of a controller via a thermocouple. The thermocouple is fitted into a slot machined into the parallel section of the nozzle directly beneath the heater band.

To minimise the transfer of heat from the heated nozzle to the mould, a circumferential clearance of at least 7 mm (¼ in) must be provided between the two parts.

13.2.2 Barb Nozzle. The conventional sprue gate (Figure 4.11) results in a sizeable blemish being left at the injection point. Now, for the majority of components this blemish is unimportant, but there are occasions where it is undesirable. To reduce it to a minimum some form of pin gate is necessary. One method of achieving this form of gate is by the use of a special nozzle in conjunction with a reverse-tapered sprue (Figure 13.4). This special nozzle is variously termed a *barb nozzle* or *Italian nozzle*.

The barb nozzle is similar to the standard nozzle except that there is a projection at the front which incorporates barbs (Y, Figure 13.4). It is this portion of the nozzle which is accommodated in the reverse-tapered sprue. Leakage of material is prevented by ensuring that the flat face of the nozzle seats on to the sprue bush at X.

In operation the plastic material flows through the nozzle, sprue, and gate, and so into the impression. After the material has solidified the sprue is pulled from the sprue bush by the barbs as the nozzle is withdrawn. The mould is opened and the moulding ejected in the normal way. To permit the removal

NOZZLE TYPES

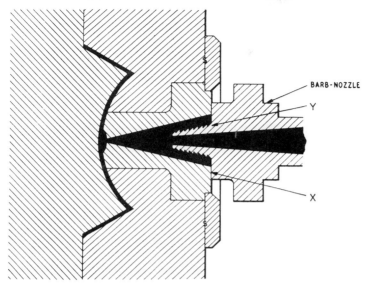

Figure 13.4—Barb nozzle

of the sprue from the nozzle a larger than normal sprue break is required and, because of this, the design is normally used on injection machines which incorporate a sliding carriage which can be withdrawn automatically as part of the machine's normal cycle of operations.

The sprue is normally removed manually from the nozzle which, of necessity, lengthens the moulding cycle. An alternative design incorporates a local stripper plate, to strip the sprue from the barbs. The barb nozzle design has largely been superseded by ante-chamber and hot sprue bush designs.

13.2.3 Ante-chamber Design. In this design, also known as the hot well design, a small mass of plastic material is retained in the ante-chamber (X, Figure 13.5). The plastic material adjacent to the mould partially insulates the central core of plastic material from the cold mould. Thus, by suitable adjustment and control of the nozzle temperature and the moulding cycle, the material remains sufficiently fluid to allow it to pass intermittently through the ante-chamber into the impression. As it is not necessary to remove the sprue from the mould, fast moulding cycles can be achieved. Note that with a standard nozzle the moulding cycle, particularly with thin-walled components, is often controlled by the waiting period for the sprue to cool sufficiently to permit its extraction.

A part-section through a mould for a thin-walled beaker which incorporates the ante-chamber design is illustrated (Figure 13.5). In this single-impression mould, the integer cavity is adopted for rigidity. A local insert forms the base of the impression, and also incorporates the ante-chamber (X). The nozzle has a seat on the insert at Y. The main body of the nozzle is accommodated in a pocket machined in the rear of the insert. The diameter

RUNNERLESS MOULDS

of this pocket must be sufficient to provide a suitable air insulation gap (at least 13 mm (½ in)) between the nozzle heater and the mould. Efficient mould temperature control is essential and in the design illustrated a cooling sleeve is fitted around the local insert. The core also incorporates a separate cooling system. (As this design is for a thin-walled container, the core is accurately located in the cavity by the angled fit location method. Ejection is by means of a stripper ring operated by a conventional ejector system.) A more detailed discussion of the ante-chamber design appears in Section 13.3.6, which deals with secondary nozzles.

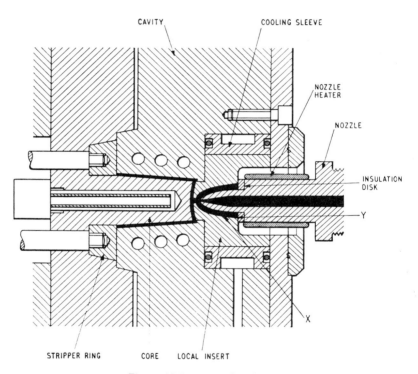

Figure 13.5—Ante-chamber design

13.2.4 Heated Sprue Systems. There are two heated sprue systems which can be used for feeding directly into an impression. In one system, in which the conventional sprue bush is replaced by an internally heated sprue bush, the concept of the injection machine's nozzle abutting onto the sprue bush is retained. In the alternative system the sprue bush is dispensed with, being replaced by an internally heated extension nozzle.

INTERNALLY HEATED SPRUE BUSH. The principle of this design is that a heating element is incorporated on the centre line of the sprue bush in the flowway between the injection machine's nozzle and the gate (entry) into the

NOZZLE TYPES

impression. By this means, the polymer material may be held at an elevated, controlled temperature up to a position relatively close to the gate. A standard injection machine nozzle, with either a flat or a radius end, is used in conjunction with this design.

One design of internally heated sprue bush is illustrated in Figure 13.6 and consists of five basic parts: (a) body-outer, (b) body-inner, (c) torpedo, (d) torpedo tip and (e) cartridge heater. The cartridge heater is housed within the torpedo assembly, as shown. The heater wires pass through one of the torpedo legs via an internal insulator and adaptor (f). The torpedo assembly is mounted within the body-inner which is then screwed into the body-outer to form the sprue bush.

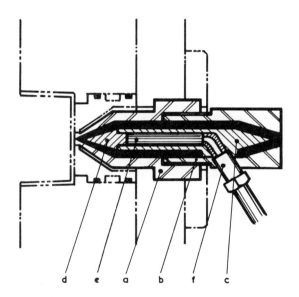

Figure 13.6—Internally heated sprue-bush (Courtesy Incoe Corporation, USA)

The front end of the body-outer is basically conical in shape, with a small parallel nose. This fits into a complementary hole machined in the cavity plate as shown. Note the position of the insulation gap between the body-outer and the cavity plate to minimise heat transfer.

The torpedo-body assembly forms an annular flow-way for the polymer melt. Thus the melt is in close contact with the heated torpedo throughout its entire passage from the nozzle to the gate. An additional band-type heating element may be fitted to the rear side of the body-inner to compensate for convection and conduction losses. When an internally heated sprue bush is incorporated in a design, careful consideration must be given to the disposition of the coolant flow-way system in the cavity plate to minimise local heating effects. In the design shown (Figure 13.6) an annulus-type cooling bush is incorporated.

RUNNERLESS MOULDS

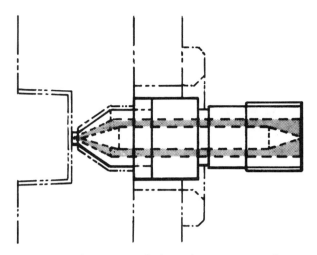

Figure 13.7—Internally heated extension nozzle

Several companies produce internally heated sprue bushes as standard parts. The one illustrated in Figure 13.6 is by the Incoe Corporation (USA) who call their assembly a 'Hot Tip Bushing'. A similar design is offered by W.J. Furse Ltd (UK) who term their heated sprue bush a 'Thermoject Nozzle'. A third design, available from DME (USA and Europe) and also called a 'Hot Tip Bushing', incorporates a torpedo (probe) which acts as a valve. During the injection stroke of the screw-plunger, injection pressure retracts the torpedo (probe) to a rear position, permitting the melt to pass through the gate into the impression. When the injection is complete, the pressure falls and a spring forces the torpedo (probe) forward to seal the gate and prevent drooling, etc.

INTERNALLY HEATED EXTENSION NOZZLE. This design is very similar to the one discussed above, the basic difference being that the assembly is attached to the injection machine cylinder and becomes an extension nozzle which protrudes into an accommodating recess directly behind the impression.

The body-inner incorporates an externally threaded section which permits the nozzle to be attached to the injection cylinder in the conventional way. A simplified drawing of the internally heated extension nozzle is shown as Figure 13.7.

13.2.5 Multi-nozzle Manifold. The direct feeding of an impression from a nozzle is not limited to the single-impression types that were discussed in the preceding sections. Many multi-impression moulds have been designed for direct feeding via a multi-nozzle manifold and, while some machine manufacturers offer these units as an optional extra for their machines, they are generally designed for a specific mould.

A multi-nozzle manifold consists of a manifold block suitably drilled to

HOT-RUNNER UNIT MOULDS

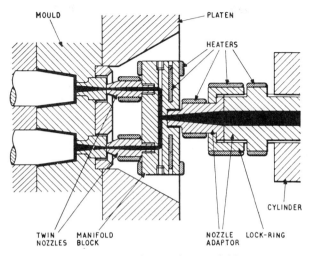

Figure 13.8—Multi-nozzle manifold

provide a flow path for the material, into which is fitted the requisite number of nozzles. The manifold block is attached to the cylinder of the machine via a nozzle adaptor and a lock ring. This latter feature permits radial adjustment to align the nozzles with the sprue bushes during the initial setting operation. Heating of the assembly is essential and this is accomplished by the provision of external band heaters, supplemented by internal cartridge heaters if necessary. To permit even filling of the impressions, separate temperature control of each nozzle is necessary. A standard mould design is used in conjunction with the multi-nozzle manifold except that each impression has its own sprue bush. Standard nozzles are shown feeding the impression via sprue gates but the barb nozzle or ante-chamber designs can equally well be used if pin gating is required.

While theoretically any number of impressions can be fed in this way, in practice, unless platen modifications are contemplated, the number is limited to the number of nozzles that can be accommodated within the aperture in the platen. In view of this and other limitations, the multi-nozzle manifold is now seldom used. It is included in this chapter because historically it formed an important step in the development of the hot-runner unit.

13.3 HOT-RUNNER UNIT MOULDS

This is the name given to a mould which contains a heated runner manifold block within its structure. The block, suitably insulated from the rest of the mould, is maintained at a closely controlled elevated temperature to keep the runner permanently as a melt. The polymer material can thereby be directed to the mould extremities without loss of heat and without the pressure loss associated with temperature variations.

To illustrate the more important features of the design a part-sectional drawing through a mould of this type is shown as Figure 13.9. The hot-runner unit is mounted adjacent to the cavity plate and accommodated in a suitably designed grid. The polymer material enters via a centrally positioned sprue bush, passes through the flow-way (x) and leaves the unit via a secondary nozzle in line with the impression. When the mould is opened the moulding is pulled from the cavity, and the sprue is broken at the small diameter end. The remainder of the feed system remains heated within the unit, ready for the next shot.

Before commencing our study of the design of hot-runner unit moulds in

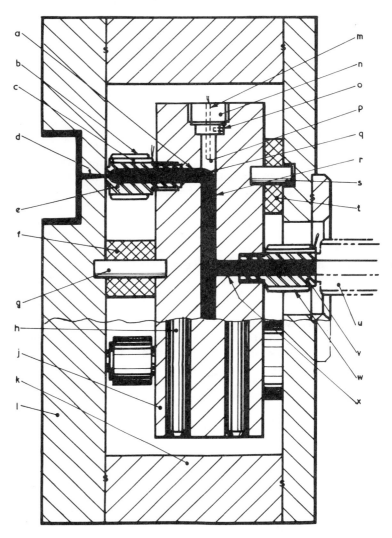

Figure 13.9—Hot-runner unit mould

HOT-RUNNER UNIT MOULDS

detail, we consider a number of aspects of the design which are relevant at this stage. This introduction is to give the novice sufficient background to be able to decide when to use a hot-runner unit design in preference to other mould types. A good starting-point, in this respect, is to consider the advantages and limitations of the system.

ADVANTAGES OF THE HOT-RUNNER UNIT TYPE OF MOULD

(1) There is no feed system for the operator to remove from the mould.

(2) On manually controlled injection machines the mould open time is reduced.

(3) The cost of separating, storing and regrinding scrap feed systems is thereby saved.

(4) As no scrap material is produced it cannot be contaminated.

(5) The moulding is automatically degated on 'direct feed' designs.

(6) All of the impressions fill at practically the same time (this feature is dependent upon impression layout).

(7) Thin wall-section mouldings fill relatively easily because the melt is at a higher temperature close to the impression.

(8) The diameter of the flow-way within the hot-runner unit can be larger than a corresponding runner machined in the parting surface of a two-part mould. There is, therefore, less pressure loss throughout the system.

(9) Mouldings are produced with less inherent strain because lower pressures are required.

(10) As there are no runners (in the conventional sense) to fill and cool, the time cycle is marginally shorter with the hot-runner unit design. This advantage is most pronounced with respect to multi-impression moulds. (The overall cycle time on a two-part mould is often controlled simply by the time taken for the feed system to solidify.)

(11) Improved moulded-part quality is obtained, particularly where large gates can be incorporated (see valve-type secondary nozzles).

(12) The hot-runner mould is not restricted to use on a particular injection machine. (Certain nozzle manifold designs necessitate the boring of holes in the injection machine platen).

(13) The location of the impression is not confined to the central region of the mould plate, as is the case when a small manifold nozzle design is adopted.

(14) On conventional runner-type moulds the weight of the feed system must be taken into account when determining the number of impressions which can be moulded on a specific size of injection machine. This is unnecessary with the hot-runner unit design, and therefore more impressions can be incorporated. The significance of this saving of weight of the runner system can be appreciated by comparing the volumes of the runner systems shown in Figures 13.10-12.

Consider the 16-impression layout shown in Figure 13.10: if the same layout is adopted for a semi-runnerless mould (Figure 13.11), a considerable saving in the volume of the feed system is achieved (see volume calculations,

RUNNERLESS MOULDS

Table 8). Here the main and secondary runners of the system shown in Figure 13.10 are replaced by a hot-runner flow-way (dotted lines) and are thus disregarded in the volume calculation.

The calculations show that the volume of the material in the runners of this second arrangement is about 30 per cent of that in the conventional system. Taking the principle one stage further and extending the hot-runner unit concept to direct feeding into the base of all the impressions (Figure 13.12), as there is now no runner system, the corresponding volume is nil.

LIMITATIONS OF THE HOT-RUNNER UNIT TYPE OF MOULD

(1) The mould setting time is generally greater than for a corresponding two-part mould.

(2) An extended period, waiting for the hot-runner unit to heat, is required

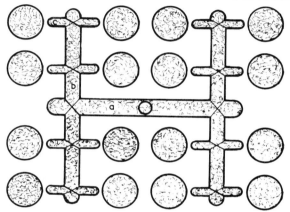

Figure 13.10—Conventional runner system for 16-impression mould

TABLE 8

COMPARATIVE CALCULATION OF FEED SYSTEM VOLUME

Runner type	No. of runners	Runner length (mm)	Runner diameter (mm)	Runner volume (mm³)	(in³)
CONVENTIONAL RUNNER SYSTEM (Figure 13.10)					
Main (a)	2	80	8	8 042	0.49
Secondary (b)	4	85	6	9 613	0.59
Branch (c)	16	15	3	1 696	0.1
			TOTAL VOLUME	19 357	1.18
SEMI-RUNNERLESS SYSTEM (Figure 13.11)					
Runner (r)	16	50	4	5 026	0.31
Sprue (s)	4	5	4	254	0.02
			TOTAL VOLUME	5 280	0.33

HOT-RUNNER UNIT MOULDS

before production can commence.

(3) The initial 'debugging' of a new hot-runner unit mould is usually more extensive than with a standard mould.

(4) The cost is higher than that of a standard mould and in some cases than that of an underfeed (three-part) mould.

(5) The area of the moulding adjacent to the gate may be blemished with surface heat marks.

(6) Polymer melt leaking from the hot-runner unit can create problems.

(7) Polymer material at the gate may solidify and interrupt production.

(8) Certain materials (e.g. nylon) have a tendency to drool (dribble) from the gate into the impression when the mould is open. This causes blemishes on the subsequent moulding.

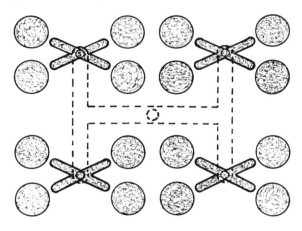

Figure 13.11—Semi-runnerless system using hot-runner mould

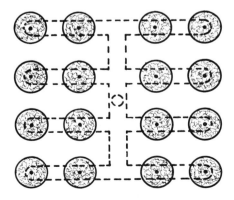

Figure 13.12—Runnerless system using hot-runner unit

RUNNERLESS MOULDS

(9) Some materials (e.g. polypropylene) have the tendency to 'string' when the moulding is extracted from the cavity. This has adverse effects similar to the above.

(10) There is a tendency for certain types of heating element to fail during production.

(11) Replacement of the heating element is sometimes difficult and can be time-consuming.

(12) With certain materials (e.g. pvc) there are degradation problems.

(13) Changing the colour and the type or grade of a material can also be a problem.

(14) Heat expansion of the hot-runner unit can create difficulties, if not allowed for.

It should be pointed out that many of these problems arise because insufficient thought has been given to the design in the planning stage. To minimise their effect, and in some cases to obviate them altogether, a number of design features should be considered at the planning stage: these are listed here for ease of reference, and are considered in detail later in the chapter.

DESIGN FEATURES OF THE HOT-RUNNER UNIT

(1) Sufficient heating elements should be incorporated so that the hot-runner unit heats quickly to the required moulding temperature from cold.

(2) Sufficient heat energy must be supplied to the hot-runner unit to replace conduction, convection and radiation heat losses.

(3) To ensure that the temperature of the melt in the flow-way is maintained without either hot or cold regions, careful location of the heating elements is essential.

(4) Considerable production time may be lost if a heater fails. Thought must be given therefore to choice of heating element, its location and the facilities for removing it should it fail.

(5) The layout of the wiring system should be neat and easily traceable.

(6) Heating element wires which are subject to heat or abrasion attack should be protected.

(7) Use manufacturers' recommended clearances when fitting heating elements of the cartridge type.

(8) Note that a cartridge heating element does not release heat over its entire length. There are end 'cold spots'.

(9) Flat-type heating elements and induction heating elements should be completely enclosed within the unit for maximum efficiency.

(10) Allowance must be made for the fact that the hot-runner unit increases in size in three dimensions when heated.

(11) As the hot-runner unit is relatively hot, consideration must be given to insulating it from the rest of the mould structure.

(12) For many materials, close control of the temperature of the melt is vital. Careful consideration as to the location of thermocouples is essential.

(13) To minimise degradation, colour-changing and material-changing problems the melt flow-way must be as streamlined as possible, without

HOT-RUNNER UNIT MOULDS

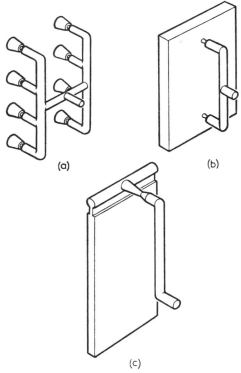

Figure 13.13—Typical applications of the hot-runner design
(a) Pin gating on multi-impression moulds
(b) Multi-point pin gating
(c) Film gating of large component

sharp corners, ledges or other stagnation points which tend to hold back the polymer melt for extended periods.

APPLICATIONS OF THE HOT-RUNNER UNIT TYPE OF MOULD

(1) It allows for the pin gating of mouldings on multi-impression types of mould (Figure 13.13a).

(2) It allows for multi-point gating on single-impression and multi-impression moulds (Figure 13.13b).

(3) It allows for side or film gating of large mouldings (Figure 13.13c).

(4) It permits the semi-runnerless design to be adopted, where small groups of impressions are fed from secondary sprues (Figure 13.11).

13.3.1 Hot-runner Unit. A hot-runner unit is an insulated heated assembly which is mounted within the structure of a mould to provide a flow-path for the polymer melt from the injection machine's nozzle to the gate entry into the impression.

RUNNERLESS MOULDS

The component parts of the hot-runner unit are as follows (see Figure 13.9): a manifold block (j) is drilled with a series of holes (x, r, a) to provide a flow-path for the polymer material. Secondary nozzles (e) are fitted to the manifold block to provide 'outlet' apertures. These apertures are coupled to the impression either directly via a small pin gate or indirectly via small sprue-gates as shown (d).

A manifold bush (v) may be fitted to the manifold block to provide the connection between the injection machine's nozzle (u) and the flow-way (x). Alternatively, an extended nozzle may be used for this purpose. The front part of the nozzle extends into a recess machined in the manifold block (see, for example, Figure 13.17).

Profiled plugs (p) are fitted at the ends of the main flow-way (r) to facilitate changes in the flow direction. The profiled plugs are held in position by an end screw (n).

In the design illustrated (Figure 13.9) the heating system comprises a number of resistance-type cartridge heating elements (h) which are fitted into holes bored parallel to the main melt flow-way drilling. In addition, band-type resistance heating elements (b, w) are fitted around both the secondary nozzle and the manifold bush.

The temperature of the hot-runner unit must be carefully controlled and it is normal practice to fit thermocouples (c, m) in a suitable position close to the main flow-way (r) and the outlet position (e).

Particular aspects of hot-runner unit design are considered in more detail below, as follows:

The manifold block (13.3.2)
The melt flow-way (13.3.3)
The inlet aperture (13.3.4)
Expansion problems (13.3.5)
Secondary nozzle design (13.3.6)
Valve systems (13.3.7)
Heating considerations (13.3.8).

Figure 13.14—Rectangular hot-runner unit manifold block

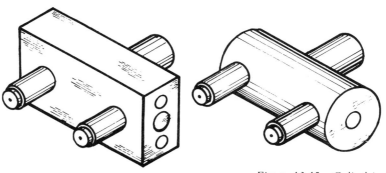

Figure 13.15—Cylindrical hot-runner unit manifold block

HOT-RUNNER UNIT MOULDS

13.3.2 The Manifold Block. The manifold block can be classified as being of one of two types, according to its cross-sectional shape: rectangular or circular. While the rectangular type is usually manufactured as a one-piece structure, the circular cross-sectional manifold block is of composite structure.

The two types are shown schematically in Figures 13.14 and 15. The greatest difference between the two is in the method adopted for heating the manifold block. In the rectangular design, cartridge, coil or flat-type heating elements are fitted into suitably shaped holes or recesses within the manifold block, whereas in the cylindrical design, band or coil-type heating elements are fitted on the outside of the manifold block. The various heating systems are discussed in later sections.

In general, the cylindrical manifold block is less bulky than its rectangular counterpart and as a result less total energy input is necessary to maintain the required temperature. The choice of design depends not only on the machining facilities available but also on the designer's preference. The novice is recommended to start with the rectangular design, as it is the simpler to draw.

THE RECTANGULAR MANIFOLD BLOCK. The rectangular cross-sectional manifold block may be one of a number of alternative shapes on plan; a few examples are shown in Figure 13.16. The 'right prism' (a) is the most common

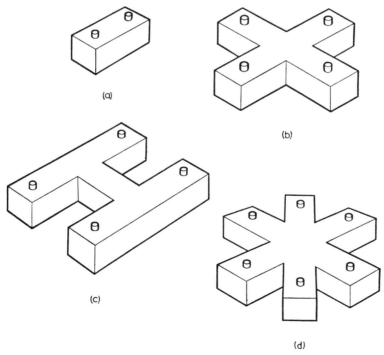

Figure 13.16—Hot-runner manifold block designs
  (a) prism (b) 'X' (c) 'H' (d) spoke-wheel

363

design encountered in practice as this shape is the simplest to machine and to accommodate within the mould structure. The right prism (usually shortened to 'prism') manifold block is manufactured from a rectangular block of steel in which holes are bored to form the flow-way system. The outlet apertures may be in one line to permit the individual holes to be interconnected by a straight drilling (shown schematically in Figure 13.20a). Where the number or layout of the impressions makes the in-line design impracticable, the outlet points, provided they are relatively close together and in some regular arrangement, can be interconnected by a series of longitudinal and transverse drillings and still be accommodated in a prism manifold block design.

However, for widely spaced outlet points or for applications where the outlet points are not symmetrical about the mould's centre line, the use of an alternative manifold block shape should be considered. Manifold blocks often approximate to a letter shape on plan and they are usually designated in this way: e.g. an 'X'-shaped unit (b) and an 'H'-shaped unit (c) are illustrated in Figure 13.16. The odd one out in this series of drawings is the 'spoke wheel' unit (d) which does not lend itself to the lettering system.

Note that all of the outlet apertures illustrated in Figure 13.16 could have been achieved using the prism design. The resulting manifold block would, however, have been bulky and heavy, and would have required considerably more heating capacity than the alternative types.

THE CIRCULAR MANIFOLD BLOCK. The simplest version of the circular cross-sectional manifold block is illustrated in Figure 13.15. It is manufactured from a cylindrical bar of steel and incorporates a central flow-way drilling. The outlet flow-ways are bored to interconnect with this central drilling. Naturally this single-drilling design is rather restrictive in that it can only be used to feed impressions which are positioned in a straight line.

To achieve varied exit-point positions to correspond to the designs discussed in the preceding section, a more complex composite design must be adopted. The composite circular cross-sectional manifold block consists of an assembly of cylindrical sections, an inlet block and outlet blocks. The principle of the design is illustrated in Figure 13.17, which shows the flow path from the nozzle to one exit point.

The inlet block (p) is located on the centre-line of the mould by the location pin (o). An extended nozzle (s) abuts onto the bottom surface of a recess at (r) machined in the inlet block. A manifold bush of the type shown in Figure 13.9 could alternatively have been fitted. The force applied by the extension nozzle is transferred to the cavity plate by the support block (n).

The outlet block (one of which is shown at (e)) incorporates a profiled plug (f) which facilitates the change in melt flow direction. The profiled plug is secured in position by the end screw (g) as shown. The secondary nozzle (a) is fitted to the outlet block to complete the flow path to the impression. Support pads (m) are incorporated directly below the outlet block.

The interconnecting melt flow-way between the inlet and outlet blocks is accommodated in the cylindrical section (l). The length of this cylindrical sec-

HOT-RUNNER UNIT MOULDS

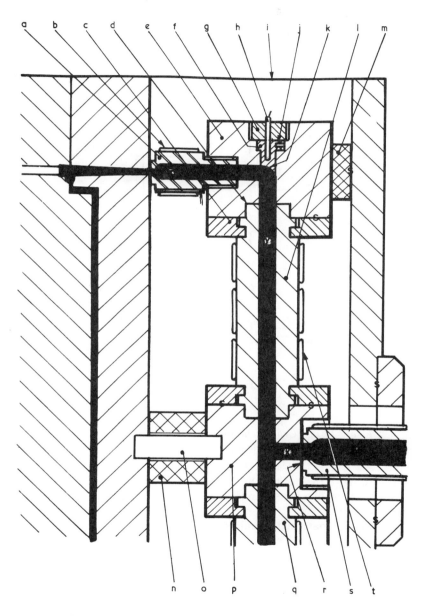

Figure 13.17—*Cylindrical manifold block, composite construction*

RUNNERLESS MOULDS

tion controls the centre distances between the individual blocks. Cylindrical projections (d) fit into accommodating recesses in the inlet and outlet blocks to ensure correct alignment. Half-shoes (c) securely clamp the cylindrical section to the respective inlet and outlet blocks to form a leak-free joint. The cylindrical section is heated either by band-type heating elements (t) or by low voltage coil elements.

The composite nature of this design permits various alternative types and sizes of unit to be constructed from relatively few 'standard' components. For example, as shown in Figure 13.18, the in-line, 'X', 'T', 'H' and 'Y' units are easily formed using this technique. Note that the shape of the inlet block

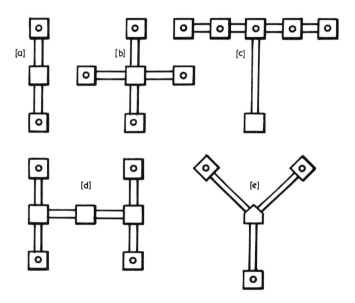

*Figure 13.18—Cylindrical hot-runner manifold block designs
(a) in-line (b) 'X' (c) 'T' (d) 'H' (e) 'Y'*

varies according to the design: the 'Y' unit, for example, requires a five-sided block, while the spoke-wheel unit (see Figure 13.20) requires a cylindrical inlet block.

Plate 14 shows a patented cylindrical manifold block hot-runner unit manufactured in the UK by Tooling Products Ltd of Langrish. This manifold block is of double-'X' configuration, having eight outlet feed points. The design incorporates telescopic joints to compensate for heat expansion of the manifold block. Note in particular the method adopted for housing the manifold block and also the way the heating element leads are fed to a series of junction boxes located at the top, centre and bottom of the assembly. This design keeps the length of the heater leads to a minimum and facilitates the removal and replacement of the heating element.

HOT-RUNNER UNIT MOULDS

The following discussion relates specifically to the rectangular manifold block, since this is the type more frequently encountered in practice. However, most of the comments apply equally well to the cylindrical manifold block design.

13.3.3 The Melt Flow-way. This is the name given to the series of drillings within the manifold block which connects the injection machine nozzle (usually via a manifold bushing) to the various secondary nozzles. Ideally, the path for the polymer melt through the manifold block should be as smooth and as streamlined as possible. Unless this aspect of the design is given careful consideration, polymer material may be retained in crevices, corners, blind alleys, etc. Any region in which the material remains within the manifold block for an extended period of time without being moved towards the outlet

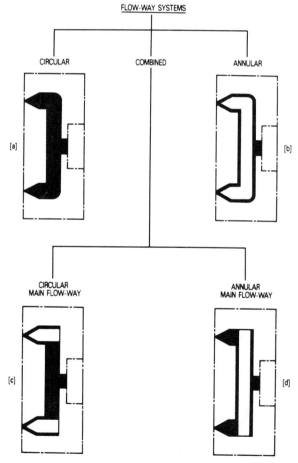

Figure 13.19—Melt flow-way systems

may create a production problem. When polymer material is held up in this way, difficulties arise when changing colour or type of material, and also degradation of the stagnant material may lead to discoloration of the moulding.

There are three possible designs for the flow-way, shown schematically in Figure 13.19. The first system, and the most commonly met in practice, is based upon a circular hole. The second is based upon an annular hole, and the primary purpose of this design is to permit an internal flow-way heating system to be incorporated. The third system is a combination of the first two, in that the circular hole and the annular hole are combined.

CIRCULAR MELT FLOW-WAY. The flow-way is formed by machining a 10-13 mm (3/8-1/2 in) hole longitudinally through the manifold block as shown in Figure 13.9. This is the 'main flow-way' (r). It is preferable to drill this hole from one end only of the manifold block, rather than attempt to drill a hole from each end and join at the centre. Admittedly the latter method results in shorter holes having to be drilled, but in practice a perfect match where the two holes meet is unlikely because of drill wander. If it is impracticable to drill from one end, an extra boring and finishing operation should be undertaken to ensure the smooth continuity of the flow-way. For the same reasons, the inlet (x) and outlet (a) flow-ways must be carefully located and drilled. Careful blending of these holes at the junction with the main flow-way is essential.

The end of the main flow-way at q is blocked up to its junction with the outlet flow-way (a) with a profiled plug. To ensure that the polymer melt has an uninterrupted passage when the flow changes direction, the face of the profiled plug should incorporate either a chamfer or a fillet (as shown). A location pin (o) on the profiled plug is accommodated in a slot machined in the manifold block. This ensures that the profiled plug is always fitted into the manifold block in the correction position, in which it is then secured by the end-screw (n). Note that a hole through both end-screw and profiled plug may be provided to accommodate a thermocouple (m) for temperature control.

The above discussion relates specifically to the in-line flow-way layout where all secondary nozzles are positioned in one line. For more complex layouts, one or more flow-way holes are drilled on the same plane as the main flow-way but at an angle to it. These additional holes are called secondary flow-ways. The same comments apply to the end of the secondary flow-way as apply to the main flow-way. Various configurations of complete flow-way systems are shown in Figure 13.20.

COMBINED CIRCULAR AND ANNULAR FLOW-WAY SYSTEM. The combined flow-way system may be of one of two possible designs. In the first (Figure 13.19d) the cross-section of the main flow-way is an annulus, while the outlet flow-ways are circular. The alternative design (Figure 13.19c) has a directly opposite arrangement: the cross-section through the main flow-way is circular, whereas the outlet flow-ways are annular in form.

Consider (d) first: this basic design is somewhat restrictive, but it has the

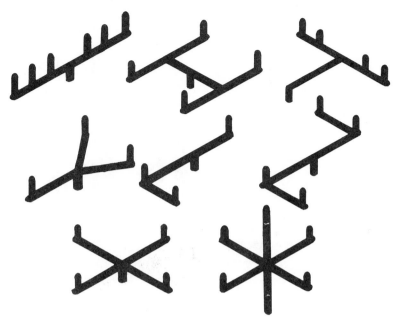

*Figure 13.20—Configurations of melt flow-way systems
In-line, 'H', 'T', 'Y', 'L', 'Z', 'X' and spoke-wheel*

advantage of providing the novice with a good introduction to the more complex complete annulus system (b). A part cross-sectional view of a mould of this type is shown in Figure 13.21. The main flow-way is formed by machining a large hole, 30 mm (1¼ in) diameter or more, longitudinally through the manifold block (a). A heater tube (d) is fitted centrally in this hole. Thus the required annular hole is formed by the inner surface of the manifold bore and the outer surface of the heater tube. The heater tube is positioned by means of end-caps (b), and these are a sliding fit in the 30 mm diameter hole. The heater tube assembly is held in position by grub screws (c) or location pins.

Heating elements (e) of the cartridge heater or low voltage coil type are fitted inside the heating tube. Some allowance should be made in the design to accommodate the expansion in length of the tube when it is heated. In the design illustrated a small gap (p) is left between the end of the heater tube and the bottom of the recessed hole in the end-cap.

As noted above, the inlet (h) and outlet (q) flow-way holes of this design are circular in cross-section and are drilled to intercept the main annular flow-way. Careful blending of the holes at these critical points is necessary. Note that the short-length entry flow-way (h) in conjunction with an extended nozzle (k) has been adopted for this design (see Section 13.3.4). The extended nozzle protrudes into a recess machined in the manifold block (a) thereby obviating the need for a manifold bush (compare with Figure 13.9). The remaining aspects of the design are identical to those discussed in the preceding section.

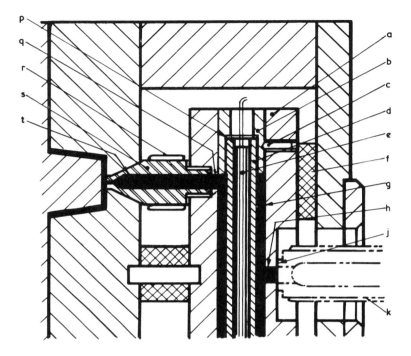

Figure 13.21—Combined circular and annular flow-way system

Heater tubes of various lengths, and matching end-caps, are available as standard parts from DME (USA and Europe) in two sizes (see Section 13.5.1).

The alternative combined flow-way system (Figure 13.19c) has a cylindrical main flow-way and an annular outlet flow-way. The primary objective of this design is close temperature control at the secondary nozzle outlets. This is achieved by incorporating an internal core of either the heat transfer rod or the internally heated torpedo type (see Section 13.3.6).

THE ANNULAR FLOW-WAY SYSTEM. In this design the manifold block incorporates an annulus for the main flow-way, secondary flow-way, and outlet flow-way as shown schematically in Figure 13.19b. The annular flow-way system is used primarily for the hot-runner plate mould (see Section 13.5).

13.3.4 The Inlet Aperture. The inlet aperture into the main flow-way may be incorporated in the manifold block design (i) via a manifold bushing, (ii) via an internally heated manifold bushing or (iii) via a short-length flow-way.

(i) THE MANIFOLD BUSHING (Figure 13.22). This corresponds to the sprue bush of a standard two-part mould, and is termed by many designers a 'sprue bush'. The manifold bushing (p) is shown in its assembled position, that is,

HOT-RUNNER UNIT MOULDS

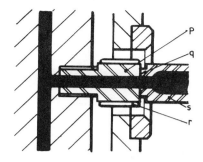

Figure 13.22—The inlet aperture: manifold bushing design

screwed into the manifold block. It is essential that there is metal-to-metal contact at the butting surface (at q) to prevent leakage and possible stagnation of polymer at this point. The manifold bush is usually heated externally via a band-type heating element (r).

A hole of at least 10 mm (3/8 in) diameter should be specified for both the manifold bushing and the machine's nozzle (s) which feeds it. There is no point in creating restriction at the entry to the manifold block by using a standard nozzle which may have a relatively small exit hole diameter. A clearance of at least 13 mm (½ in) should be provided around the manifold bushing to minimise the transfer of heat to this plate. Note that in this design the register ring (locating ring) is accommodated in a recess in the back plate for the purpose of centering, and is not connected to the manifold bushing.

(ii) INTERNALLY HEATED MANIFOLD BUSHING (Figure 13.23). This is basically an assembly of two parts: a manifold bushing (t), and an internally heated probe (u). The design of the manifold bushing is an extension of that discussed above. The general features, including the fitting arrangements, are identical, but the diameter of the melt flow-way is enlarged in order to accommodate a probe.

The probe is fitted to the manifold block as shown, being secured either by a backing plate or a grub screw. The centre of the probe is hollow to accommodate a cartridge heating element (w) or a low voltage coil-type element. The

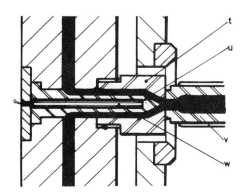

Figure 13.23—The inlet aperture: internally heated manifold bushing design

RUNNERLESS MOULDS

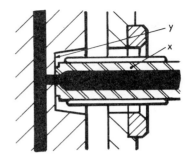

Figure 13.24—The inlet aperture: direct entry via short-length flow-way

wires for the element are fed through a suitable channel or hole at the base of the manifold block to an adjacent junction box.

This is an example of an annular entry aperture where heat energy is added to the polymer melt during its passage from the heated machine nozzle (v) to the main flow-way.

(iii) THE SHORT-LENGTH FLOW-WAY (Figure 13.24). In this design the manifold bushing is dispensed with. A nozzle of the extended type (x) protrudes into a recess machined into the face of the manifold block (y). The length of the inlet aperture is thereby drastically reduced, minimising the distance the melt has to travel from the heated nozzle to the flow-way. Some designers prefer to incorporate a hardened plate at this butting surface to prevent the extended nozzle from bedding into the relatively soft manifold block.

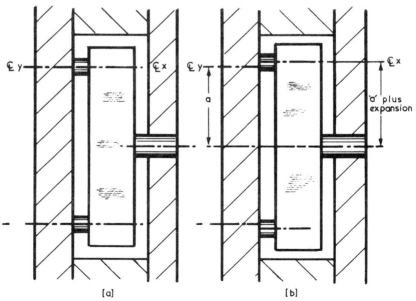

Figure 13.25—Expansion of the manifold block (length) (a) manifold block cold (b) manifold block heated

HOT-RUNNER UNIT MOULDS

13.3.5 Expansion Problems. One important consideration which must not be overlooked when designing a hot-runner unit mould is that metals expand when heated. The manifold block increases in all three dimensions when the temperature is increased. This distance between the centres of the secondary nozzles will increase with respect to the 'fixed' distance between the centres of the impressions machined in the cavity plate (Figure 13.25). At (a) the manifold block is cold and the centre lines of the secondary nozzles 'x' are in line with the corresponding impression centre lines 'y'. However, when the manifold block is heated, the distance between the secondary nozzle centres is increased, and the respective centre lines 'x' and 'y' are no longer in line. When designing the hot-runner mould, allowance must be made for this expansion to ensure that the centre lines are in line during production.

Expansion in depth is generally not a major design problem. Figure 13.26 shows a detail of the manifold block secondary nozzle assembly. When cold

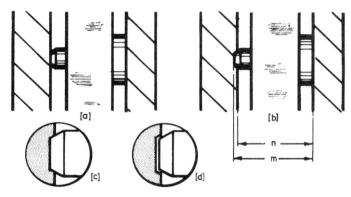

Figure 13.26—Expansion of the manifold block (depth)
(a) manifold block cold (b) manifold block heated

the depth (m) of this assembly corresponds to the depth (n) of accommodating space provided in the hot-runner unit grid. However, when the hot-runner unit is heated, the depth of the assembly increases as shown in Figure 13.26b, so that the secondary nozzles tend to bed into the front face of the cavity plate, as shown. This bedding-in is usually small enough to be negligible and, in fact, provides a leak-free union between the two mould members. However, if the hot-runner unit is used for more than one material whereby different manifold temperatures are required, differential expansion can be expected and this may create difficulties.

Scrap views (c) and (d) illustrate this point: at (c), the assembly has expanded when used for material A, and the secondary nozzles have bedded into the cavity plate. Material B, however, requires a lower manifold block temperature than material A, therefore the expansion of the assembly is less and, as shown at (d), a gap appears between the face of the secondary nozzle and the indented cavity plate. Polymer melt may creep through this gap and flood the hot-runner unit.

RUNNERLESS MOULDS

EXPANSION THEORY. The value for the thermal expansion of steel is in the region of $12\text{-}13 \times 10^{-6}$ mm/mm°C ($12\text{-}13 \times 10^{-6}$ in/in°C), the precise value depending upon the specific type of steel used, in this case, in the manufacture of the manifold block. The general equation for calculating expansion is as follows:

$$e = L \times C \times dT \tag{13.1}$$

where e = expansion (mm or in)
L = length dimension (mm or in)
C = coefficient of thermal expansion (mm/mm°C or in/in°C)
dT = increase in temperature (°C).

Example. A manifold block is to be heated from 20°C to 230°C. Calculate the increase in dimensions between the secondary nozzles situated 635 mm (25 in) apart. Use a value for the thermal expansion of steel of 13×10^{-6} mm/mm°C (in/in°C).

Solution

SI units	Imperial units
$e = (635)(13 \times 10^{-6})(230 - 20)$	$e = (25)(13 \times 10^{-6})(230 - 20)$
$= 1.73$ mm	$= 0.068$ in

Such a large increase in the dimension between the secondary nozzle centres and the impression centres could not be ignored. However, when the centres of the hot runner unit are contained within a pitch circle of radius less than 75 mm (3 in) the maximum expansion of the secondary nozzle from the central point is about 0.2 mm (0.008 in); this degree of expansion is usually ignored and no allowance is made in the hot-runner unit design. The designer should, however, when designing a nozzle of the direct-feed type, ensure that the strength at the front end of the nozzle is sufficient to withstand the shearing forces produced by the manifold block expansion.

The various ways in which expansion is accommodated in designs where the pitch circle radius is greater than 75 mm (3 in) are discussed below with respect to each secondary nozzle.

13.3.6 Secondary Nozzles. A secondary nozzle is defined as that part of the hot-runner unit which provides a connecting flow path from the manifold block to the cavity plate entry into the impression.

There are many possible designs for secondary nozzles and to cover this aspect of mould design completely would require a separate volume. However, we can look at the more commonly used types in order to classify them and to lay down a few general design principles. In the following discussion, the various types described are numbered as sn1, sn2, etc, for ease of reference.

Secondary nozzles are cylindrical in shape and usually incorporate a

HOT-RUNNER UNIT MOULDS

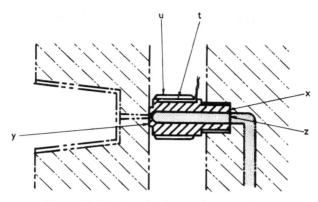

Figure 13.27—Standard secondary nozzle (sn1)

threaded section which screws into the manifold block (Figure 13.27). A leak-free joint between the two parts is essential, therefore the mating faces of both the secondary nozzle and the manifold block recess (x) must be flat and at right angles to the centre line.

The nose of the secondary nozzle may simply abut onto the face of the cavity plate directly behind the impression or it may protrude into the cavity plate in order to feed the impression directly with polymer melt. The alternative designs are shown in Figure 13.27 and Figure 13.28 respectively.

It is normal practice to provide a separate heating and control system for the secondary nozzle in order to achieve close temperature control of the melt in the outlet flow-way. Various types of heating element are used, including band, cartridge, and low voltage coil types. Whether the heater is fitted internally or externally depends primarily on the secondary nozzle design. The thermocouple, which is connected to the external control equipment, should be located relatively close to the outlet aperture. There are various possible arrangements, which include:

(i) incorporating the thermocouple (t) in a slot machined directly below an external heating element (u) (Figure 13.27);

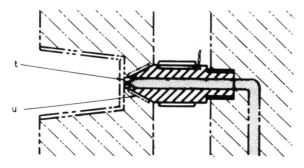

Figure 13.28—Direct feed secondary nozzle (sn2)

(ii) accommodating the thermocouple (t) in a hole in the centre of the core-rod, when fitted (Figure 13.30);

(iii) fitting the thermocouple (t) in a hole bored at an angle to the melt flow-way (Figure 13.29).

Note that certain types of heater include a thermocouple as part of the assembly. This can be useful, particularly where space is limited.

To avoid gaps and crevices which may create stagnation problems, and also to avoid possible restriction of the melt flow, it is preferable for the flow-way in the secondary nozzle to be of the same diameter as the outlet flow-way in the manifold block ((z), Figure 13.27).

STANDARD SECONDARY NOZZLE (sn1). This is virtually identical to a standard injection machine nozzle. Recall that the hot-runner unit evolved from the manifold nozzle, which originally incorporated standard nozzles. It was logical, therefore, that the pioneers in hot-runner unit design (c. 1940) should adopt a similar design for secondary nozzles.

A typical standard secondary nozzle is shown in Figure 13.27. Its front face is shown to be in sliding contact with the cavity plate at (y). Thus, when expansion of the manifold block occurs in the heating-up phase, the secondary nozzle simply slides across the surface of the cavity plate. Naturally, allowance must be made for this expansion when calculating the position of the secondary nozzles in the manifold block so that, when this is at working temperature, the corresponding holes in the secondary nozzles and mould plate are in line.

A conventional sprue gate is adopted to connect the secondary nozzle outlet aperture to the impression. Alternatively, when the hot-runner unit is being used as a semi-runnerless system, a conventional sprue is used to connect the secondary nozzle outlet apertures with the runner system.

DIRECT FEED SECONDARY NOZZLE (sn2). As the title implies, this type of secondary nozzle is used to feed the impression with melt directly from the manifold block flow-ways. There is no intermediate sprue gate as in the standard (sn1) design. In direct feeding, the nose of the secondary nozzle, which is basically conical in shape, protrudes into the cavity plate as shown in Figure 13.28.

The front face (t) of the secondary nozzle nose forms an integral part of the impression and must have the same surface finish as the rest of the impression. Since the nose of the secondary nozzle is in direct contact with the cavity plate immediately adjacent to the impression, problems arise as a result of heat transfer. Heat flows from the 'hot' nose of the secondary nozzle to the cooler cavity plate, with two adverse effects with regard to moulding:

(i) the temperature at the tip of the secondary nozzle decreases, with the result that the polymer melt tends to solidify and block the outlet aperture;

(ii) The temperature of the steel forming the cavity impression adjacent to the secondary nozzle nose increases. This is likely to produce 'heat-marked' mouldings and, in addition, may necessitate an extended production cycle.

HOT-RUNNER UNIT MOULDS

To minimise these heat transfer effects certain design features should be noted:

(i) the area of the front face (t) of the secondary nozzle should be kept to a minimum;

(ii) the surface contact between the nose of the secondary nozzle and the cavity plate should be reduced to a minimum by providing relief regions (u). (An alternative method for relieving this region is the double cone method illustrated in Figure 13.33.)

(iii) Careful thought must be given at the design stage to the location and type of water flow-way system in the locality of these secondary nozzles.

The direct feed secondary nozzle design does not allow for expansion of the manifold block. With reference to Figure 13.28, note that the nose of the secondary nozzle fits into a complementary recess in the cavity plate, while the rear end is screwed into the manifold block. Thus, when the manifold block is heated and expands, the secondary nozzle nose is subject to shearing forces, and this must be taken into account in the design of this type of secondary nozzle. The application of the direct feed secondary nozzle is therefore restricted to use where impressions are closely spaced and where calculated expansion of the manifold block is minimal.

The external conical shape of this type of secondary nozzle necessitates a more restrictive internal flow-way, compared with design sn1. To minimise the adverse effect of such a restriction a 'rocket nose shape' is adopted for the outlet region of the flow-way.

DIRECT FEED — SLIDING-TYPE SECONDARY NOZZLE (sn3). The preceding secondary nozzle (sn2) suffers from a disadvantage in that the layout of the impressions is restricted by the inability of the design to accommodate any extensive expansion of the manifold block. The direct feed — sliding-type secondary nozzle (Figure 13.29) is a variation of sn2 in which, as it is not physically attached to the manifold block, any amount of expansion can be catered for.

The nose end of the secondary nozzle is conical and fits into a complementary recess in the cavity plate behind the impression, just as in sn2. An additional shoulder section (m) is provided to give stability. The rear end (n) of the

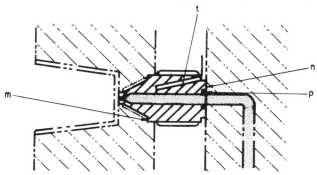

Figure 13.29—Direct feed—sliding-type secondary nozzle (sn3)

secondary nozzle is in sliding contact with the face of the manifold block. Pressure pads (not shown) mounted directly below the manifold block in this region hold the two faces together.

This secondary nozzle can be compared directly with sn1, in which the secondary nozzle is attached to the manifold block and is in sliding surface contact with the face of the cavity plate (Figure 13.27). In sn3 the secondary nozzle is fitted to the cavity plate and is in sliding surface contact with the face of the manifold block. Thus in both designs the manifold is free to expand without applying stress to the secondary nozzle.

With sn3 there is the danger that the polymer melt may leak from between the butting surfaces of the secondary nozzle and the manifold block. If this occurs the whole of the hot-run unit may be flooded and production interrupted. To provide a more effective leak seal than that achieved with the basic metal-to-metal contact, stainless steel O-rings (p) may be fitted. The O-ring is incorporated in a shallow groove machined in the sliding surface of the secondary nozzle (Figure 13.29). The O-rings, which protrude slightly, are compressed by the heat expansion of the manifold block to form an effective barrier against melt seepage.

Compared with sn2, more heat is transferred to the cavity plate from the secondary nozzle with sn3 because of the increased metal-to-metal contact. This loss of heat makes it difficult to maintain close temperature control at the secondary nozzle exit point. To overcome this, the secondary nozzle can be internally heated (see sn9).

DIRECT FEED — ANNULUS-TYPE SECONDARY NOZZLE (sn4 and 5). Production problems may be encountered with secondary nozzles sn1-3 because it is relatively difficult to control the temperature at the small outlet apertures sufficiently precisely to avoid either drooling or solidification of the polymer melt.

To improve temperature control in the outlet aperture region, secondary nozzles have been developed which incorporate a central metal core. This means that the polymer melt in the secondary nozzle is held in annular form. The object of incorporating this central member is twofold:

(i) The central rod, being manufactured from a good heat-conducting material (e.g. beryllium-copper), permits rapid heat transfer from the rear part of the secondary nozzle to the tip.

(ii) The control of temperature in the outlet region is greatly improved because the response to temperature change is quicker with an annular cross-section than with a cylindrical cross-section.

Two specific designs of secondary nozzle in this category have been developed; in one, a metal core-rod extends right through the secondary nozzle, as shown in Figure 13.30; in the other, a 'torpedo' is incorporated in the central region only (Figure 13.31).

(i) *Core-rod design* (sn4 — Figure 13.30). A conical-ended core-rod (p) extends through the centre of a correspondingly shaped bore-hole in the secondary nozzle body (q). The core-rod is secured in position by a grub screw,

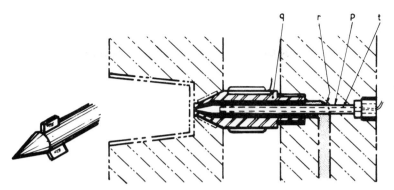

Figure 13.30—Direct feed—annulus-type secondary nozzle, core-rod design (sn4)

as shown. The tip of the core-rod extends into the outlet aperture and is approximately level with the gate entry into the impression.

Again, streamlining of the flow path is essential: profiled plugs (not shown) must be fitted at the ends of flow-ways and careful finishing of the flow-way surface is required. Additional attention must be given to the region (r) where the core-rod passes through the main flow-way. There should not be any crevices or gaps through which the polymer melt can creep and thereby cause subsequent problems of stagnation.

Note that this design permits a thermocouple to be fitted in the centre of the core-rod, and as this is closer to the melt (compared with the surface-mounted thermocouple), the response to temperature fluctuation is quicker. With the centrally mounted thermocouple, a modified grub screw is required to accommodate the thermocouple wires.

One limitation of the core-rod design should be noted: if the core-rod is relatively long compared with its diameter it has a tendency to deflect as a result of unequal pressures exerted from either side. Movement of the core-rod tip obstructs the melt flow through the gate entry into the impression. This problem can be overcome by incorporating fins (Figure 13.30 detail) which fit into the secondary nozzle bore-hole. Note that the manifold block end of the core-rod must be increased in diameter to permit the fins to be incorporated. An alternative nose-fin design is shown in Figure 13.58.

(ii) *Torpedo design* (sn5 — Figure 13.31). This secondary nozzle incorporates a torpedo-shaped element in the centre of the outlet flow-way. The general shape of this secondary nozzle is the same as that of previous designs, except that it is slightly larger in diameter in order to accommodate the torpedo.

The criterion for an efficient design is that the torpedo is held centrally, while the melt flow obstruction effect of the supports is minimal. There are several alternate designs; Figure 13.32 (isometric view) shows the 'spider' type. The holes, which are shown parallel to the periphery of the torpedo, provide a flow path for the polymer melt, while the steel webs between the holes give the

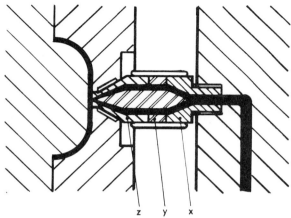

Figure 13.31—Direct feed-annulus-type secondary nozzle, torpedo design (sn5)

required support. Careful blending and shaping of the steel at the entrance to and exit from the holes is essential to avoid melt stagnation points. The spider torpedo may be of one-part (Figure 13.31) or multi-part (Figure 13.32) construction. The latter design, which is adopted to facilitate manufacture, has three components: front cone (a), spider block (b), and torpedo tip (c).

The complete secondary nozzle assembly, shown schematically in Figure 13.31, consists of three main parts; the body (x), the spider torpedo (y) and the nose section (z). In practice, the body and nose sections incorporate male and female threads respectively, so that the two parts can be securely attached together. The spider torpedo is housed in a suitable recess in the nose-section. The internal bore of both body and nose-section follows the contour of the torpedo, thereby providing a smooth, constant flow path for the polymer melt. The tip of torpedo extends into the outlet aperture and is approximately level with the gate.

If the two annular designs, the core-rod and the torpedo, are critically compared we note that whereas the former suffers from problems of possible deflection of core, the latter has inherent problems of flow-restriction.

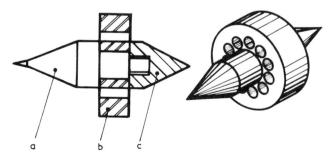

Figure 13.32—The 'spider' torpedo

HOT-RUNNER UNIT MOULDS

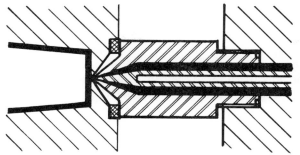

Figure 13.33—
ICI hot-runner secondary nozzle incorporating double-cone seating system

A further secondary nozzle seating method which is applicable to either of the annular designs is illustrated in Figure 13.33. The design is known as the ICI hot-runner nozzle and is credited to Mr E. Billington. In this design both the entry through the cavity plate and the front nose of the secondary nozzle are conical in form. However, metal-to-metal contact is only made at the extreme tip, as the two included angles differ as shown. This reduces to a minimum the flow of heat from the secondary nozzle to the cooler mould. To protect the delicate tip of the secondary nozzle, the 'body' abuts onto the lower face of a recess provided in the cavity plate directly behind the impression. Heat loss at this point is minimised by the incorporation of an insulation ring.

ANTE-CHAMBER SECONDARY NOZZLE (sn6 — Figure 13.34). In this design the nose of the secondary nozzle is bell-shaped and it protrudes into a well, machined into the cavity plate directly behind the impression. The body of the secondary nozzle, which is larger in diameter than the well, abuts against the face of the cavity plate. A ring, manufactured from an insulating material, is usually incorporated at the junction between the two parts for two reasons:

(i) the transfer of heat from the secondary nozzle to the cavity plate is thereby minimised;

(ii) as the insulating material is slightly compressed due to the expansion of the manifold block, a good leak-free material seal is achieved.

The well in the cavity plate is parallel-sided and terminates with a complementary bell-shaped entry into the impression. The ante-chamber is formed

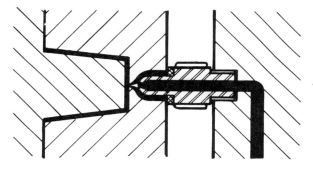

Figure 13.34—
Ante-chamber secondary nozzle (sn6)

by the outer surface of the secondary nozzle nose and by the well machined in the cavity plate. When functioning, this ante-chamber is filled with a small volume of polymer melt.

In operation, the polymer melt passes from the manifold block, through the central hole of the secondary nozzle and into the ante-chamber via two or three small apertures machined at an acute angle in the bell-shaped nose. The melt then flows into the impression via the gate. Note that this secondary nozzle design is basically the same as the ante-chamber design which is used for single-impression moulds.

The principle of the ante-chamber design is that, because polymers are poor conductors of heat, the material adjacent to the wall of the secondary nozzle nose remains sufficiently fluid to allow melt to pass intermittently into the impression via the gate. The layer of material adjacent to the cavity plate acts as an insulator, preventing undue loss of heat from the secondary nozzle to the cavity plate.

This design is also known as the 'hot well' design and the secondary nozzle is often called the 'Distrene nozzle'. Note that the secondary nozzle tip extends into the gate region. Thus, heat is transferred from the body of the secondary nozzle to the tip, allowing reasonable control of the melt temperature in this critical region. Expansion of the manifold block does not present any problems with this design. Provided the well is of sufficient diameter, the expansion of the secondary nozzle nose relative to the cavity plate can be accommodated.

INTERNALLY HEATED SECONDARY NOZZLES. An internally heated secondary nozzle is defined as a secondary nozzle which incorporates a heating element within the assembly. Recall that in all of the previously discussed designs the

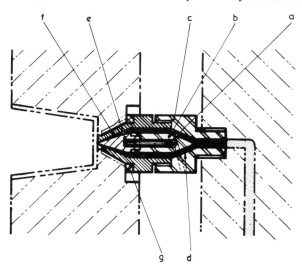

Figure 13.35—*Internally heated, direct feed—torpedo-type secondary nozzle (sn7)*

HOT-RUNNER UNIT MOULDS

heating element is mounted on the outside surface of the secondary nozzle body. The primary advantage of internal heating is that the material in the flow-way is heated from the inside outwards instead of the other way round. This permits closer temperature control of the gate-entry region. As the outer surface operates at a temperature lower than that of the corresponding externally heated secondary nozzle, less heat has to be removed from the vicinity of the impression by the mould's fluid circulation system.

Internally heated secondary nozzle designs are similar to, and extensions of, designs discussed in previous sections. The only difference between the designs is that space must be provided in the interior of the secondary nozzle to accommodate a heating element of either the cartridge or low-voltage coil type.

The internally heated secondary nozzle designs which are discussed below are based upon the direct feed — annulus designs (sn4 and sn5) and the antechamber design (sn6). As the principal features of each of these designs have already been covered, only those aspects of the design relating to internal heating are discussed.

INTERNALLY HEATED, DIRECT FEED — TORPEDO-TYPE SECONDARY NOZZLE (sn7 — Figure 13.35). This design is similar to the direct feed — torpedo type (sn5) except that a heating element (a) is incorporated in the central region.

The main problem associated with this design is that of providing suitable accommodation for the heating element wires. It will be noted from the illustration that the wires from the heating element must pass through the 'melt' region, in order to be coupled to the electrical supply. By making the entry part (b) of the torpedo integral with the secondary nozzle body (c) and causing the melt to flow through angled drillings (d), the heating wires can be accommodated in an intermediate hole (shown dotted) which is bored between the flow-way holes. (See also Figure 13.39 which illustrates a similar design.)

The heating element is secured in position by a removable torpedo tip (e). The secondary nozzle nose (f) screws into the body (c) to complete the assembly. In the design illustrated an insulating disc (g) is incorporated at the seating surface of the secondary nozzle to minimise heat loss.

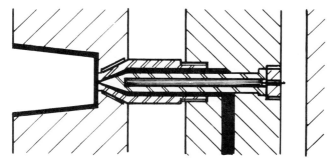

Figure 13.36—Internally heated, direct feed-core-rod-type secondary nozzle (sn8)

RUNNERLESS MOULDS

INTERNALLY HEATED, DIRECT FEED — CORE-ROD-TYPE SECONDARY NOZZLE (sn8 — Figure 13.36). The heating element is fitted into a recess bored down the centre of the core-rod. As the core-rod passes completely through the flowway there are no problems with respect to accommodating the heating element wires.

The only difference between the internally heated and the unheated core-rod secondary nozzle design is that the core-rod of the former must be larger in diameter to accommodate the heating element. This means that the overall diameter of the secondary nozzle must be correspondingly increased in size. (Compare Figures 13.30 and 13.35.)

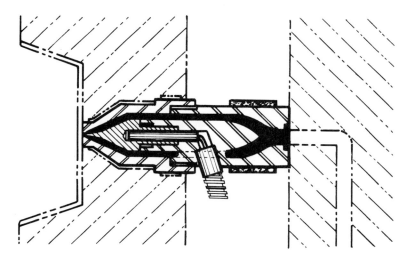

Figure 13.37—Internally heated, direct feed—sliding-type secondary nozzle (sn9) (Courtesy W.J. Furse Ltd, UK)

INTERNALLY HEATED, DIRECT FEED — SLIDING-TYPE SECONDARY NOZZLE (sn9 — Figure 13.37). The heating element is fitted inside the torpedo section in a similar manner to that described for sn7 and the same observations apply. The primary object of the sliding secondary nozzle design is to permit direct feeding of the impression with polymer melt while making allowance for heat expansion of the manifold block. A general description of the sliding type of secondary nozzle is given under sn3 above.

The main limitation of the use of the sn3 design is that close temperature control at the secondary nozzle exit point is difficult to achieve. By incorporating the heating element internally, and providing a torpedo tip which extends into the gate-entry region, this limitation is overcome.

Various manufacturers offer this type of secondary nozzle as a standard part. For example, W.J. Furse (UK) market a design of this type as the 'Thermoject nozzle', while the Incoe Corporation in the USA call their secondary nozzle a 'Hot-tip bushing'. Both these commercial designs incorporate a stainless steel O-ring at the mating surface between the secondary nozzle and

HOT-RUNNER UNIT MOULDS

the manifold block. These O-rings are hollow and are compressed slightly when installed. Polymer melt enters the centre of the O-ring via a few small holes incorporated in the inner surface and this causes the O-ring to expand and to create a positive seal.

INTERNALLY HEATED, PROJECTING CORE-ROD SECONDARY NOZZLE (sn10 — Figure 13.38). This system is similar to the ante-chamber design (sn6) in that a well is provided in the cavity plate directly behind the impression. However, the similarity between the two designs ends at that point. In sn10 the secondary nozzle has a flat face which abuts against the surface of the cavity plate. A disc of insulating material is incorporated at the interface, as shown, to minimise heat loss. The major diameter of the conical well is the same as the exit diameter of the flow-way from the secondary nozzle. Thus, provided

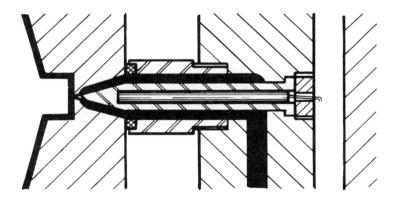

Figure 13.38—Internally heated, projecting core-rod secondary nozzle (sn10)

the correct expansion of the manifold block has been allowed for when specifying the respective centre distances of the wells and secondary nozzles, the two sets of holes will be in alignment during production.

A core-rod, which incorporates an internal heating element, is fitted to the manifold block as discussed under sn4. The core-rod extends through the secondary nozzle body and protrudes into the cavity plate well (Figure 13.38). As the core-rod is relatively long it is advisable to incorporate locating fins (see Figure 13.58) to bear against the sides of the well to prevent core-rod deflection.

Note that because the front end of the core-rod is fixed, and provided locating fins are incorporated, the flow path is not changed when the manifold block expands. Naturally the core-rod is bowed in the process and this point must not be overlooked when adopting this design. The bowing must not be excessive if a cartridge heater is used. There is no restriction, however, with the low voltage coil heating element as this acts in the same way as a spring when it is deformed.

RUNNERLESS MOULDS

It is interesting to note that the DME standard 'fixed-probe' design incorporates the essential features of the core-rod design discussed above. While the DME 'fixed-probe' is designed for another purpose (see Section 13.5) it can be also be adapted for this use.

INTERNALLY HEATED, PROJECTING TORPEDO SECONDARY NOZZLE (sn11 — Figure 13.39). The general principle of this system is the same as that described for sn10. The main difference between the two systems is that in the first a core-rod is used as the central element and in the second a torpedo is adopted.

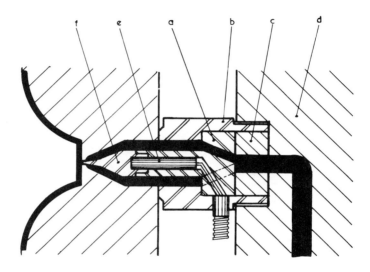

Figure 13.39— Internally heated, projecting torpedo secondary nozzle (sn11)

The face of the projecting secondary nozzle abuts onto the surface of the cavity plate. The projecting nose of the torpedo extends into a well machined directly behind the impression, forming an annular entry system.

This design illustrates an alternative method of mounting the torpedo in the secondary nozzle. The spider torpedo (a) is mounted in a recess in the secondary nozzle body (b) and supported by a disc (c). The hole in this disc forms a continuation of the manifold outlet hole. The secondary nozzle body is screwed into the manifold block (d), which secures the torpedo and disc in the required position. By manufacturing the torpedo as shown, the machining and finishing of the angled flow-ways is greatly simplified compared with the alternative design shown in Figure 13.35.

The heating element (e) is secured in position by the torpedo tip (f) in a similar manner to sn7.

SECONDARY NOZZLES — CONCLUDING REMARKS. In order to simplify the drawings in this section as much as possible, fluid circulation systems have

HOT-RUNNER UNIT MOULDS

not been shown except in a couple of illustrations. In practice, careful consideration must be given to the type, location and arrangement of the coolant fluid circulation system in the immediate vicinity of the secondary nozzle in order to control the temperature in this region as closely as possible.

Figure 13.40 shows the interrelationship between the various secondary nozzles discussed in this section, and to some degree it also serves to illustrate the evolution of secondary nozzle design. However, as in the case of the hot-runner unit, later designs have not necessarily superseded the older and more well-established designs for certain purposes.

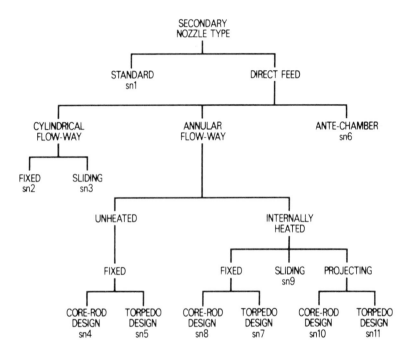

Figure 13.40—'Family tree' of secondary nozzles

13.3.7 Valve Systems. Various types of valve system have been developed for use with nozzles, manifold secondary nozzles, hot-runner secondary nozzles and hot-runner plate systems. The valve systems for all of the above designs are similar, and valve systems in general are discussed in this section.

Valve systems are of two types: (i) the automatic shut-off valve and (ii) the externally operated shut-off valve. The primary purpose of a shut-off valve in any feed assembly is to permit the segregation of the melt in the mould from the melt in the injection cylinder at some specified time in the operating cycle. The precise moment when the valve is operated depends upon whether it is of type (i) or type (ii) above, and upon the production technique in use.

RUNNERLESS MOULDS

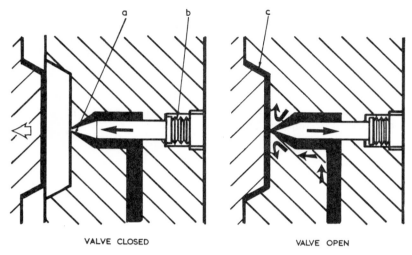

Figure 13.41—Principle of the automatic valve

AUTOMATIC VALVE SYSTEMS. The automatic shut-off valve is based upon a spring-loaded system. One design is illustrated schematically in Figure 13.41: the valve nose (a) is held onto a seating by bell-view spring washers (b), preventing the material from flowing through the secondary nozzle into the impression (c). When the injection pressure builds up to the setting of the spring, the valve opens, permitting the melt to flow through the gate into the impression. Note that the force applied to operate the valve is the material pressure, multiplied by the effective projecting area of the valve stem. Immediately the material pressure falls below the setting of the valve spring, the valve gate closes once again.

The reasons for fitting an automatic valve are as follows:

(i) A larger diameter gate can be used with the valve system. This permits faster filling and less inherent strain in the mouldings.

(ii) The valve prevents drooling (dribbling) or stringing when the mould is open.

(iii) The valve effectively prevents 'suck-back' of material from the impression when the screw plunger is withdrawn. This reduces the waiting period for the melt to solidify before the screw plunger is withdrawn. It must be pointed out that many mould designers feel that the automatic valve creates more problems than it solves. With modern injection machine control methods the advantages listed above are not so apparent.

Figure 13.42 shows the automatic valve applied to a hot-runner design.

EXTERNALLY OPERATED VALVE SYSTEMS. The principle of operation of this system is shown in Figure 13.43. The needle valve seats at the outlet aperture of the secondary nozzle as in the automatic valve. However, in this case the valve stem passes through the hot-runner manifold block (or other system)

HOT-RUNNER UNIT MOULDS

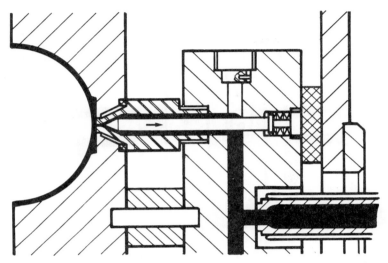

Figure 13.42—Direct-feed secondary nozzle fitted with automatic valve

and is attached to an actuating device. The valve has two positions: either open, when material is free to pass through the gate into the impression, or closed, when melt flow ceases.

The main consideration, therefore, is the method to be adopted for actuating the valve. Various methods have been developed and four examples are given below to illustrate the general principles.

(i) *Actuator/linkage system.* Figure 13.44 shows the valve stem (c) coupled to a hydraulic (or pneumatic) actuator (a) via a linkage system (b). The pivot

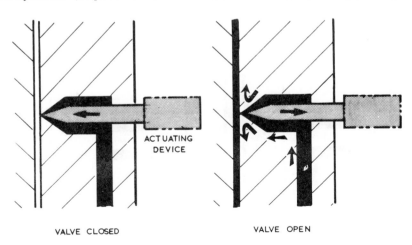

Figure 13.43—Principle of the externally operated valve system

389

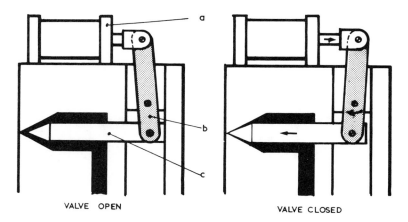

Figure 13.44—Actuator/linkage system for valve operation

points are so arranged that the needle valve is opened and closed by a relatively short movement of the actuator ram.

(ii) *Actuator/wedge system* (Figure 13.45). A wedge-shaped steel block (c) is attached to the base of the valve stem (a). The wedge block is held in contact with a complementary wedge-shaped operating block (e) by a spring (b). The operating wedge block is attached to the ram of a short-stroke hydraulic or pneumatic actuator (d). When the ram is actuated, the operating wedge block is moved downwards (as drawn) or upwards and the needle valve is correspondingly moved inwards (to close the gate entry) or outwards (to open the gate entry), respectively.

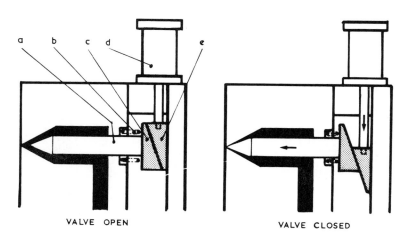

Figure 13.45—Actuator/wedge system for valve operation

HOT-RUNNER UNIT MOULDS

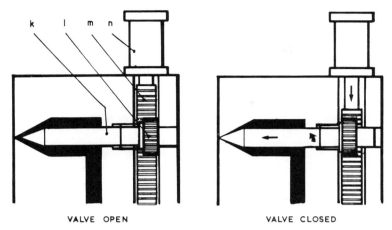

Figure 13.46—Actuator/rack and pinion system for valve operation

(iii) *Actuator/rack and pinion system* (Figure 13.46). A hydraulically (or pneumatically) operated rack (m) is in engagement with a pinion (l), which is attached to the valve stem (k). When the rack is actuated, the pinion and the valve stem rotate. Because the valve stem is in threaded engagement within the manifold block, it is caused to move backward or forward (opening and closing the gate) according to the direction of rack movement.

(iv) *Direct acting actuator system* (Figure 13.47). The valve stems (u) are fitted to an operating plate (t) mounted directly behind the manifold block. The ram of a parallel mounted hydraulic or pneumatic actuator (s) is attached to this operating plate. When the ram is actuated, the operating plate moves the needle valves in unison to open or close the gates.

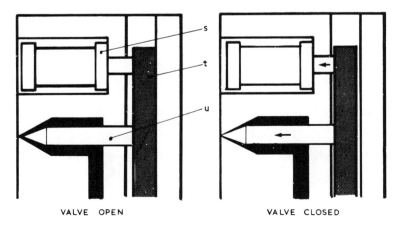

Figure 13.47—Direct-acting actuator system for valve operation

RUNNERLESS MOULDS

The advantage of the externally operated valve system is that the valve can be actuated exactly when required. Recall that the automatic valve is actuated by the injection pressure and that the precise moment at which the valve opens is controlled by a spring. The main disadvantage of the externally operated valve system is cost: the overall cost of the mould is increased substantially if this system is incorporated.

The externally operated valve system forms the basis of several specific moulding techniques, of which two are described below.

(i) *Precompressed moulding.* This technique, which is also known as 'explosion moulding', was developed in the USA by Columbus Plastics Products Inc. and W.R. Grace and Co. In operation, the valves are kept closed until the injection screw-plunger has moved part-way forward, compressing the melt in the injection cylinder and manifold block flow-way. A relatively high pressure within the injection cylinder and hot-runner unit is therefore developed. The needle valves are opened at the required time by the hydraulic actuator and the melt enters the impression at high velocity. Once the impressions are filled, the valves are closed, allowing the screw-plunger to be withdrawn. The cycle of operations is continued.

(ii) *Sequential impact moulding.* Another technique which can be used with a valve-gate hot-runner unit is sequential impact moulding. This was developed by G.D. Gilmore and J. Decker of the Bopp-Decker Corporation, USA. The basic principle is the same as for the previous method, that is the valves are kept closed until a relatively high pressure is developed within the injection cylinder and unit. With this method, however, the valves to the individual impressions are opened and closed in sequence instead of simultaneously, the impressions being filled separately in rapid succession.

This system has an additional advantage over the precompressed moulding technique in that, as the full pressure is only applied to one impression at a time, a smaller clamping force is required to keep the mould closed.

13.3.8 Heating the Manifold Block. The purpose of the hot-runner unit is to provide a flow path for the polymer melt from the injection machine's nozzle to the entry point (sprue or gate) in the cavity plate. The polymer material in the flow-way must be kept at an elevated temperature so that it remains in the melt condition during its passage to the impression.

The polymer is initially heated by the injection cylinder heating elements, and the melt is transferred through the flow-way of the manifold block by the movement of the screw-plunger. Unfortunately the polymer melt loses some of its heat during its passage through the hot-runner unit because heat is continuously transferred to cooler regions. Therefore heat energy must be added to the manifold block to replace the energy lost in this way.

When heating the hot-runner unit from cold, the material in the flow-ways must reach the required melt temperature before production can commence. This must be taken into account when calculating the total power requirements for a particular manifold block. It is beneficial to have a reserve of power for this purpose and a loading value of 2 watts per cm^3 (30 W/in^3) of manifold block volume is suggested.

Some designers prefer to double up on the number of heaters incorporated so that, should a heater fail in production, the reserve heating system can be switched on. The point should be borne in mind here that extra heaters take extra space, and therefore the manifold block may become inconveniently large.

The temperature range over which the manifold block can be operated is dependent upon the particular type and grade of material being processed. The temperature of the manifold block must not be so high that degradation of the material occurs, nor must it be so low that high pressures are required to force the highly viscous melt through relatively small flow-ways. Close control of the temperature of the melt in the flow-ways is essential.

HEATING METHODS. In practice, electrical devices are normally used for heating the manifold block although other systems such as circulating hot oil, high-pressure hot water and steam have been reported. The following notes are restricted to the electrical systems: (i) standard voltage resistance heating; (ii) low voltage resistance heating; (iii) induction heating.

(i) *Standard voltage resistance heating.* This method of heating the manifold block is based upon commercially available mains voltage resistance heating elements. The heating element is embedded in a refractory material and enclosed in a metal casing. Many types of resistance heater may be utilised in manifold block design, but in practice the cartridge heating element, the flat heating element and the band heating element are the types most commonly encountered. The method of fitting the heating element to the manifold block depends upon its shape (see below).

(ii) *Low voltage resistance heating.* The heating element in this case is a length of relatively thick wire which is manufactured from a material which has a high resistivity value (eg 80/20 nickel-chromium alloy). The wire is formed into a shape such as a coil or a zigzag to facilitate its incorporation into the manifold block. When a current is passed through the wire, it heats as a result of the resistance to flow. (Note that a material which has a low resistivity value will pass a current with little heating effect.)

(iii) *Induction heating.* When a conductor is placed within a magnetic field carrying an alternating current, eddy currents are induced into the conductor, causing it to heat. The heating effect is most intense at the surface, but if the current is maintained the heat is conducted right through the body. Induction coils are manufactured from heavy gauge copper wire and bound with tape.

Figures 13.48 to 13.53 illustrate how each of the above heating methods is incorporated in the manifold block design. Note that in all cases the melt flow-way is located at some finite distance from the heating element and that heat is transferred to the polymer material by conduction.

Two alternative heating arrangements are adopted in the manifold block design: the temperature of the melt is maintained by conducted heat from heating elements situated either (i) externally or (ii) internally with respect to the flow-way. As the design principles of each method are different, they are discussed separately below.

(i) THE EXTERNAL FLOW-WAY HEATING SYSTEM. In the diagrams illustrating this heating system, the cylindrical flow-way is shown, purely for convenience, in the centre of the rectangular (or cylindrical) manifold block cross-section. In fact, the flow-way is not necessarily central in the manifold block.

The conduction heat flow path from the heating element to the flow-way is indicated by arrows. The various illustrations should be compared and note taken of the relative distances between heating element and flow-way.

The methods adopted for external flow-way heating include (a) cartridge, (b) low voltage coil (internal), (c) flat, (d) low voltage zigzag, (e) induction, (f) band, (g) low voltage coil (external). The last two methods relate to the heating of cylindrical manifold blocks.

(a) *Cartridge heating element.* This is a main voltage heating element which is commercially available as a standard part in various diameters, lengths and wattages. Watlow Firerod (USA), for example, produce a series of cartridge heaters, the diameters of which range between ¼ and ¾ in (6.35-19mm) in steps of ⅛ in (3 mm approx). The ⅜ in (10 mm) and ½ in (13 mm) diameter cartridge heating elements are the most commonly used sizes for manifold block heating applications. (Smaller diameters are used for the internal heating of secondary nozzles.)

The cylindrical cartridge heating element is fitted into a hole bored and reamed through the manifold block. The relevant manufacturer's catalogue

Figure 13.48—*(a) Cartridge and (b) low voltage coil heating elements*

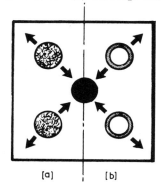

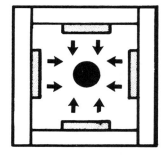

Figure 13.49—*Resistance-type strip heating elements*

should be consulted before specifying the diameter of the hole, to achieve the correct fit and thereby obtain optimum heater life under specific conditions.

Wherever practicable the cartridge heaters are mounted parallel to the melt flow-way and in close proximity to it, as illustrated in Figure 13.48a. Naturally, for a complex melt flow-way system consisting of main and secondary flow-ways this idealised arrangement is not possible and a compromise must be accepted.

(b) *Low voltage coil heater.* A length of relatively thick wire is wound onto a former to form a spring-like coil. Suitable terminals are attached to the coil to permit it to be connected to a low voltage supply. As noted previously, the wire chosen for this purpose must have a high resistivity value in order to obtain the desired heating effect as the current is passed.

The coil heaters are fitted into holes bored through the manifold block adjacent to the flow-way as illustrated in Figure 13.48b. Mr E. Billington, in his paper 'The design of hot-runner systems for moulds',* makes the point that when a torque is applied to a spring it becomes smaller in diameter, and when the torque is released the spring returns to its original size. This characteristic is made use of when fitting and removing coil-type heaters. It is of interest to note that ICI (UK) adopt the low voltage coil heating element in their basic hot-runner unit design.

(c) *Flat heating element.* This is a relatively shallow heating element, of rectangular cross-section, which is commercially available as a standard part in a range of sizes and wattages. The flat heating element is accommodated in a slot machined into the side of the manifold block, and is clamped in position by a cover plate (Figure 13.49). The cover plate ensures that a good surface-to-surface contact between heating element and manifold block is achieved for the purpose of heat conduction.

Flat heating elements are commonly fitted on all four sides of the manifold block, as illustrated. The available surface on two of the faces is, of course, restricted because of secondary nozzle and manifold bushing fitting arrangements.

(d) *Low voltage zigzag heating element.* A length of high-resistivity wire is formed into a zigzag configuration, the adjacent parallel lengths being relatively close together. Terminals are fitted to either end of the wire to facilitate its connection to the low voltage supply. The overall length of the wire required depends upon a number of factors including the resistivity value of the wire, the voltage used and the power input required.

The zigzag element is fitted into a recess in the side wall of the manifold block and secured by a cover plate in a similar manner to that described for the flat heating element (Figure 13.50).

(e) *Induction heating coil.* Heavy gauge copper wire is wound onto a former

---

*Paper presented to the Plastics and Rubber Institute's Symposium on Mould Design, January 1976.

RUNNERLESS MOULDS

Figure 13.50—Low voltage zigzag heating elements

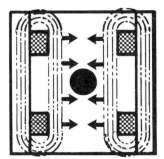

Figure 13.51—Induction heating system

and then bound with tape to form a coil of convenient size. The induction heating coil for this application is normally made to specification and is not available as a standard part.

The coil is fitted into an accommodating recess machined into the sides of the manifold block. A cover plate secures the coil in position (Figure 13.51). As noted previously, when a current is passed through the coil, eddy currents are induced into the surface of the manifold block (see chain-dotted lines) and heat is transferred from the immediate vicinity of the coil to the melt flow-way by conduction.

All of the above external flow-way heating techniques relate specifically to the rectangular cross-section manifold block. We now consider the methods adopted for heating cylindrical manifold blocks. Basically, two types of heater are applicable, the band heating element (and variations) and the low voltage coil heater.

(f) *Band heating element.* The mains voltage resistance type of element is enclosed within a casing which is in the form of split hollow cylinder. This is mounted on the external surface of the cylindrical manifold block and secured in position by clamp screws (Figure 13.52a).

The band heater is available as a standard part in a wide variety of diameters, widths and wattages. In many composite cylindrical manifold block designs, use of the conventional band heater necessitates dismantling the unit each time a heater has to be replaced. To overcome this limitation, 'half-shoe' heating elements may be used. These half-shoes can be clamped around the manifold block to form what is, effectively, one continuous heating element.

The band heating element has other applications in hot-runner unit design, such as the heating of secondary nozzles, manifold bushings, etc.

(g) *Low voltage coil heater.* This is the same type of heater as that described under (b) above. The flexible nature of the coil permits the heater to be fitted just as easily to the outer surface of the cylindrical manifold block as internally within a bored hole (Figure 13.52b). A limitation of this heating method

HOT-RUNNER UNIT MOULDS

is that the composite cylindrical manifold block must be disassembled when replacing a coil heater. This is not a major restriction to its use, however, as the coil heater has a particularly long production life.

The low voltage coil heater may also be used for heating secondary nozzles and manifold bushings.

(ii) THE INTERNAL FLOW-WAY HEATING SYSTEM. The annular flow-way system is adopted for this design. The annulus is formed by the inner surface of a relatively large central bore-hole and the outer surface of a heater tube. A heating element is fitted down the centre of this tube (Figure 13.53). Note that the heating element is in close proximity to the relatively thin shell of polymer melt.

This design makes use of the excellent insulating properties of polymer materials. The low thermal conductivity value reduces the flow of heat from the polymer to the cooler manifold block. The outer layers of the melt stream act as a most efficient insulator. Recall that the purpose of the manifold block is to maintain the temperature of the melt in the flow-way. Thus, if heat is effectively prevented from being conducted from the flow-way, a more efficient system results. In practice, the outer wall of the manifold block is relatively cool compared with a corresponding externally heated manifold block.

Figure 13.52—(a) Mains voltage band heating element
(b) Low voltage coil heating element

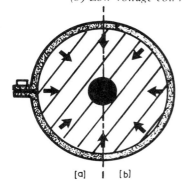

Figure 13.53—Mains voltage cartridge heating element

The internal heating system utilising an annulus is used very effectively in hot-runner plate design (Section 13.5).

Two types of heating element are used in conjunction with the annulus system, the cartridge heating element and the low voltage coil heater, already discussed under (a) and (b) above.

Table 9 indicates the relative merits, on a scale of 5, of the various heating systems with regard to ease of fitting, ease of removal, effective distance from melt flow-way, life expectancy and availability as standard parts. A rating of

RUNNERLESS MOULDS

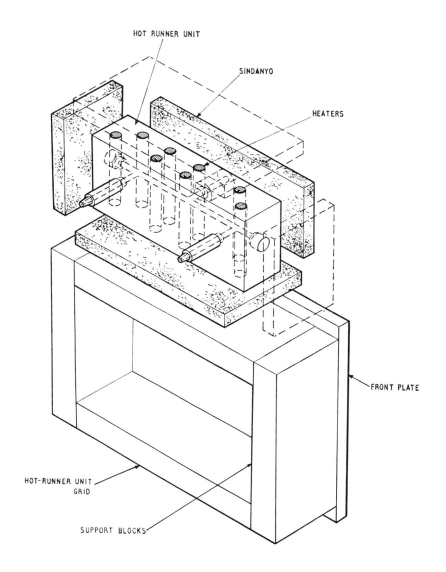

Figure 13.54—Exploded view of hot-runner unit and associated grid

HOT-RUNNER UNIT MOULDS

TABLE 9
HEATING SYSTEMS — A COMPARATIVE SURVEY

Heating element	Ease of fitting	Ease of removal	Effective distance from flow-way	Life expectancy	Available as standard part
Band	4*	4	2	4	5
Cartridge	5	1	5*	3*	5
Strip	4	4	2	4	5
Low voltage coil	5*	4*	5*	5	1
Low voltage zigzag	4	4	2	5	1
Induction coil	4	4	2	5	1

*Value dependent upon application.

1 indicates, with respect to these five criteria,
(a) great difficulty in fitting the element
(b) great difficulty in removing the element
(c) considerable distance between element and flow-way
(d) very poor life expectancy
(e) non-availability as a standard part.

The opposite is indicated, of course, by a rating of 5.

Note that an additional consideration with respect to the low voltage system is that a voltage reducer is necessary, which increases the cost of the initial installation.

WIRING ARRANGEMENTS. One feature that all electrical systems have in common is that the heating elements must be wired to a power source. The more heaters that are incorporated in the manifold block, the greater the complexity of the wiring system. The designer should aim to simplify the wiring system as much as possible by connecting batches of heating elements to readily accessible junction boxes. A confused wiring system makes fault-finding difficult and the replacement and rewiring of the heating element a lengthy operation.

INSULATING THE MANIFOLD BLOCK. The fundamental principle of the injection moulding of thermoplastics is that the mould must be relatively cold to permit rapid solidification of the melt. We are now considering incorporating a heated runner manifold block within the structure of the mould. It follows that some form of insulation must be provided between the manifold block and the rest of the mould.

Two methods are in general use. In one the manifold block is almost entirely surrounded with compressed asbestos board. Small blocks of this insulating material are fitted to all sides of the manifold block by screws. Suitable apertures are provided in the insulation for the secondary nozzles, manifold bushings, etc (Figure 13.54). The second method is to provide an insulating air gap of at least 16 mm (5/8 in) all round the manifold block except where contact with the rest of the mould is essential — even in these places contact is reduced to a minimum.

RUNNERLESS MOULDS

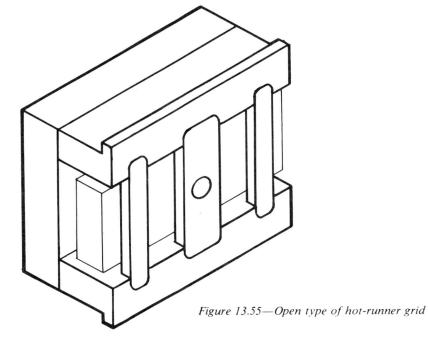

Figure 13.55—Open type of hot-runner grid

13.3.9 Hot-runner Unit Grid. The hot-runner unit is accommodated in a pocket behind the mould plate. This pocket may be formed in a similar fashion to the ejector grids discussed in Chapter 3. The frame type of hot-runner unit grid (Figure 13.54) completely encloses the unit and provides maximum support for the mould plate. The in-line hot-runner unit grid (similar to Figure 3.1) is favoured by many designers in that it allows access to the heating elements without stripping the mould. Thin sheet metal end-plates can be fitted to provide protection for the electrical wiring and mounting space for electric sockets.

An alternative hot-runner grid design is the 'open grid' system which is illustrated schematically in Figure 13.55. Here again, either a frame or an in-line support system is used to support the mould plates and provide a pocket into which the hot-runner unit can be fitted. However, in this design a back plate is not used. Ledges or recesses are incorporated in the support block so that mould half can be clamped to the machine platen. The hot-runner unit is secured in the required position by clamp plates which fit into accommodating recesses machined in the support blocks on either side of the hot-runner unit. The advantage of this arrangement is that the entire hot-runner unit may be opened for inspection by moving the complete injection mould half forward with the moving half (being suitably supported). This facilitates the servicing and replacement of an element, should this become necessary. The ease with which this can be done with respect to a cylindrical manifold block is illustrated in Plate 14.

INSULATED RUNNER MOULDS

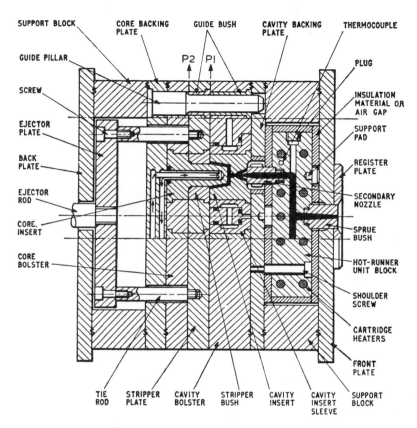

Figure 13.56—Two-impression hot-runner unit mould

13.3.10 Worked Example. Sketch a section through a multi-impression injection mould for a beaker. Incorporate the following features:
  (i) Insert-bolster assembly.
  (ii) Stripper plate ejection.
  (iii) Ante-chamber feed with a pin gate.
  (iv) Hot-runner unit.
  (v) Frame-type hot-runner unit grid.
  (vi) Central cooling of the core inserts.
  (vii) Annular-groove sleeve-cooling for the cavity insert.
A mould incorporating the above features is illustrated in Figure 13.56.

13.4 INSULATED RUNNER MOULDS

An insulated runner mould is one in which the melt flows through a large-diameter runner machined in the butting surfaces of the cavity plate (b) and the feed plate (c). The two plates are attached together by quick-release swing

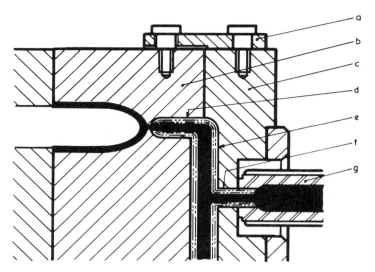

Figure 13.57—Insulated runner mould

latches (a) during the normal moulding operation. A part cross-sectional drawing of this type of mould is shown in Figure 13.57.

Polymer melt flows from a standard or extended nozzle (g) into a large-diameter runner (e), and finally into the impression via the reverse tapered sprue (d) and gate. The mould is operated as a simple two-part mould in that the runner system is not removed during the normal cycle of operations. As the mould opens, the moulding is extracted from the cavity and the gate is broken at its junction with the reverse tapered sprue.

This technique is only practicable because thermoplastics have such good insulating properties. The outer layer of the melt (shown chain-dotted) solidifies against the cold runner wall (metal) which forms an insulating shell, while the central core of the material remains in the melt state. Provided that the moulding cycle permits the melt to pass intermittently through the runner system without extended hold-ups, the mould can operate continuously in this fashion. Naturally, solidification of the runner system will occur if the production cycle is interrupted for any reason. When this happens the latches are released, the respective plates are separated, and the 'solidified' runner is removed.

The insulated runner mould is similar in construction to the underfeed (three-part) mould. Basically, the only difference between the two is that the former design incorporates a large-diameter runner, a large-diameter reverse tapered sprue, and a swing latch system. The diameter adopted for the runner is in the range 13-25mm (½-1 in), and a correspondingly large diameter is chosen for the reverse tapered sprue. From the above comments it will be apparent that an underfeed mould can be readily modified to an insulated runner design, if required.

The development of the insulated runner mould is credited to Phillips

INSULATED RUNNER MOULDS

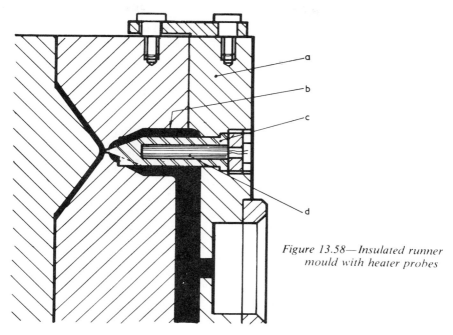

Figure 13.58—Insulated runner mould with heater probes

Chemical Company (USA) who developed the system for polyethylene. The conditions required for moulding this material are not critical, but the coolant flow-way holes should be positioned close to the runner so that reasonably accurate temperature control can be achieved in this region. The mould needs to be fairly warm when starting up, but to achieve an economic cycle the temperature is then progressively reduced until the required conditions are reached.

Note that in this basic design of insulated runner mould no heat is introduced to the material in the mould. The only heat supplied is by the injection cylinder heating elements. It follows that, even allowing for the excellent insulation provided by the polymer shell within the runner, the temperature of the melt at the outlet point (the gate) is considerably lower than the temperature of the melt which leaves the nozzle. There is always the possibility, therefore, that the melt will freeze at the gate and thereby interrupt production. To achieve better control of temperature in the outlet region, an insulated runner mould with heater probes has been developed.

13.4.1 Insulated Runner Mould with Heater Probes. As noted above, the most sensitive part of the basic insulated runner design is the gate region. The gate is, of course, at the end of the flow path and thus at the greatest distance from the heat source. By incorporating a heated probe in the outlet region, an additional input of energy can be added to the melt exactly where it is required (Figure 13.58).

The probe (c) is a hollow cylindrical pin which is fitted perpendicular to the feed plate (a). It is centrally disposed with respect to the outlet flow-way (b) (previously called reverse tapered sprue, but note modified shape). The probe incorporates a conical nose which is often shaped to fit into the gate, as shown. Note that the incorporation of the probe results in an annular outlet system.

A cartridge heating element (d), often combined with a thermocouple tip, is accommodated in the central bore of the probe. A slot is provided in the feed plate to accommodate the cartridge heater wires.

The system permits the temperature in the exit region of the melt to be controlled within fairly close limits. It will be noted that the melt acts as insulation to the outside of the outlet flow-way (metal) to retain the heat in this region.

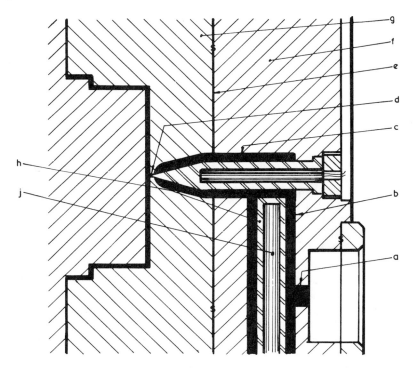

Figure 13.59—*Hot-runner plate mould — simplified*

13.5 HOT-RUNNER PLATE MOULDS

The hot-runner plate mould evolved from the insulated runner design and it similarly utilises the polymer melt as an insulating barrier to heat flow. However, unlike the insulated runner design, the flow-ways are heated by a system of internal heaters using the annular flow-way system.

Figure 13.59 shows a simplified drawing of the flow-way arrangement. The melt enters the hot-runner plate at (a) through a short cylindrical aperture

HOT-RUNNER PLATE MOULDS

from an extended nozzle. It flows through the main flow-way (b), outlet flow-way (c) and finally enters the impression via the gate (d).

The shell of polymer melt in the main and outlet annular flow-ways is heated from the inside outwards, which means that the outer layers act as an insulating film preventing undue loss of heat from the melt to the surface (e) of the hot-runner plate. This surface is therefore relatively cool and does not need to be insulated from the rest of the fixed half. Thus the hot-runner plate (f) is attached directly to the cavity plate (g) and becomes an integral part of the fixed mould half-assembly.

The mould thereby functions as a simple two-part mould, the melt passing intermittently through the flow-ways of the hot-runner plate directly into the impression. Close control of the temperature of the melt in both the flow-ways and the gate region is possible with this design.

13.5.1 Hot-runner Plate. This plate is normally manufactured to the same length and width dimensions as the cavity plate. It must be deep enough, however, to accommodate a relatively large flow-way hole plus the flow-way inlet and outlet system.

The flow-way shown in Figure 13.59 is not practicable because no support is provided for the heater tube (h). This part is shown 'floating' in mid-stream of the melt. While spider supports could be incorporated to make the design a more practical proposition (the heater (j) wires being fed through one of the spider legs), it is usual practice to support the heater tube at either end of the flow-way hole. To accommodate this it is necessary to stagger the various flow-ways and outlet holes. However, before proceeding to discuss this aspect of the design let us recapitulate the principal features of the annular design, which were discussed in Section 13.3.3 with respect to the hot-runner unit.

With reference to Figure 13.21, a hole (g) 30 mm (1¼ in) diameter or above, is machined through the plate. A heater tube assembly is fitted centrally into this hole and is held in position by a grub screw (c) or location pin. The required annular hole is thereby formed by the outer surface of the heater tube and the inner surface of the bored hole. The heater tube (d) is positioned by means of end-caps (b) and these are a sliding fit in the hole. A heating element (e) of the cartridge heater or low voltage type is fitted inside the heater tube. Some allowance for expansion should be made in the design to accommodate the expansion in the length of the tube when it is heated. In the design illustrated a small gap (p) is left at the end of the heater tube and the bottom of the recessed hole in the end-cap.

We now turn to consider the aspect of the design which novices find difficult to comprehend: the flow-way system is staggered and is not on one plane. This is a three-dimensional problem and is difficult to visualise without practice.

A schematic view of the flow-way for a four-impression mould is shown in Figure 13.60. Note that the outlet flow-way (n) is offset with respect to the secondary flow-way (s) and that this flow-way, in turn, is offset with respect to

RUNNERLESS MOULDS

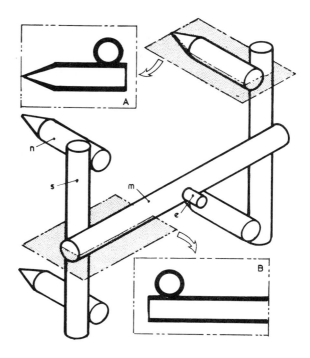

Figure 13.60—Schematic view of hot-runner mould flow-way system

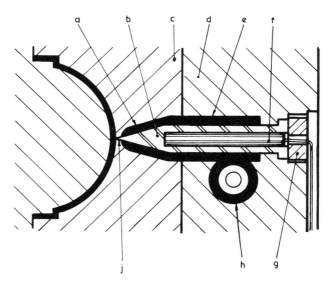

Figure 13.61—Hot-runner plate outlet flow-way

HOT-RUNNER PLATE MOULDS

the main flow-way (m). Had branch flow-ways been required, these again would have been offset. The amount of offset required is indicated in the two details, A and B. The length of the cylindrical entry aperture (e) should be as short as possible to reduce the distance the melt has to travel from the heated secondary nozzle to the heated main flow-way.

Referring to Figure 13.61: a 'well' (a) is machined in the cavity plate (c) directly behind the impression. The diameter chosen for this well is 25 mm (1 in) or more, depending upon the diameter of heater probe used. A corresponding sized hole (e) is adopted for the outlet flow-way from the hot-runner plate (d), so that when assembled (as shown) they provide an uninterrupted flow-path for the melt from the main flow-way (h) to the gate entry (j) into the impression.

The heater probe (b) is centrally disposed in the outlet flow-way and is secured in the hot-runner plate. Locating fins may be incorporated at the front end of the heater probe, to fit into an accommodating recess in the well (see Figure 13.58), and by this means the tip of the heater probe is held concentric with the gate entry. A small-diameter cartridge heating element (f) is fitted in the centre of the heater probe and can incorporate a thermocouple tip for precise temperature control.

The complete heater tube assembly and heater probes for this type of mould are available as standard parts from DME (USA and Europe). Two sizes of heater tube are available, diameter $1\frac{1}{4}$ and 2 in (31.75 and 50.8 mm) respectively.

Two types of heater probe are also available: a fixed probe (similar to that illustrated in Figure 13.58) and the 'auto-shut probe', which is a spring-loaded shut-off valve system. A three-dimensional cutaway drawing which incorporates the DME heater tube assembly and auto-shut probes is included as Plate 15.

14

Standard Mould Units

14.1 GENERAL CONSIDERATIONS

The reader who has progressively worked through this book will have noted the similarity in the structural build-up of the mould plates in many of the designs. The majority of the designs discussed in the text fall within the category of *two-part* (two-plate) moulds. The other designs are of either three-part or four-part construction.

Because of this similarity in the structural build-up of the majority of moulds it is desirable that some form of standardisation be adopted to permit mould units to be produced in quantity and thereby reduce manufacturing costs. Logically it is advantageous for the mould-maker to purchase a mould unit at reasonable cost, rather than to be actively engaged on this relatively unimportant aspect of mould manufacture.

A mould unit may be defined as an assembly of parts which conforms to an accepted structural shape and size. The mould unit is purchased with the mould plates suitably attached together and a guidance system incorporated. Naturally the mould unit does not contain the impression form and this aspect of mould manufacture must be left to a specialist mould-maker.

The two-part mould is adopted as the 'standard mould unit' by mould unit manufacturers because this particular mould construction is the most widely used design in industrial practice. The unit comprises two mould plates (a cavity plate and a core plate) plus an ejector system, as illustrated in Figure 14.1.

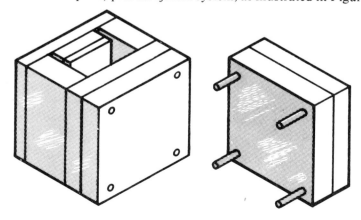

Figure 14.1—Standard mould unit

GENERAL CONSIDERATIONS

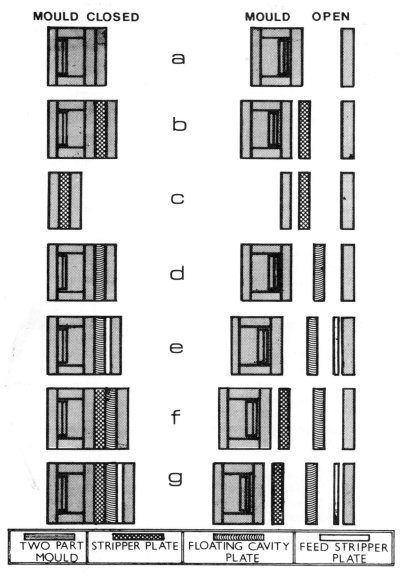

Figure 14.2—Mould units shown in the closed and open positions
    (a) Standard mould unit
    (b) Stripper plate mould unit (standard type)
    (c) Stripper plate mould unit (basic type)
    (d) Underfeed mould unit
    (e) Underfeed mould unit incorporating a feed stripper plate
    (f) Underfeed mould unit incorporating stripper plate ejection
    (g) Underfeed mould unit incorporating both a feed stripper plate and stripper plate ejection

STANDARD MOULD UNITS

The progressive build-up of various other mould types is shown in Figure 14.2, using the two-part mould unit (a) as a structural base. Extra plates are added to (or subtracted from) this standard mould unit to form alternative structures. The majority of the conventional mould types can be obtained in this way.

Figure 14.2 illustrates seven types of mould unit in the closed and open positions. Designs (b)-(g) are discussed in Section 14.3. Note that the names given to these mould types, used here for consistency within this book, are not necessarily those used by the manufacturer.

It is important for the beginner to appreciate that the majority of mould designs can be broadly classified within one of these mould types: for example, a splits-type mould can be considered as a two-part mould with the addition of a splits assembly.

Even more complex mould structures can be envisaged with the addition of further extra plates but this type of structure is beyond the scope of this book. The novice should aim to be conversant with the range of standard mould units that are available so that he is in a position to use mould units whenever the opportunity occurs.

The advantages and disadvantages of using standard mould units may be compared to those of buying a ready-made suit from a reputable shop. We know that there are many disadvantages in buying a suit in this way. For example, the fit may not be perfect; there is a limit to the number of styles and cloths available; some alteration to the suit may be necessary. Considered overall, however, good value is obtained and the suit is satisfactory for most occasions. Nevertheless, for the special occasion a 'made-to-measure' suit may allow the choice of a particular style or cloth, or a perfect fit.

The above comments apply equally to the standard mould unit. The designer may have to accept a compromise in mould size; the number of mould types and the range of sizes is limited; for certain moulds some modification to the mould unit is necessary to accommodate a specific feature. Considered overall, however, the advantages of using standard mould units outweigh the disadvantages and for a large number of mould designs the standard mould unit can be beneficially used by both the mould designer and mould-maker. Nevertheless, on many occasions it is necessary to have a 'made-to-measure' mould in order that a special feature may be incorporated or a specific coolant flow-way system adopted, or simply because a standard mould unit of a suitable shape or size is not available.

14.1.1 Advantages and Limitations of Standard Mould Units

ADVANTAGES

(1) Drawing sheets for individual unit sizes are available. This reduces drawing time.
(2) Less steel needs to be carried in stock, therefore investment is reduced.
(3) Buying and stock control are simplified.
(4) The cost of the mould unit is known, therefore estimating is easier.

GENERAL CONSIDERATIONS

(5) Waiting time for steel blanks, etc., is avoided.
(6) Shaping, planing and drilling of steel plates and blocks is avoided.
(7) Turning, grinding and fitting of guide pillars and bushes is similarly avoided.
(8) The ejector plate is pre-positioned and located.
(9) The individual mould plates are screwed and dowelled together.
(10) Machine time is saved as a result of 6 and 7 above.
(11) Labour time is saved as a result of 6-9 above.
(12) Work on the impression can usually begin immediately.
(13) The individual mould unit components are standard: if damage occurs during manufacture or in production, a part can be quickly replaced.
(14) In small moulding companies, one mould unit can be used for several similar jobs. Only the impression inserts need to be changed.
(15) The overall time the mould is in the toolroom is reduced.
(16) Mould delivery time is reduced.
(17) Highly paid mould-makers may be employed on important impression work rather than on relatively minor 'bolster-work'.

LIMITATIONS

(1) The number of sizes available is limited.
(2) Maximum size available is relatively small (see Figure 14.4).
(3) Maximum depth of mould plate is also relatively small.
(4) The ejector stroke may be larger than is actually required.
(5) The positioning of coolant flow-way holes is often made difficult by pre-positioning of guide pillars, guide bushes, screws and dowels.
(6) The support blocks are positioned relatively far apart. Extra support blocks may have to be fitted if deflection of the mould plate is to be avoided.
(7) The back plate cannot be unscrewed independently of the support block and backing plate. To expose the ejector assembly, the mould half must be disassembled.

14.1.2 Manufacturers of Mould Units. The first standard mould units became available in the UK in the early 1960s. Naturally in this early period only a relatively few sizes of mould unit were available and, because of this, their acceptance by the mould-making industry was not immediate.

Mould designers were reluctant to specify mould units for their designs for several reasons, which included:
 (i) the range of sizes was limited;
 (ii) delivery of units was often unpredictable;
 (iii) mould-makers would not use mould units because of their unfamiliarity.

However, over the ensuing years the range of sizes has increased, delivery has improved, and the mould-maker has slowly, but progressively, begun to accept this innovation to his technology.

Several companies now provide a major service in standard mould units in the UK at the present time. Between them, these companies offer about 12 000 possible variations of the standard two-part mould unit assembly alone.

Brief details of the mould unit suppliers are as follows:

(i) Diemould Service Company Ltd, commonly known as DMS. This is a British company with registered offices in London and a factory in Devon.

(ii) DME Europe, formally the Detroit Mold Engineering Co. of the USA (hence DME). This company's main plant and offices are located in Mechelen (Belgium), and the UK information centre is at High Wycombe, Bucks.

(iii) Hasco is a West German Company, manufacturing standard mould units in Ludenscheid. Hasco's mould units and other standard parts are available in the UK from Hasco-Internorm Ltd, Daventry.

(iv) Uddholm-Sustan manufacture standard mould units in Sweden, West Germany and Japan. The units are available in the UK from Uddholm Ltd, Birmingham.

Each company produces a large catalogue containing the complete range of standard mould units that are available. To specify a particular unit is simply a matter of quoting the catalogue number. The catalogue also list many other mould components which are produced as standards.

14.2 STANDARD TWO-PART MOULD UNITS

The two-part mould unit is adopted by the manufacturers as the primary unit upon which more complex mould unit systems may be constructed. It forms, therefore, the *standard mould unit assembly* (Figure 14.1).

As mentioned earlier, the two-part mould is the most widely used design in current practice. It is therefore essential for the beginner to become conversant with the constructional details and availability of two-part mould units as quickly as possible.

The two-part mould unit consists of two mould plates (a cavity plate and a core plate) plus an ejector system. A typical cross-section through a two-part mould unit assembly is illustrated in Figure 14.3.

14.2.1 Two-part Mould Unit Variables. To be useful to the practising mould designer and mould-maker, the mould unit must be available in a large number of sizes. Basically, the size of mould unit varies in three dimensions, i.e. in length, in width and in overall height (Figure 14.3). Note that overall height is directly dependent upon the thicknesses of the individual mould plates and, in addition, on the length of the ejector stroke. Thus there are a considerable number of variables in a mould structure to be considered.

The mould unit designer has the difficult task of rationalising these variables to allow an economic range of sizes to be produced while providing the mould designer with a reasonable choice of mould units for specific requirements.

With reference to Figure 14.3, the variables may be divided into two types:

STANDARD TWO-PART MOULD UNITS

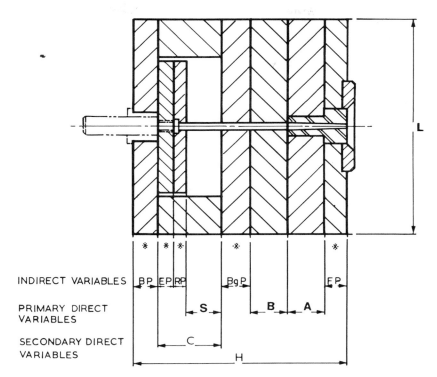

Figure 14.3—Cross-sectional view of a standard mould unit

direct and indirect; the direct variables may be further subdivided into primary and secondary direct variables. The primary direct variables are:

Length of mould plate	L
Width of mould plate	W
Depth of cavity plate	A
Depth of core plate	B
Length of ejector stroke	S

The secondary direct variables, dependent upon the primary direct variables, are designated thus:

Overall mould height	H
Depth of ejector space	C

PRIMARY DIRECT VARIABLES W AND L. Width x length (W x L) is used as the designatory title for the individual mould unit series by both manufacturers. For example, a 150 x 250 unit has a mould plate 150 mm wide x 250 mm long. Thus the overall size of mould plate, established in the drawing office, is the mould unit size, and the relevant page of the catalogue can be consulted.

Figure 14.4 shows a chart giving the various width x length combinations currently available from the three companies. Note that in this chart all of the values are rounded to the nearest centimetre.

STANDARD MOULD UNITS

The method of reading the chart is to locate the box which is relevant to the particular mould unit required. The symbol inside the box indicate which company manufactures a standard mould unit of that size. In the diagram each company has been allocated a symbol. Refer to Figure 14.4 for the code. For example, suppose a small 15 cm by 15 cm mould unit is required. The relevant box contains a small triangle which indicates that a unit of that size is available from DMS.

While the box in the above example only contains one supplier, in the majority of cases a particular size of mould is available from more than one source. For reasons of simplicity, a larger symbol is adopted when more than two companies produce a similar size of unit.

PRIMARY DIRECT VARIABLES A, B and S. We now turn to the other direct variables, depth of cavity plate (A), depth of core plate (B) and length of ejector stroke (S). In practice, the depths of the cavity and core plates are a function

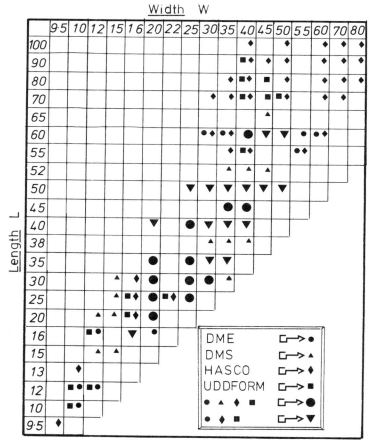

Figure 14.4—Width and length combinations of standard mould units available

STANDARD TWO-PART MOULD UNITS

A	B	S	Cat No	A	B	S	Cat No	A	B	S	Cat No
A_1	B_1	S_1	1	A_2	B_1	S_1	10	A_3	B_1	S_1	19
		S_2	2			S_2	11			S_2	20
		S_3	3			S_3	12			S_3	21
	B_2	S_1	4		B_2	S_1	13		B_2	S_1	22
		S_2	5			S_2	14			S_2	23
		S_3	6			S_3	15			S_3	24
	B_3	S_1	7		B_3	S_1	16		B_3	S_1	25
		S_2	8			S_2	17			S_2	26
		S_3	9			S_3	18			S_3	27

Figure 14.5—Typical layout of mould unit variables to permit a catalogue number to be specified. In this 3 x 3 x 3 arrangement A = depth of cavity plate, B = depth of core plate, S = ejector stroke

of a number of factors including the following:
 (i) the shape of the moulding;
 (ii) the dimensions of the moulding;
 (iii) the method of impression construction;
 (iv) the number of impressions required.

As mouldings are required in all shapes and sizes, and a large variation in the number of impressions is required in practice, the mould unit manufacturer must provide the mould designer with a range of mould plate thicknesses. This variation in thickness applies to both the cavity plate and the core plate, therefore a combination of various mould plate thicknesses is required.

As different depths of moulding occur in practice it follows that various lengths of ejector stroke, too, are required. This adds another possible variable to the two discussed above.

Consider the problem mathematically. We have three variables termed A, B and S. The total number of alternative arrangements which can be obtained is the product of the number of values (sizes) for each variable. For example, if three depths for each mould plate are available in association with three ejector stroke lengths, we have a product of three terms, that is, $3 \times 3 \times 3$. Thus a range of 27 mould units is necessary if the above variation in plate and ejector stroke size is to be offered by the manufacturer.

The above combination of variable is known as a $3 \times 3 \times 3$ arrangement or 'matrix', and can be laid out as shown in Figure 14.5. This type of layout permits a specific catalogue number to be given to each mould unit. If the number of alternatives for each variable is increased to four, a $4 \times 4 \times 4$ arrangement results, requiring 64 units, and so on.

STANDARD MOULD UNITS

INDIRECT VARIABLES. The indirect variables of a standard mould unit are the thicknesses of those plates which are marked with an asterisk in Figure 14.3. These plates include the back plate (BP), ejector plate (EP), retaining plate (RP), backing plate (BgP), and the front plate (FP).

The thicknesses of the above plates are primarily dependent upon the overall size of the mould plates ($W \times L$) and are independent of the size and shape of the component. The mould unit manufacturer treats these thicknesses as a controlled variable, that is, the thickness of each plate is kept constant for a particular unit series. The actual plate thicknesses for a specific series may be ascertained from the diagram associated with that series. This diagram also shows relevant details regarding the position and dimensions of guide pillars, guide bushes, screws, dowels, sprue pin, etc.

Typical examples of the diagrams covering specific unit series are shown in Figures 14.7 and 14.8, reproduced from the DMS and DME catalogues.

14.2.2 DMS Standard Mould Unit. The DMS system for the standard two-part mould unit (termed 'standard mould base unit' by DMS) is basically similar to the design discussed above in Section 14.2.1. A summary of the currently available DMS two-part mould unit range is given in Table 10. (Note that the terms listed under 'Mould unit size' are relative and apply only to this mould unit series. For example, the size range listed as 'very large' is not in fact very large in relation to the size of injection moulds currently in use.) The total number of units of a particular mould size range is obtained by multiplying the product of the $A \times B \times S$ arrangement by the number of sizes in the range.

TABLE 10
DMS TWO-PART MOULD UNIT RANGE

Mould unit size	Unit size range $W \times L$ (mm)	No. of sizes in range	A, B, and S arrangement	Total
Small	115 × 146 – 115 × 196	2	3 × 3 × 3 = 27	54
Intermediate	146 × 146 – 296 × 446	15	4 × 4 × 3 = 48	720
Large	346 × 306 – 396 × 586	8	3 × 3 × 3 = 27	216
Very large	446 × 516 – 446 × 656	2	3 × 3 × 3 = 27	54

The DMS catalogue page is arranged slightly differently from the basic matrix layout shown in Figure 14.5. In the DMS system the ejector stroke S is made the primary variable which must be established immediately following that of the specified mould plate ($W \times L$).

Each value of the ejector stroke S is contained in a separate rectangle or 'box'. Once the correct box has been found, the various combinations of mould plate thickness (A and B) can be inspected to permit the selection of the required unit size. A typical DMS catalogue page is shown in Figure 14.6.

Example. Suppose a two-part mould unit is required having the following dimensions: width: 146 mm; length: 246 mm; mould plate A thickness: 56 mm; mould plate B thickness: 36 mm; ejector stroke: 45 mm.

STANDARD TWO-PART MOULD UNITS

146 x 246 Series
STANDARD MOULD BASE UNIT

ORDERING PROCEDURE:

Catalogue Number
Details this page.

Clamping Slots
Details section DMS 6 pages 58 & 59. If required specify Catalogue Number followed by the letter **W** to denote unit supplied with clamping slots.

Oversize Clamp Plates
Can be fitted on this series of Mould Base Unit. Details section DMS 6 page 60. If required specify Catalogue Number followed by the letter **X** to denote unit supplied with oversize clamp plates.

Location Spigot
Details section DMS 3 pages 2 & 3. For this series use Code Numbers **LRM10** to **LRM17**.

Sprue Bush
Details section DMS 3 pages 4, 5, 6 & 7. For this series use Code Numbers **M60** to **M64**, **M70** to **M74**, **R60** to **R62**, **R70** to **R72**.

Ejector Rod
Details section DMS 3 pages 8 & 9.
B for Bushed.
UB for Unbushed.

Return Pins
For position refer to drawing note 8.
P for fitted.
O for omitted.

Example: 146 x 246 series unit, catalogue number 4259, supplied with clamping slots, having 125 mm. location spigot, with 3,5 mm. by 2° inclusive bore flat face sprue bush, ejector rod bushed and return pins fitted.

Order:
4259W/LRM13/M63/B/P.

Dim. A	Dim. B	Dim. S	Dim. C	Mould HT	Fitted with GP	Fitted with GB	Cat. No.
26	26	15	43	161	604	601	4250
	36			171	605	602	4251
	46			181	605	603	4252
	56			191	606	604	4253
36	26			171	608	601	4254
	36			181	608	602	4255
	46			191	609	603	4256
	56			201	610	604	4257
46	26			181	611	601	4258
	36			191	612	602	4259
	46			201	613	603	4260
	56			211	613	604	4261
56	26			191	615	601	4262
	36			201	615	602	4263
	46			211	616	603	4264
	56			221	616	604	4265
26	26	30	58	176	604	601	4266
	36			186	605	602	4267
	46			196	605	603	4268
	56			206	606	604	4269
36	26			186	608	601	4270
	36			196	608	602	4271
	46			206	609	603	4272
	56			216	610	604	4273
46	26			196	611	601	4274
	36			206	612	602	4275
	46			216	613	603	4276
	56			226	613	604	4277
56	26			206	615	601	4278
	36			216	615	602	4279
	46			226	616	603	4280
	56			236	616	604	4281
26	26	45	73	191	604	601	4282
	36			201	605	602	4283
	46			211	605	603	4284
	56			221	606	604	4285
36	26			201	608	601	4286
	36			211	608	602	4287
	46			221	609	603	4288
	56			231	610	604	4289
46	26			211	611	601	4290
	36			221	612	602	4291
	46			231	613	603	4292
	56			241	613	604	4293
56	26			221	615	601	4294
	36			231	615	602	4295
	46			241	616	603	4296
	56			251	616	604	4297

Figure 14.6—DMS catalogue: layout of variables for a 146 mm x 246 mm standard mould unit

STANDARD MOULD UNITS

146 x 246 Series
STANDARD MOULD BASE UNIT

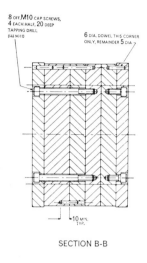

SECTION B-B

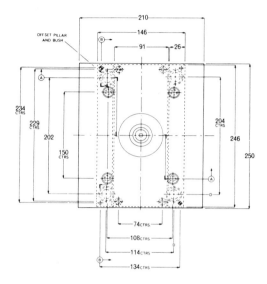

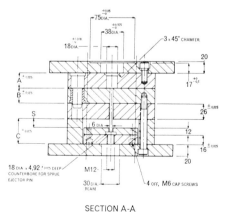

SECTION A-A

DRAWING NOTES

1. 16 mm. diameter Guide Pillars and Guide Bushes fitted.
2. 18 mm. diameter clearance holes for Guide Pillars in Backing Plate.
3. 6,5 mm. diameter clearance hole for the Sprue Ejector Pin in Backing Plate.
4. Effective length of the Sprue Ejector Pin in the Moving Cavity Plate is four times its diameter.
5. The Sprue Ejector Pin is supplied with the unit but not fitted.
6. 30 mm. deep tapping drill on 36 mm. cavity plates upwards.
7. 'D' denotes datum holes.
8. If required 4 Return Pins of 10 mm. diameter can be fitted on 44 mm. x 222 mm. centres.
9. Material specification see section DMS 5 page 4.

Figure 14.7—DMS catalogue: typical diagram of a DMS standard mould unit

STANDARD TWO-PART MOULD UNITS

Procedure. (i) Consult the catalogue and find the page relating to the 146 × 246 series (see Figure 14.6).

(ii) Inspect each box labelled 'DIM S' to find the required ejector stroke (45 mm). This box will be found in the lower third of the page.

(iii) Inspect each box in this section under the heading 'DIM A' to find the required mould plate A thickness (56 mm). This dimension is located in the bottom left-hand corner of the section.

(iv) Adjacent to this dimension are four possible variations in mould plate B thickness. (Located under heading 'DIM B'. The required dimension (36 mm) is the second possibility listed.

(v) Reading across the quarter-page line with the required $A \times B \times S$ combination, the mould unit height (231 mm) will be found, and the catalogue number, 4295, in the right-hand column.

The columns 'Fitted with GP and GB' refer to the specific guide pillars and guide bushes which are fitted in the standard mould unit.

The mould unit diagram associated with the above DMS catalogue page is shown in Figure 14.7. There are two points which the beginner should carefully note:

(i) DMS adopt the third-angle projection drawing method.
(ii) This diagram does not follow normally accepted mould drawing practice. The conventional engineering orthographic approach has been adopted for this diagram, and hence the plan view shows the *rear face of the fixed mould half* and not the view on the parting surface as would normally be expected.

When ordering a mould unit, several other features have to be decided upon in addition to the basic mould unit size. These items include the following: the register ring (location spigot); the sprue bush bore dimensions; whether a flat or radius entry to the sprue bush is required; whether the ejector bush is required to be bushed or unbushed; whether push-back pins (return pins) are required. To determine the code numbers to give to these items, various pages of the manufacturer's catalogue must be consulted. The ordering procedure is given at the left of the catalogue page (Figure 14.6).

A number of variations on the standard mould unit can be obtained, for example:

(i) The unit can be ordered without a backing plate being fitted to either mould plate A or B.
(ii) The back and front plates can be of the same dimensions ($W \times L$) as the other mould plates. A clamping slot is provided.

14.2.3 DME Standard Mould Unit. The DME standard mould unit (termed 'standard mould base' by DME) is similar to the design discussed in Section 14.2.1. However, instead of five, only four direct variables are used in this system: width (W), length (L) mould plate thickness (A) and mould plate thickness (B). The stroke length is a constant for a particular mould plate size, and therefore is a controlled variable in this system.

STANDARD MOULD UNITS

In order to accommodate the range of ejector strokes likely to be required in practice, a relatively long ejector stroke is provided for each mould unit. (The manufacturer will offer alternative specific ejector strokes if requested.)

A summary of the available DME standard mould unit range is given in Table 11. The total number of units of a particular mould unit size is obtained by multiplying the product of the $A \times B$ arrangement by the number of sizes in the range.

TABLE 11
DME TWO-PART MOULD UNIT RANGE

Mould unit size	Unit size range $W \times L$ (mm)	No. of sizes in range	A, B and S arrangement	Total
Very small	$125 \times 125 - 125 \times 156$	2	$5 \times 5 \times 1 = 25$	50
Small	$156 \times 156 - 196 \times 396$	8	$6 \times 6 \times 1 = 36$	288
Intermediate	$246 \times 246 - 246 \times 496$	5	$7 \times 7 \times 1 = 49$	245
Large	$296 \times 296 - 346 \times 596$	10	$8 \times 8 \times 1 = 64$	640
Very large	$396 \times 396 - 594 \times 594$	12	$10 \times 10 \times 1 = 100$	1200

The DME catalogue page is arranged as a conventional matrix. The thickness of the cavity plate A is established immediately following the size of specific mould plate ($W \times L$). Subsequently the required thickness of core plate B is located. Finally the values for ejector space height and overall mould height and the catalogue number are read off the table.

A typical page from the DME catalogue is given in Figure 14.9. The series number given at top right of the page indicates the *nominal* width × length dimension in *centimetres*. For example, a 156 mm × 196 mm mould unit becomes Series 1620.

This particular mould unit series adopts a $6 \times 6 \times 1$ arrangement resulting in 36 possible variations of mould unit size.

The mould unit associated with the above catalogue page is shown in Figure 14.8. The diagram is in first-angle projection and follows normal mould drawing practice.

A number of variations on the standard mould unit can be obtained from DME, as follows:

(i) The back and front plates are available as overhanging clamping plates. This feature is indicated in Figure 14.8 by dotted lines.
(ii) The backing plates to mould plates A and B can be omitted, if required.
(iii) Long ejector plates can be specified.

14.2.4 Uddform Sustan-Standard Mould Unit. This company gives the term 'mould sets' to their standard mould unit range. The basic design is similar to the design discussed in Section 14.2.1.

There are four direct variables in this system: width (W), length (L), mould plate thickness (A), and mould plate thickness (B). The ejector

STANDARD TWO-PART MOULD UNITS

SERIES 1620
156 mm × 196 mm

DME STANDARD MOULD BASE
(USA Pat. Nr 2 874 409)

Available in D M E Nr 1, 2 and Nr 3 Steel
See page 9 for steel instructions

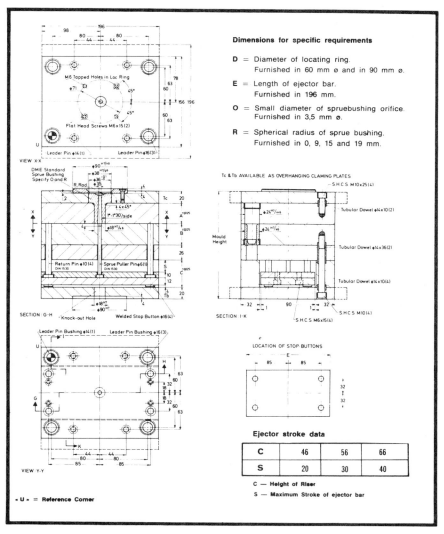

Figure 14.8—DME catalogue: typical diagram of a DME standard mould unit

421

STANDARD MOULD UNITS

SERIES 1620

156 mm × 196 mm

DME STANDARD MOULD BASE
(USA Pat. Nr 2 874 409)

Available in D M E Nr 1, 2 and Nr 3 steel
See page 9 for steel instructions

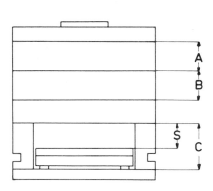

A	B	C	MOULD HEIGHT	CATALOGUE NUMBER	NET WEIGHT
26	26	46	164	1620SF-26-26	38
	36	46	174	1620SF-26-36	41
	46	56	194	1620SF-26-46	43
	56	56	204	1620SF-26-56	46
	66	66	224	1620SF-26-66	49
	76	66	234	1620SF-26-76	51
36	26	46	174	1620SF-36-26	41
	36	46	184	1620SF-36-36	43
	46	56	204	1620SF-36-46	47
	56	56	214	1620SF-36-56	49
	66	66	234	1620SF-36-66	52
	76	66	244	1620SF-36-76	54
46	26	46	184	1620SF-46-26	43
	36	56	204	1620SF-46-36	45
	46	56	214	1620SF-46-46	48
	56	66	234	1620SF-46-56	50
	66	66	244	1620SF-46-66	53
	76	66	254	1620SF-46-76	56
56	26	46	194	1620SF-56-26	45
	36	56	214	1620SF-56-36	47
	46	56	224	1620SF-56-46	50
	56	66	244	1620SF-56-56	52
	66	66	254	1620SF-56-66	55
	76	66	264	1620SF-56-76	58
66	26	46	204	1620SF-66-26	48
	36	56	224	1620SF-66-36	51
	46	56	234	1620SF-66-46	54
	56	66	254	1620SF-66-56	56
	66	66	264	1620SF-66-66	60
	76	66	274	1620SF-66-76	62
76	26	56	224	1620SF-76-26	50
	36	56	234	1620SF-76-36	53
	46	66	254	1620SF-76-46	56
	56	66	264	1620SF-76-56	58
	66	66	274	1620SF-76-66	61
	76	66	284	1620SF-76-76	64

When ordering please specify:
- Catalogue Number
- D M E steel quality
- D, E, O, R and C dimensions
- Preferred routing

NOTE

Standard Mould Base Ordered	#1	#2	#3
Plate Tc is Steel N?	1	2	2
A	1	2	3
B	1	2	3
G	1	2	2
Tb	1	1	1

SEE SECTION 4 FOR ADDITIONAL PLATE THICKNESSES

Figure 14.9—DME catalogue: layout of variables for a Series 1620 mould unit

STANDARD TWO-PART MOULD UNITS

space (C) and ejector plate dimensions are constant for a particular mould size, and the stroke is, therefore, a controlled variable in this system.

Each catalogue page is arranged as a conventional matrix covering a specific $W \times L$ size rounded to the nearest centimetre. For example, a 395 mm × 396 mm unit is designated the number 4040 and this number is given at the top of each page.

Seven depths of mould plate are available, plus seven depths of plate B, thereby giving an ABS matrix of $7 \times 7 \times 1$, resulting in 49 possible variations for each mould unit size.

In addition to the standard series (designated S) an alternative design is available for the small and intermediate size ranges. This design eliminates the space taken up by screws and therefore allows more space for impressions, coolant channels, etc. This latter system is called 'Europa' and these alternative mould units are designated E.

The major mould plates are also available in a number of different Uddeholm steels. In addition to the mould units, Uddform-Sustan also produce a standard splits type mould (termed 'sliding core' mould) and a variety of mould component parts.

A summary of the available Uddform-Sustan standard mould unit range is given in Table 12.

TABLE 12
UDDHOLM-SUSTAN TWO-PART MOULD UNIT RANGE

Mould unit size	Unit size range $W \times L$ (mm)	No. of sizes in range	A, B and S arrangement	Total
Very small	100 × 100–126 × 126	4	49	196
Small	156 × 156–196 × 396	8	49	392
Intermediate	218 × 246–246 × 496	6	49	294
Large	296 × 296–346 × 496	9	49	421
Very large	396 × 396–396 × 896	15	49	735

14.2.5 Hasco Standard Mould Unit. A completely modular system is adopted by Hasco, and the term 'Standard Elements' is used to designate their standard system. The Hasco catalogue page is interestingly different from those previously discussed in that there is not a commentary to explain the ordering procedure. This is achieved entirely by diagrams and which, cunningly, overcomes language problems.

A 'thumb nail' size drawing of a basic three part mould is repeated nine times on each double catalogue page. Each drawing identifies a specific plate (or plates) by the use of colour. A correspondingly small dimensioned detail drawing is shown adjacent to the assembly drawing, together with a list of alternative depths which are available for that specific plate.

The catalogue page is identified by a number which represents the width and length dimensions (in mm) of the mould plate (i.e. $W \times L$). A letter-number combination is used to identify a specific plate in the assembly.

The mould plates (A & B) are available in a number of thicknesses and this number varies between six and nine depending upon the unit. The ejector space (C), too, is a variable in this system, and either three or four depths are available. In addition, the back plate and backing plate of certain size assemblies are available in two thicknesses. Alternative steels for the mould plates and certain other plates are also available.

The ordering procedure for this system is to initially establish the specific mould plate size ($W \times L$) and then consult the relevant catalogue page. Decide upon the mould plate thickness (A & B), together with the ejector space (C) required. Finally, if more than one depth of backing plate is available—a decision as to the most suitable depth must be decided upon. The order consists of a list of plates plus appropriate screws, guide pillars, guide bushes etc.

There are 63 basic width-length combinations available with the basic system, but as the number of variables for each combination does not follow any set pattern it is impracticable to table the range. However an example will suffice to indicate the very large number of variations which are possible with this modular system.

Consider a mould plate 296 mm wide by 296 mm long. Under this heading there are eight mould plate depths, three ejector space depths, and two alternative depths of backing plate. Thus the number of variations for a basic mould of this size is $8 \times 8 \times 3 \times 2$ making 284 in total.

Hasco also offer a range of 16 alternative circular mould units which can be extremely useful for certain applications.

14.2.6 Summary. From the preceding discussion it is evident that the standard mould unit (SMU) may be used for most of the designs given in Part One — Elementary Mould Design. The only limiting factor is that of size of SMU available. At the time of writing the largest unit is 1000 mm × 800 mm and the maximum plate depth if 150 mm.

Thus, provided the size of the mould falls within specified limits, the following design features can be accommodated in SMU: pin ejection (Figure 3.22); D-pin ejection (Figure 3.33); blade ejection (Figure 3.40); valve ejection (Figure 3.44); stripper bar ejection (Figure 3.49); stripper ring ejection (Figure 3.58). In the last two designs the ejector retaining plate is not necessarily used.

There are two items in the elementary design section which necessitate some deviation from the SMU: the sleeve ejection design (Figure 3.36) (this involves only a minor deviation) and the stripper plate design (Figure 3.51). As the latter is basically of a three-part (plate) construction, obviously a considerable deviation from the SMU is necessary.

While some of the mould designs discussed in Part Two, Intermediate Mould Design, can be accommodated in a SMU, in the majority of cases some modification is required.

14.3 DEVIATIONS FROM THE STANDARD MOULD UNIT

Any deviation from the SMU simply involves the addition or subtraction of plates to achieve the required assembly. Guide pillars and guide bushes

DEVIATIONS FROM THE STANDARD MOULD UNIT

may have to be reversed and, in addition, the sprue puller may have to be removed.

Mould designs necessitating a deviation from the SMU are discussed in turn below. In each illustration the SMU is shown as a reference base, and the required mould unit is shown below. Added plates are shown cross-hatched. Mould plates added to or subtracted from the SMU are marked 'IN' or 'OUT', respectively.

14.3.1 Sleeve Ejection Mould Unit. In the sleeve ejection design (Section 3.4.4) the ejector element is incorporated in the ejector assembly instead of, or in addition to, ejector pins. An extra plate is required to secure the core pin which passes through the sleeve ejector element (Figure 3.36). This extra plate may be local, as shown, in which case it can be incorporated by the mould-maker in a suitable recess in the back plate of the SMU. Alternatively, an additional plate, the core-retaining plate (CRP), may be added between the ejector assembly and the back plate as shown in Figure 14.10. As the width of this plate can be the same as that of the ejector plate assembly, deviation from the SMU is minimal.

14.3.2 Stripper Plate Mould Unit. There are basically two types of stripper plate mould which may be classified as the standard design and the basic design. In the standard design, the stripper plate is incorporated between the fixed and moving halves of the mould.

STANDARD STRIPPER PLATE DESIGN. This mould unit assembly is achieved, quite simply, by adding an extra plate (the stripper plate) between the two primary plates A and B. The stripper plate may be positively coupled to the ejector plate assembly by means of tie-rods (Figure 3.51) (the retaining plate being therefore unnecessary), or 'operating pins' (basically ejector pins) may be incorporated in the ejector plate assembly (Figure 3.52).

Note that the mould unit manufacturer fits the guide pillar in plate A of the SMU, and the guide bush in plate B. It is therefore necessary to reverse the guide pillar/guide bush arrangement for a stripper plate design. The sprue puller, too, is omitted. (Note that in stripper plate moulds the sprue puller is fitted in the core plate — Figure 7.6). The necessary SMU adaption and the final unit assembly are shown in Figure 14.11.

BASIC STRIPPER PLATE DESIGN. The directly operated stripper plate assembly is achieved by: (i) removing the ejector plate assembly and ejector grid; (ii) fitting an extra plate (the stripper plate) between the two primary plates, A and B. Actuation of the stripper plate may be either by means of the injection machine actuator rods (Figure 3.57) or by some other means, such as length bolts (Figure 3.54) or chains (Figure 3.56), etc.

Whether or not the guide pillar/guide bush arrangement is reversed depends upon whether the stripper plate is required to be operated from the 'injection' or the 'ejection' side. In the former case the guide pillar assembly can be left as standard, but in the latter case the guide pillar/guide bush assem-

STANDARD MOULD UNITS

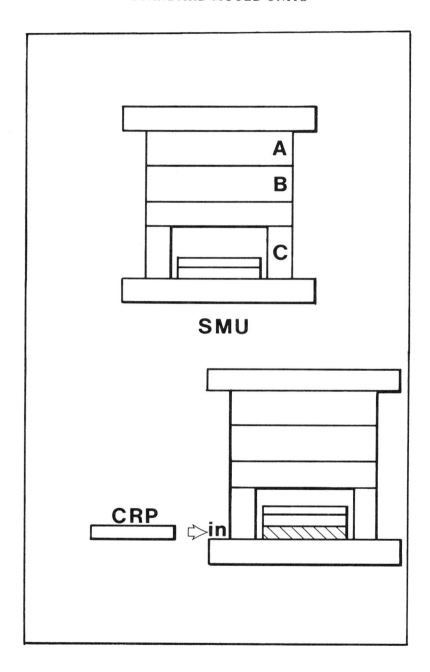

Figure 14.10—Sleeve ejection mould unit. SMU with additional core-retaining plate (CRP)

DEVIATIONS FROM THE STANDARD MOULD UNIT

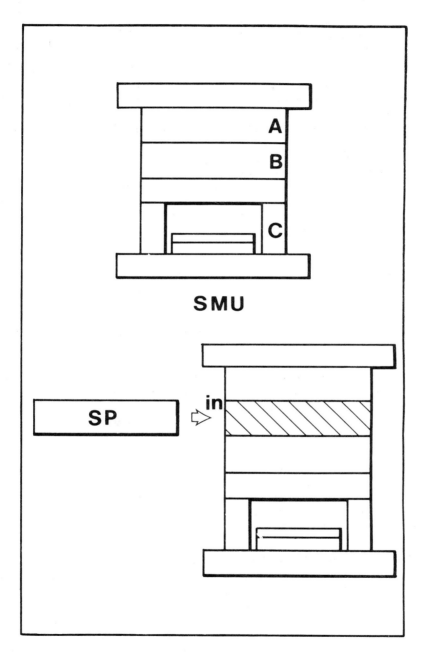

*Figure 14.11—Stripper plate mould unit (standard type).
SMU with additional 'stripper plate' (SP) mounted between plates A
and B. Reverse the guide pillar/guide bush assembly (not shown)*

STANDARD MOULD UNITS

bly must be reversed as discussed above. The main deviations from the SMU are shown in Figure 14.12.

14.3.3 Splits Mould Unit. Mouldings which incorporate external undercuts usually require a splits-type mould so that the undercut portion of the component may be relieved before ejection takes place (refer to Chapter 8).

Sliding splits (Section 8.2) are accommodated in a recess machined into mould plate A (Figure 8.8) and the mould plate therefore needs to be relatively deep. The problem can be overcome to a limited extent by incorporating separate 'heel blocks' in the mould plate, as illustrated in Figure 8.25.

For 'angled-lift' splits designs (Section 8.3), a relatively deep recess is required in mould plate B so here again a deep mould plate is necessary.

In conclusion, therefore, while the SMU can be used for a splits-type mould in general, it may be necessary to specify an extra-deep (non-standard) plate for either mould plate A or B.

14.3.4 Side-core Mould Unit. When a moulding incorporates a hole or recess in a side wall, some form of side-coring is required. This necessitates incorporating a movable element within the structure of the mould which can be actuated prior to ejection (refer to Chapter 9).

In general, the SMU can be used for mould designs which incorporate side cores, the only modification being that heel blocks may have to be fitted by the mould-maker as shown in Figure 9.14.

14.3.5 Internal Undercut Mould Unit. Mouldings which incorporate projections or recesses on an inside surface usually require a movable core element to be incorporated within the structure of the mould. The internal undercut can often be accommodated on a 'form pin' or 'split core', which is actuated by the movement of the ejector system (refer to Chapter 10).

Thus the SMU is suitable for internal undercut designs although, to achieve the required actuation movement for the movable element, it may be necessary to specify an extra-deep support block system. If relative movement is required between the operating member (for example, the guide pin in Figure 10.10) and the ejector assembly, an alternative steel may have to be specified for the ejector retaining plate.

14.3.6 Unscrewing Mould Unit. Many component designs incorporate threaded sections; while the external thread can usually be produced using the splits mould, internally threaded components usually necessitate some form of unscrewing action as described in Chapter 11. In general, some adaptation of the SMU is necessary when it is used for an unscrewing mould in order to accommodate the unscrewing mechanism and drive system.

As discussed in Chapter 11, there are many alternative designs of unscrewing mould and each requires a slightly different mould structure. It would be impracticable to cover in this section every possible variation of assembly met with in industrial practice, and discussion is therefore restricted to two typical

DEVIATIONS FROM THE STANDARD MOULD UNIT

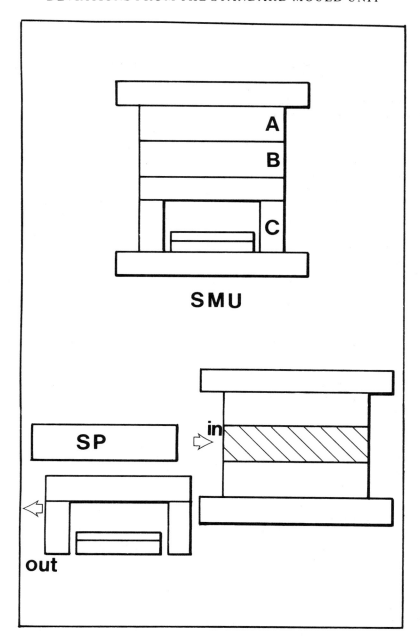

*Figure 14.12—Stripper plate mould unit (basic type).
SMU with additional 'stripper plate' (SP) mounted between plates A
and B. Reverse the guide pillar/guide bush assembly. Remove the ejector grid and
ejector assembly*

STANDARD MOULD UNITS

designs, which illustrate what can be accomplished using the SMU as a structural base.

One simple example of the mould construction for an in-line layout of cores is shown in Figure 11.41. Note that the moving half only is shown in this illustration. Thus, from the standardisation viewpoint, the only deviation from the SMU is that the support blocks and ejector assembly are replaced by a transmission plate (TP). This is illustrated in Figure 14.13.

Another example of the construction of an unscrewing type mould is illustrated in Figure 11.40. This design is for a pitch circle layout of the cores. Once again, only the moving half of the assembly is illustrated. The structure is slightly more complex than the previous example in that three extra plates are required, namely the transmission plate (TP), the core-shaft plate (CP) and the gear plate (GP). These three plates replace the support blocks and ejector assembly of the SMU (Figure 14.14). Note, with reference to Figure 11.40, that an extra-large back plate may be required to accommodate the drive mechanism. This latter plate would most likely have to be a 'special' in view of its unusual proportions.

On some unscrewing designs an ejector system is required in addition to the unscrewing mechanism. Naturally in these cases the ejector plate assembly and support block system is retained.

14.3.7 Multi-daylight Mould Unit. The purpose of most multi-daylight mould designs is to underfeed components, using pin gates. While the following discussion has been restricted to the applications of SMUs in the intermediate level mould designs discussed in Chapter 12, the principles can be extended to more complex structural assemblies. The discussion is therefore restricted to the following mould types:

(a) Basic underfeed mould (three-part mould).
(b) Underfeed mould with feed stripper plate (four-part mould).
(c) Underfeed mould with stripper plate ejection (four-part mould).
(d) Underfeed mould with feed stripper plate and stripper plate ejection (five-part mould).

It is a matter of choice whether the feed plate (first plate) is of one-part or two-part construction. Note that the SMU utilises two-part construction for the cavity plate which, in effect, becomes the feed plate of a multi-part mould. The backing plate of this construction can be dispensed with for this particular application and plate A used alone. When this alternative design is adopted, it is usual to call off for a deeper (oversize) plate. Note that the only disadvantage in using the two-part construction is that the sprue length is unnecessarily long.

For reasons of simplicity and uniformity the drawings illustrating this mould show the two-plate construction for the feed plate.

BASIC UNDERFEED MOULD. The only difference between a basic underfeed mould unit and a standard mould unit is that an extra plate is required, sandwiched between the two primary plates A and B. The primary plate assembly

DEVIATIONS FROM THE STANDARD MOULD UNIT

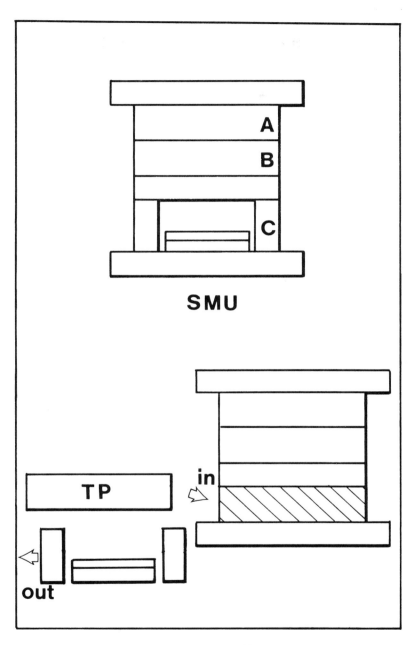

*Figure 14.13—Unscrewing mould unit (example 1).
SMU with support blocks and ejector assembly replaced by a
transmission plate (TP)*

STANDARD MOULD UNITS

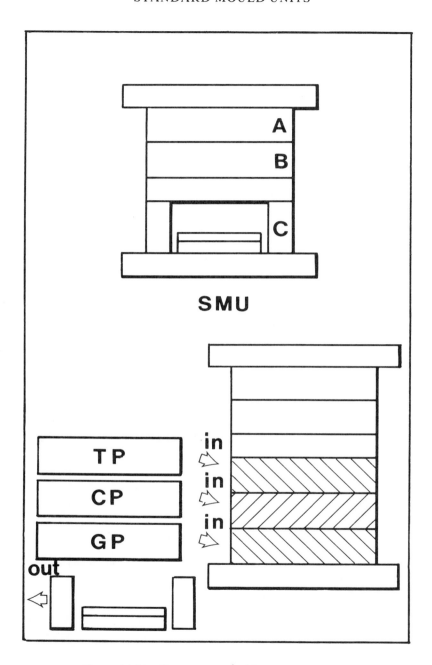

Figure 14.14—Unscrewing mould unit (example 2).
SMU with support blocks and ejector assembly replaced by three plates: transmission plate (TP), core-shaft plate (CP) and gear plate (GP)

DEVIATIONS FROM THE STANDARD MOULD UNIT

of the SMU becomes the feed plate in this design (see notes above). The additional plate is the 'floating cavity plate' (FCP), into which the cavity impression is incorporated by the mould-maker.

The floating cavity plate may be of one-plate or two-plate construction, depending upon whether the impression is machined directly into the plate or whether a backing plate for cavity inserts is required. The modifications necessary to the SMU in order that it may be used as a basic underfeed mould unit are illustrated in Figure 14.15. Note that the floating cavity plate in this design consists of one plate only.

UNDERFEED MOULD WITH FEED STRIPPER PLATE. The majority of underfeed designs necessitate the incorporation of an extra plate to the basic underfeed mould: the purpose of this plate is to strip the feed system from a secondary puller arrangement. A typical operational sequence for this type of mould is shown in Figure 12.9. The extra plate, in this system, is known as the 'feed stripper plate' (FSP) or 'runner stripper plate'.

There are two alternative feed stripper plate designs; in one, the plate is relatively narrow in width and is local to the feed system; in the second, the plate is the same size as the other mould plates of the unit.

For the first design a basic underfeed mould unit is used. The mould-maker machines a suitable recess in plate A to accommodate the feed stripper plate, as shown in Figure 14.16.

The modifications necessary to the SMU to achieve the second design are illustrated in Figure 14.17. In this case, between the feed plate A and the core plate B, two extra plates are required: the 'feed stripper plate' (FSP) and the 'floating cavity plate' (FCP).

BASIC UNDERFEED MOULD WITH STRIPPER PLATE EJECTION. This design combines the principles of the basic underfeed design and the stripper plate design. Thus from a mould unit viewpoint the only deviation required is the addition, between the primary plates A and B, of two extra plates, the 'floating cavity plate' (FCP) and the 'stripper plate' (SP).

Either the standard stripper plate design (as shown in Figure 14.18) or the basic stripper plate design can be adopted. In the latter case the ejector plate assembly is removed (see chain-dotted lines in Figure 14.18).

In combining these two categories of mould when using a standard mould unit as the structural base, a problem arises in the guide pillar/guide bush arrangement. Recall that the SMU incorporates the guide pillar in plate A and the guide bush in plate B, which is satisfactory for the underfeed design. However, as mentioned earlier, for the stripper plate design the guide pillar/guide bush arrangement needs to be reversed. Thus, ideally, two sets of guide pillars are required. It is therefore necessary to inform the mould unit manufacturer that extra guide pillars and guide bushes are required, and to specify their precise position.

UNDERFEED MOULD WITH FEED STRIPPER PLATE AND STRIPPER PLATE EJECTION. This design incorporates both a feed stripper plate for ejecting

STANDARD MOULD UNITS

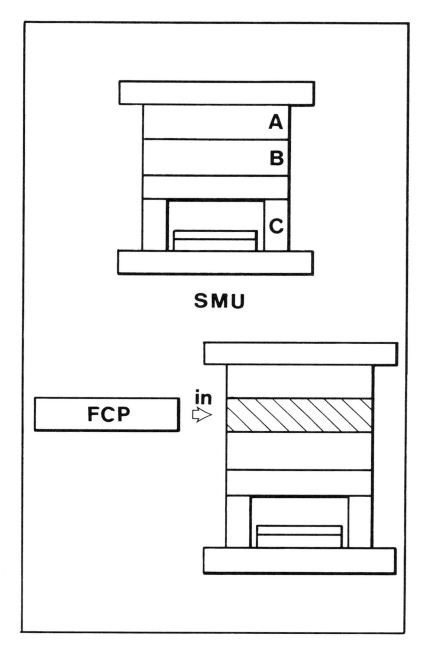

*Figure 14.15—Basic underfeed mould unit.
SMU with additional floating cavity plate (FCP) mounted between
plates A and B*

DEVIATIONS FROM THE STANDARD MOULD UNIT

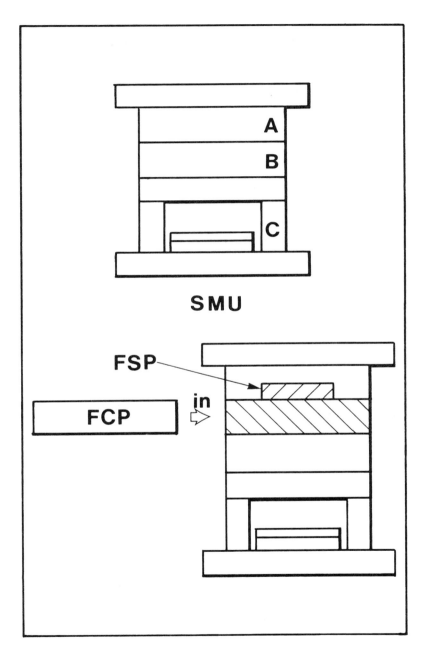

Figure 14.16—Underfeed mould unit incorporating feed stripper plate (Design 1). Basic underfeed mould unit (see Figure 14.15) with an additional feed

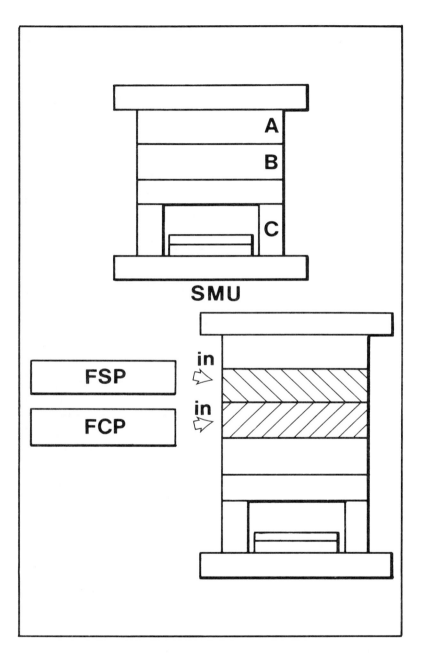

Figure 14.17—Underfeed mould unit incorporating a feed stripper plate (Design 2). SMU with two plates, feed plate (FP) and floating cavity plate (FCP), mounted between plates A *and* B

DEVIATIONS FROM THE STANDARD MOULD UNIT

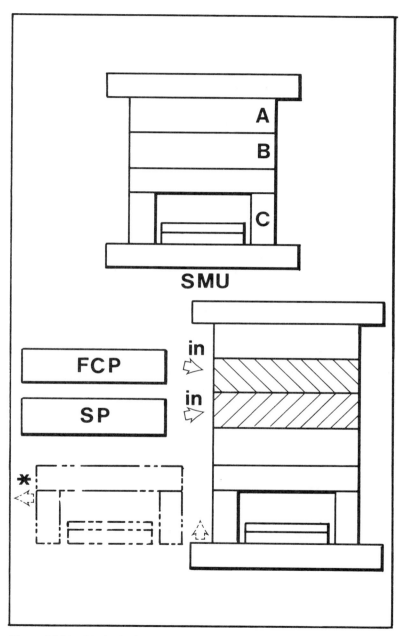

Figure 14.18—Underfeed mould unit incorporating stripper plate ejection. SMU with two plates, floating cavity plate (FCP) and stripper plate (SP), mounted between plates A and B (see text reference to basic stripper design)

STANDARD MOULD UNITS

the feed system, and a stripper plate for ejecting the mouldings, in addition to the basic concept of the underfeeding of the component. There are several possible variations which utilise the SMU as a structural base:

(a) Three plates may be incorporated between the primary plates A and B of the SMU, as shown in Figure 14.19, to act as the 'stripper plate' (SP), 'floating cavity plate' (FCP) and 'feed stripper plate' (FSP).

(b) Two plates may be incorporated between the primary plates A and B, to act as the stripper plate (SP) and floating cavity plate (FCP). The feed stripper plate (shown chain-dotted and denoted by an asterisk (1)) is incorporated by the mould-maker in a recess.

(c) A design may be adopted as either (a) or (b) but having the ejector grid and ejector plate assembly removed (shown chain-dotted and denoted in Figure 14.19 by an asterisk (2)). The mould can thus be operated on the basic stripper plate principle.

Note that, as in the preceding design, an extra set of guide pillars and guide bushes is required.

14.3.8 Hot-runner Mould. The principle of the hot-runner design is that a heated manifold block (hot-runner unit) or a heated plate (hot-runner plate) is incorporated within the structure of the mould to permit the plastics material to be maintained at an elevated temperature during its passage from the nozzle to the impression (see Chapter 13).

(i) The hot-runner unit is incorporated in a suitable recess in the cavity plate assembly. The recess can be provided in one of two ways when utilising a standard mould unit as the base.

(a) An ejector grid assembly consisting of support blocks and plate is added in front of plate A of the SMU, as shown in Figure 14.20. Note that for this application the addition is termed a 'hot-runner unit grid' (HRUG). A space is thereby provided behind the cavity plate (A) into which the hot-runner unit may be fitted.

(b) In the alternative design, illustrated in Figure 14.21, an extra plate is fitted between the front plate and the cavity plate A. This 'hot-runner grid plate' (HRP) is machined by the mould-maker to accommodate the hot-runner unit (see chain-dotted lines).

(ii) The hot-runner plate is added to the front of plate A of the SMU, as shown in Figure 14.21. The chain-dotted lines should be ignored for this application. The hot-runner plate is bored to provide a flow path for the melt and to accommodate heating elements.

14.3.9 Insulated Runner Design. From the mould unit viewpoint, this design is identical to the basic underfeed mould discussed in Section 14.3.7. An extra plate is incorporated between the primary plates A and B (Figure 14.15). The insulated runner is machined between plate A and the extra plate. In production, these two plates are fastened together to form what is basically a two-part mould. For further details of this design refer to Section 13.3.3.

DEVIATIONS FROM THE STANDARD MOULD UNIT

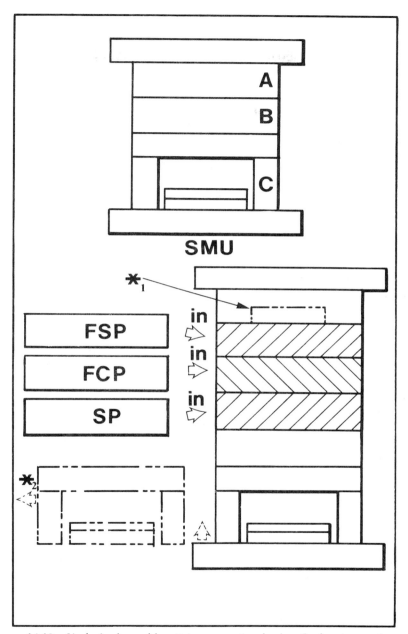

Figure 14.19—Underfeed mould unit incorporating both a feed stripper plate and stripper plate ejection.
SMU with three plates, feed stripper plate (FSP), floating cavity plate (FCP) and stripper plate (SP), mounted between plates A and B (see text reference to alternative designs)

STANDARD MOULD UNITS

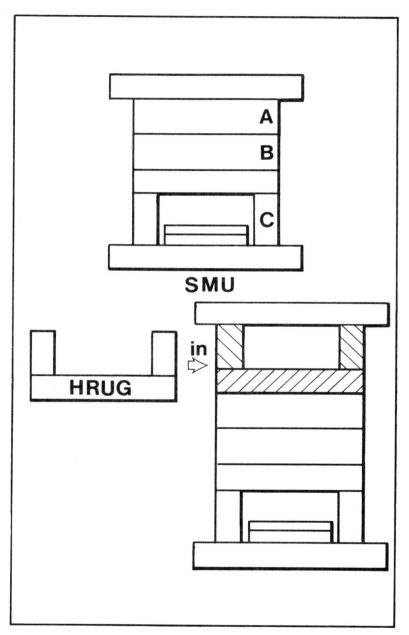

Figure 14.20—Hot-runner mould (Design 1)
SMU with additional hot-runner unit grid (HRUG) mounted in front of plate A

DEVIATIONS FROM THE STANDARD MOULD UNIT

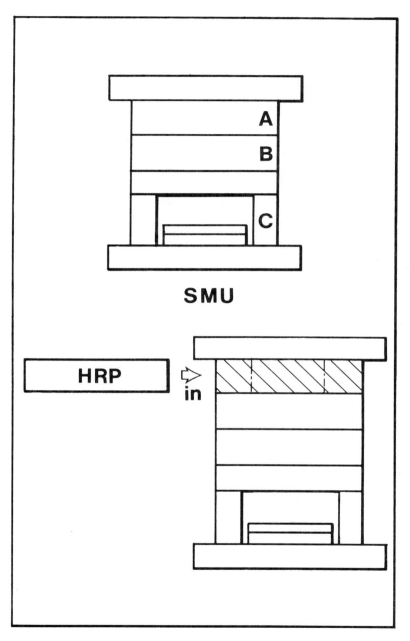

Figure 14.21—Hot-runner mould (Design 2)
SMU with additional plate mounted in front of plate A

STANDARD MOULD UNITS

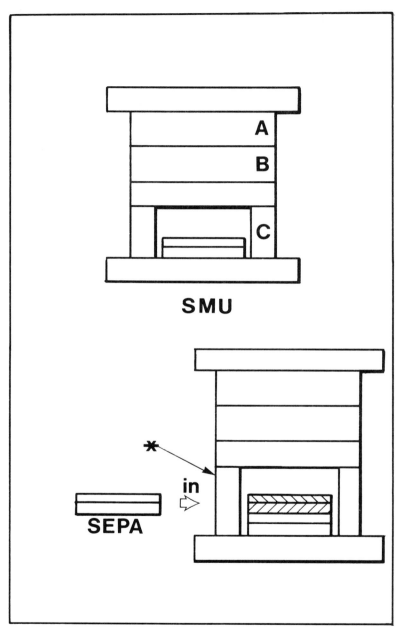

Figure 14.22—*Double ejection mould unit.*
SMU with additional ejector plate assembly mounted within the ejector grid.
*Note that extended support blocks may be required

COMPARATIVE TERMINOLOGY

14.3.10 Two-stage Ejection Mould Unit. To provide for two-stage ejector systems an additional ejector plate assembly, known as the 'secondary ejector plate assembly' (SEPA), is incorporated in the ejector space of the SMU. Note that extra-deep, non-standard support blocks may have to be specified (Figure 14.22).

14.4 COMPARATIVE TERMINOLOGY

The terminology used throughout this book for the various mould parts does not necessarily accord with that of either of the mould unit manufacturers. To avoid confusion, the following glossary is provided.

	DMS	DME	U-S
Back and front plate	Clamp plate	Top and bottom Clamping plate	Clamping plate
Backing plate	Backing plate	Support plate	Support plate
Cavity and core plate	Cavity plate	Mould plate	Cavity plate
Ejector plate	Ejector plate	Ejector plate	Ejector plate
Guide pillar	Guide pillar	Leader pin	Guide pillar
Guide bush	Guide bush	Leader pin bush	Guide bush
Push-back pin	Return pin	Return pin	—
Register ring	Location spigot	Location ring	Locating ring
Retaining plate	Ejector pin retaining plate	Ejector retaining plate	—
Sleeve	Sleeve ejector	Ejector sleeve	Sleeve ejector pin
Sprue puller (pin)	Sprue ejector pin	Sprue puller pin	—
Standard mould unit	Standard mould base unit	Standard mould base	Mould set
Stripper plate	Stripper plate	Stripper plate	Stripper plate
Sprue bush	Sprue bush	Sprue bushing	Sprue bush
Support blocks	Spacer blocks	Risers	Riser
Support pillars	Support pillars	Support pillars	Support pillars

15

Checking Mould Drawings

15.1 GENERAL

The final stage which must be undertaken when designing an injection mould is that of *checking* the completed drawing. During this most important phase the design is scrutinised in a critical manner with the objects of discovering and correcting errors which may have occurred. The errors may be in design, or in drawing or in dimensioning.

When the drawing is complete the draughtsman has the unique opportunity of being able to re-examine the overall design to ensure that the individual features combine to form a workable and efficient mould unit.

For the mould draughtsman, the process of checking is essential to ensure that errors incorporated in the drawing are not produced in the manufactured mould. Naturally if an error remains undetected, and if the mould is manufactured to specification, considerable expense and delay will be incurred in making the necessary corrections. It is sound practice to have the mould drawing checked by an independent designer (a checker) who, not having been connected with the design previously, will be able to check the drawing quite impartially. However, the fact that a drawing is subsequently scrutinised by a checker must not prevent the draughtsman from checking the design thoroughly himself.

15.1.1 Mould Checking Problems. One way by which a novice may increase his proficiency in the art of checking mould drawings is to take the opportunity, whenever it occurs, to check drawings produced by others. It is easier to observe an error in someone else's drawing than to note an error in one's own design. There is a tendency for beginners to have a 'blind spot' with regard to their own designs and to miss the most obvious errors during the checking phase.

Perhaps this book is being used on a self-teaching basis, in which case it is probable that the reader will not have an opportunity to peruse drawings produced by others. To overcome this difficulty a number of drawings have been prepared in which design errors have been included. Some of the errors will be obvious to the 'checker' but detection of the more subtle errors will require a very careful scrutiny of the design.

Naturally the absolute novice in the mould design field will not be able to check a specific mould design before acquiring a fundamental knowledge of the relevant design. There is little point in the beginner attempting to check a design for a 'splits' mould, for example, until the relevant chapter on splits moulds has been thoroughly understood.

PIN EJECTION MOULD

In general the drawings included in this chapter show cross-sectional views, and therefore design errors only are required. The exception is Figure 15.10, for which a plan view (Figure 15.8) and cross-section (Figure 15.9) are also given. In this case, in addition to design errors the correctness of the projection should also be checked.

Unless a comment is made to the contrary in the wording of the problem, all relevant component parts for a particular design should appear in the cross-sectional drawing. Therefore the 'checker' should note, as an error, any component part (e.g. a guide pillar) that is not shown in this view.

Every noted design error should be recorded, and the list subsequently checked against the recorded design errors. The reader is advised to refrain from reading through listed design errors before attempting to check the design.

15.2 PIN EJECTION MOULD

Problem: The drawing in Figure 15.1 gives a side-sectional view of a single-impression mould for a narrow box-shaped component. Pin ejection is adopted as the ejection method and the component is fed via a sprue gate. Check the drawing and note all design errors.

15.2.1 Design Errors
(1) A sprue bush has not been incorporated which means that the melt can seep between the mould insert (C5) and the bolster plate (C6).
(2) The register plate (R3) has not been located, and therefore alignment between the injection machine's nozzle and the sprue entry will be difficult to maintain.
(3) Guide pillars and guide bushes are omitted from this design, therefore alignment between the two mould halves is not controlled.
(4) A gap is shown between the rear side of the ejector plate (E5) and the front face of the back plate (B1). When the melt is injected into the impression, the material pressure on the ejector pins (E4) will force the ejector plate assembly to be moved rearwards.
(5) The ejector plate assembly (E7, E5, R4) is not supported or guided. An ejector rod bush could have been adopted for this purpose.
(6) The clearance hole for the ejector pins (E4) apparently extends completely through the mould plate (C12) and core insert (C11). The melt will creep through this annulus and flood the ejector space.
(7) The push-back pins (P9) are not shown to be in physical contact with the cavity plate (C6). They are therefore inoperative and do not perform any useful function.
(8) The flow-ways are bored on the joint line between the mould insert (C5) and the bolster plate (C6). Fluid leakage will occur.
(9) The temperature control flow-way system appears from the sectional view to be inadequate. However, for the flow-way system, the plan view (not given) is important as the precise configuration of the drilled holes will be observable.

CHECKING MOULD DRAWINGS

(10) The diameter of the sprue appears to be rather excessive in relation to the thickness of the moulding.
(11) The two mould halves have inadvertently been shown to be attached together by screws. (Note the letter 's' on the mating faces of inserts C11 and C5.)

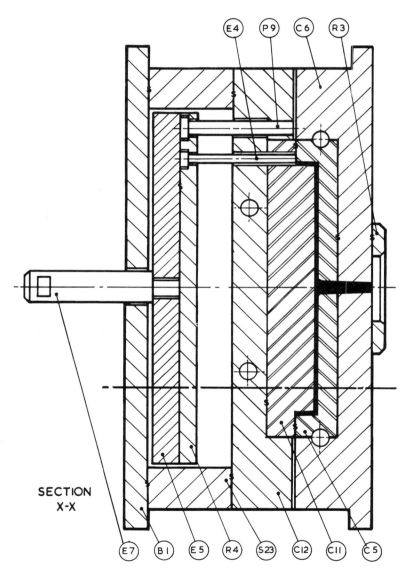

Figure 15.1—Mould checking problem 1. Pin ejection type mould. (Note that this drawing includes a number of design errors)

SLEEVE EJECTION MOULD

15.3 SLEEVE EJECTION MOULD

Problem: The drawing in Figure 15.2 gives a side-sectional view of a four-impression injection mould for a circular knob. The knob is subsurface gated and is sleeve ejected. Check the drawing and note all design errors.

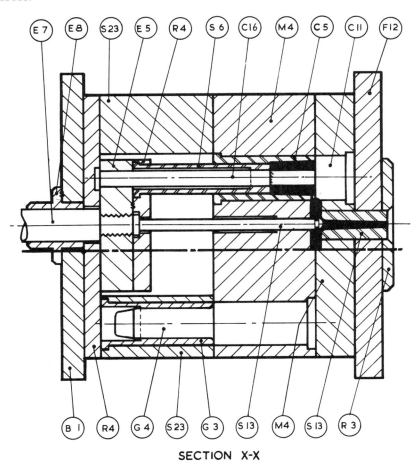

Figure 15.2—Mould checking problem 2. Sleeve ejection type mould. (Note that this drawing includes a number of design errors)

15.3.1 Design Errors
(1) The sprue bush (S13) design is unconventional in that it does not incorporate a shoulder. The nozzle seating force is thereby applied directly to the core plate.
(2) The insert (C5) is unsupported and will be displaced into the ejector space when the melt is injected into the impression.

(3) The guide pillar (G4) and guide bush (G3) assembly have been incorporated in the wrong plates. There is, therefore, no alignment facilities between the two mould halves.
(4) The various structural parts of the mould are not shown to be attached together by screws. Note that the letter 's' is often used to signify this feature on general arrangement drawings.
(5) The core, which forms the inside of the moulding, is shown to be a part of the insert (C11) mounted in the fixed mould plate (M4). Note that the short hatch lines between the cores indicates the position of the butting surface. This means that the moulding will remain in the fixed half when the mould is opened. From the ejection viewpoint the moulding should be retained in the moving half by extending the core pin (C16) to the parting surface.
(6) The core pin (C16) is not secured by the core retaining plate (R4) and could, therefore, be pulled from this plate by the movement of the sleeve ejector (S6).
(7) When the sleeve ejector (S6) is actuated, the inner locating diameter at the front end will lose contact with the core pin (C16). It is likely, therefore, that the extreme edge of the core pin will be damaged in operation.
(8) Radial clearance has not been provided for the sleeve ejector (S6) in the sleeve retaining plate (R4). The sleeve ejector, therefore, cannot 'float' to adjust for possible misalignments.
(9) The wrong type of sprue puller (S13) has been incorporated in this design. The mushroom-headed type (as shown) will not permit automatic extraction of the feed system.
(10) The sprue puller (S13) is extended so far forward as to restrict the melt flow into the runner system.
(11) No method for returning the ejector plate assembly to its rear moulding position is indicated (i.e. push-back pins or springs).
(12) The subsurface gate (incorporated in the cavity bush, C5) has an extremely shallow angle. There is the likelihood that the gate will become blocked with material.
(13) A coolant flow-way system has not been indicated, therefore the temperature of the mould cannot be controlled.
(14) The parting surface has not been relieved.

15.4 STRIPPER PLATE MOULD

Problem: The drawing in Figure 15.3 gives a side-sectional view of a two-impression injection mould for a deep circular box. Check this design based upon the premise that a ring-type runner and gate are suitable for feeding this component.

15.4.1 Design Errors
(1) Sprue bushes are subject to a high intensity of force by the injection unit. The corner of the sprue bush (S14), at the shoulder, should

STRIPPER PLATE MOULD

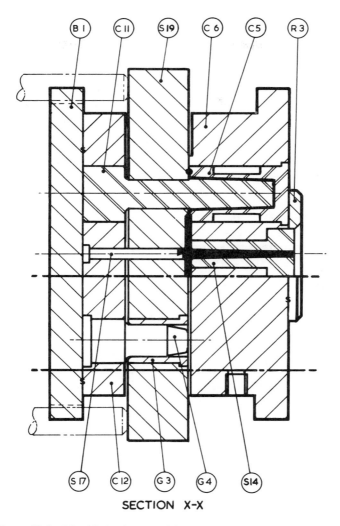

Figure 15.3—Mould checking problem 3. Stripper plate type mould. (Note that this drawing includes a number of design errors)

incorporate a suitable fillet to minimise stress concentrations in this region.
(2) The cavity insert (C5) is not supported by a backing plate. The cavity insert could be displaced by the applied injection force.
(3) A circular component which is fed by a ring runner and gate system will create moulding problems unless a suitable venting system is incorporated. In this design the base of the cavity should have been 'vented' by incorporating a local insert (with suitable vent slots) into the base of the cavity insert (C5).

CHECKING MOULD DRAWINGS

(4) A gap is indicated between the rear side of the stripper plate (S19) and the front side of the core plate (C12). Thus the mould is drawn (in effect) slightly opened. When closed, therefore, the core will abut onto the bottom of the cavity.
(5) The thickness of the steel shown between the cavity and the cooling annulus in the cavity insert (C5) is extremely small. This wall would probably collapse under the applied internal melt pressure.
(6) The guide pillars (G4) are shown to be in engagement with the guide bushes (G3) in the stripper plate only. No corresponding guide bush has been provided in the cavity plate (C6) and, hence, general alignment between the two mould halves is not controlled.
(7) In order to eject the moulding completely, the stripper plate (S19) must be moved off the guide pillar (G4)—i.e. the guide pillars are too short.
(8) The stripper plate (S19) does not incorporate a bush to provide a sliding surface for the core insert (C11). This omission complicates the manufacture of the mould and also creates maintenance problems.
(9) The shank of the core insert (C11) appears to be smaller in diameter than the 'core' itself. Note that the shank should be larger in diameter than the core to prevent the stripper plate from scoring the side walls during ejection. The mould could be neither manufactured nor operated as drawn.
(10) Some designers prefer to incorporate stop-bolts in their design to arrest, positively, the movement of the stripper plate.
(11) A Z-type sprue puller (S17) is impracticable for stripper plate designs. The feed system cannot be removed from the undercut.
(12) No method for controlling the temperature of either the stripper plate (S19) or the core plate (C12) has been indicated.
(13) O-rings have not been incorporated in the cavity insert (C5) design and seepage of fluid could, therefore, be a major problem in production.
(14) The main runner appears to be rather small in relation to the diameter of ring runner which it feeds.

15.5 SPLITS TYPE MOULD

Problem: The drawing shown in Figure 15.4 gives a side-sectional view of a single-impression mould for an egg-cup. Check the drawing and note all design errors.

15.5.1 Design Errors
(1) The base of the moulding is formed by the front face of the sprue bush (S14). In these circumstances it is desirable to support the sprue bush to prevent it from being displaced by the applied injection force within the impression.
(2) The guide bush (G3) is mounted in the guide strip (G6) instead of the core plate (C12). Direct alignment between the two mould halves is,

SPLITS TYPE MOULD

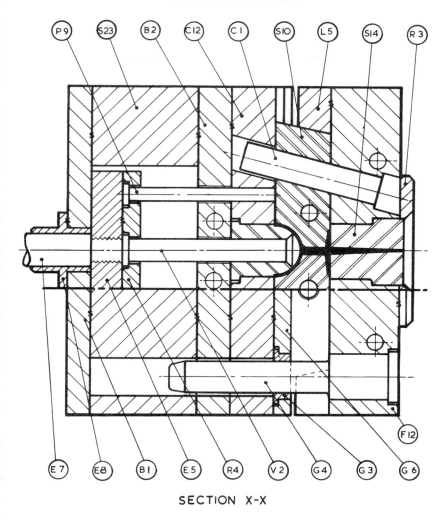

SECTION X-X

Figure 15.4—Mould checking problem 4. Splits type mould. (Note that this drawing includes a number of design errors)

therefore, not achieved. Note that this error also results in the guide bush having a very unfavourable diameter to depth ratio.
(3) The finger cam (C1) does not incorporate a lead-in at its front end. In addition, the finger cam appears to be a sliding fit in the complementary hole machined in the split. The slightest movement of the split, when in the open position, could result in a serious mishap occurring when the mould is subsequently closed.
(4) Neither a spring detent nor a spring-loaded mechanism has been incorporated in this design to hold the splits in the 'actuated' position.

CHECKING MOULD DRAWINGS

(5) A positive stop to limit the outward movement of the split has not been incorporated in this design.

(6) The opening movement of the split halves appears to be just sufficient to permit the moulding to be extracted. Always check the splits opening movement very carefully.

(7) During the closing stroke, the splits (S10) will foul the ejector valve (V2)—i.e. unless the injection machine has facilities for withdrawing the ejector system while the mould is open.

(8) Note that the clearance hole in the core plate (C12) follows the same angle as that of the finger cam (C1). This is incorrect as the finger cam cannot be withdrawn from the hole in a horizontal plane.

(9) The locking heels (L5) are shown to be bolted directly to the mould plate (F12). These locking heels are likely to be displaced by the applied injection force acting within the impression. If this design is to be adopted, it is usual for the locking heel to incorporate a lozenge-shaped projection on the underside which fits into a complementary shaped slot in the mould plate.

(10) Wear plates have not been incorporated on the locking heels (L5). These wear plates are required for maintenance and for adjustment purposes.

(11) The positioning of the push-back pins (P9) is impracticable. In the position shown, they will foul the splits (S10) when the ejector assembly is operated. It is normal practice to position the push-back pins so that they pass through the guide strips (G6).

(12) The valve ejector (V2), as designed, cannot be assembled in the mould. Note that the element has a projection at each end which will not pass through the holes bored in the retaining plate (R4), backing plate (B2) and core insert (C11), respectively.

(13) The coolant flow-way in the fixed mould plate (F12) is shown to be relatively close to the shank of the guide pillar. Some fluid seepage is possible in this region.

15.6 SIDE CORE TYPE MOULD

Problem: The drawing shown in Figure 15.5 is a side-sectional view of a single-impression mould for a box-shaped component, which incorporates an internally operated side core assembly. The side core is required to form the transverse hole in the component. Check this mould design and note as many errors as possible.

15.6.1 Design Errors

(1) This is an impracticable design from the side core design viewpoint. As the mould opens, that part of the side core element (S5) which passes through the hole in the fixed mould plate (C6) will fracture. Neglect this impracticability when considering other design errors.

(2) Sprue bushes are subject to high forces by the injection unit. The ring of steel situated in the fixed mould plate (C6) between the sprue bush

SIDE CORE TYPE MOULD

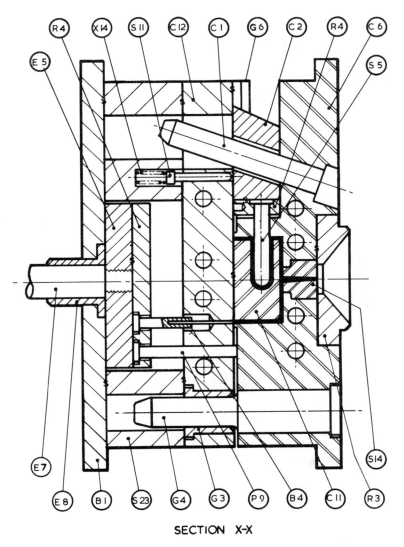

Figure 15.5—Mould checking problem 5. Side core type mould. (Note that this drawing includes a number of design errors)

shoulder (S14) and the impression is slender and could easily be distorted by this applied force.
(3) The core insert (C11) is shown to be bolted directly onto the front face of the core plate (C12). In addition to the likelihood that plastics material will creep under the insert, there is the probability that the core insert will be laterally displaced by the pressurised melt.
(4) A stop has not been provided to arrest positively the forward movement of the side core assembly (C2/R4). Thus the side core

CHECKING MOULD DRAWINGS

element (S5) could foul the base of the recess machined in the core insert (C11).
(5) The withdrawal movement of the side core assembly (C2/R4) is insufficient to extract the entire side core element (S5) from the moulding.
(6) A spring detent (S11) has been provided to nominally hold the side core assembly (C2/R4) in the 'withdrawn' position while the mould is open. The accommodating recess in the side core retaining plate (R4) is too deep, and will result in the side core assembly being 'locked' in the withdrawn position by the spring-loaded plunger. As the mould is closed the plunger will be sheared and the finger cam could possibly be damaged.
(7) A locking heel has not been provided in the design; therefore, the melt pressure will displace the side core assembly outwards. The amount of movement is dependent upon the clearance provided by the angled hole in the carriage (C2).
(8) Blade-type ejection (B4) is not appropriate for ejecting this type of moulding. Pin ejection should have been adopted.
(9) The 'ejector space', which is situated directly behind the core plate (C12), does not allow for sufficient ejector plate assembly (E5/R4) movement to completely extract the moulding from the core insert (C11).
(10) The method of attaching the blade to the pin in the ejector blade assembly (B4) is not shown.
(11) The ejector rod bush (E8) is not attached to the back plate (B1). This bush could be displaced by the movement of the ejector rod (E7).
(12) The land provided round the impression is too narrow. In addition, no extra parting surface support appears to have been provided. It is likely, therefore, that the locking force of the machine will distort this narrow land and create moulding extraction problems.
(13) The lower coolant flow-way situated in the core plate (C12) is too close to the push-back pin hole. Fluid leakage will occur.
(14) Finally, the design is quite impracticable due to component design. The internal projection formed in the core insert will not permit the moulding to be extracted from the core.

15.7 UNDERFEED TYPE MOULD

Problem: The drawing in Figure 15.6 gives a side-sectional view of an eight-impression mould for a pen cap. Check this design on the premise that an underfeed mould is required and that the impressions are to be positioned on a pitch-circle diameter.
(1) The problem specifies that an eight-impression mould is required and that the impressions are to be positioned on a PCD. It would be impossible to achieve this number of impressions with the size of cavity insert and pitch circle diameter as drawn.
(2) The cavity insert (C5) is not secured in the floating cavity plate (C6), and this insert could be displaced as the mould is opened.

UNDERFEED TYPE MOULD

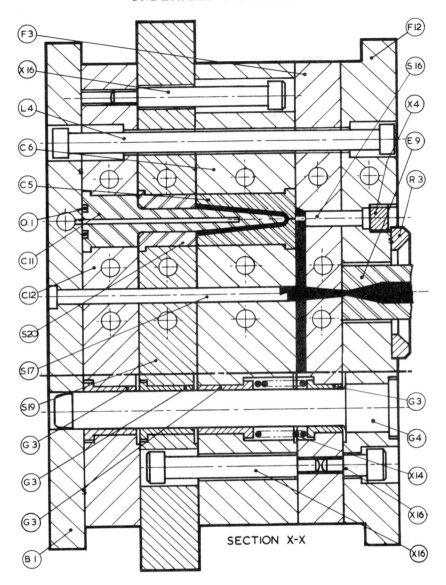

Figure 15.6—Mould checking problem 6. Underfeed type mould. (Note that this drawing includes a number of design errors)

(3) The small diameter portion of the feed puller (S16) is too long as it effectively blocks the gate entry.
(4) The shape of the component does not justify the need for a feed stripper plate design. A more simple design, similar to that shown in Figure 12.3, would have been suitable.

455

CHECKING MOULD DRAWINGS

(5) The runner should not be incorporated in the feed stripper plate (F3). As drawn, the feed stripper plate cannot perform its required function of stripping the runner from the feed pullers (S16).

(6) The length of travel permitted to the feed stripper plate (F3) by the shoulder-headed screws (X16) is insufficient to strip the feed system from the feed pullers. (This point assumed a corrected runner system—see (5).)

(7) A Z-type sprue puller (S17) is impracticable for an underfeed design.

(8) The sprue puller (S17) extends through the back plate (B1), and the puller is likely to be displaced through the aperture of the injection machine platen when the plastic melt is injected into the mould.

(9) The length of travel permitted to the floating cavity plate (C6) by the shoulder-headed screws (X16) is insufficient to permit the feed system to be extracted from the mould.

(10) A sprue bush has not been incorporated in this design. The sprue region of the mould is subject to high, localised forces by the injection unit. In this design the nozzle (E9) abuts directly onto the feed stripper plate (F3) and the region adjacent to the sprue could be damaged. A replaceable sprue bush would, therefore, be desirable. Alternatively, an extended nozzle could have been used—see Figure 12.10.

(11) The radial clearance between the nozzle (E9) and the aperture in the front plate (F12) is too small. As drawn, a considerable amount of heat will be transmitted from the nozzle to the front plate.

(12) The length of travel permitted to the stripper plate (S19) by the shoulder-headed screws (X16) is insufficient to allow for the complete ejection of the moulding.

(13) On this type of mould it is usual to have two sets of guide pillars. Only one set has been indicated in this design. Thus, when the mould is fully open, the ejector stripper plate (S19) is not supported and it is prevented from falling only by the shoulder-headed screws (X16) and length bolts (L4).

(14) The hole in the back plate (B1) to accommodate the front end of the guide pillar (G4) should be larger to provide a clearance for this component.

(15) The length bolt (L4) is incorporated for the purpose of pulling the floating cavity plate (C6) to the open position should the initial impetus given to the plate by the springs be insufficient. In this drawing the length bolt is shown to interconnect the front plate (F12) and the core plate (C12), instead of the floating cavity plate (C6) and the core plate. Thus, when the mould is opened either the mould will be pulled off the platen or the length bolts will break.

(16) The above-mentioned length bolt (L14) appears to have two heads. The bolt cannot, therefore, be assembled in the mould as drawn.

(17) The fluid circulation system consisting of a number of drilled holes is not an efficient method of controlling the temperature of the mould plate for this type of component. An annulus method of circulation is preferred.

HOT-RUNNER TYPE MOULD

(18) It is difficult to see how the coolant holes, which are positioned on either side of the centre-line, could have been drilled and yet miss passing through the cavity inserts (C5), core inserts (C11), stripper bushes (S20), and feed puller (S16). Note that the eight impressions are on a pitch-circle diameter.

(19) The cooling system for the core insert consists of a slender drilled hole which is fed via a hole bored in the back plate. This method is ineffective. The hole is too small and no method of ensuring that the fluid passes up the hole has been indicated in the design.

(20) The stripper plate bush (S20) is incorrectly fitted. The shoulder should be at the opposite end to prevent the bush being displaced during the ejection phase.

15.8 HOT-RUNNER TYPE MOULD

Problem: Figure 15.7 shows a side-sectional view of an elementary hot-runner unit design for a two-impression mould. The component, a beaker, incorporates a bead at the open end which creates an 'undercut' in the draw direction. Check the design and note design errors based upon the premise that a hot-runner unit had been specified in order to achieve a minimal gate witness mark at the base of the component.

15.8.1 Design Errors. The design errors (1)–(14) relate specifically to the hot-runner unit design. Many of these errors are the result of insufficient attention being given to the design of the melt flow-path. In an efficient hot-runner unit design it is essential that the flow-path be as streamlined as possible, without crevices, recesses, chambers, etc., into which the melt could accumulate and create colour change and degradation problems during production.

(1) There should not be a gap between the bottom of the sprue bush (S14) and the recess in the hot-runner manifold block (M1).

(2) The central hole should be drilled completely through the manifold block to facilitate streamlining and cleaning functions.

(3) The central hole (as (2) above) should not extend beyond the horizontal exit hole which feeds the secondary nozzle (S3). Shaped plugs must be fitted into the drilling to avoid material-trap chambers being formed.

(4) The diameter of the horizontal exit holes of both the manifold block (M1) and the secondary nozzle (S3) should be the same. By adopting different diameters, crevices are formed which create hold-up points.

(5) The secondary nozzles (S3) should not seat on the face of the manifold block (M1). During manufacture of these parts a gap could inadvertently occur between the bottom of the secondary nozzle and the accommodation recess in the manifold block.

(6) The cartridge heating elements (C3) are incorporated in the manifold block in the wrong direction. For narrow manifold blocks (such as this one for a two-impression mould) it is preferable to incorporate the heaters in the same direction as that of the main manifold.

CHECKING MOULD DRAWINGS

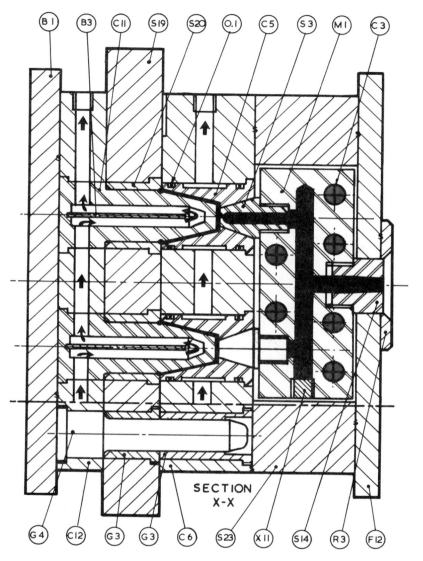

Figure 15.7—Mould checking problem 7. Hot-runner type mould. (Note that this drawing includes a number of design errors)

(7) No thermocouples have been shown in the hot-runner unit design for temperature control purposes.
(8) The insulating air gap, which surrounds the hot-runner unit, is too small. The normal minimum is about 16 mm ($\frac{5}{8}$ in).
(9) The two secondary nozzles (S3) are shown seated in the cavity insert recesses. Care must be exercised in such designs to ensure that the

STEPPED PARTING SURFACE MOULD DESIGN

expansion of the manifold block (M1) will not subject the secondary nozzles to undue strain.

(10) The permanent contact between the secondary nozzles (S3) and the cavity insert (C5) will cause part of the impression to function at an elevated temperature. Without close temperature control this could result in the distortion of the moulded product, and an extension of the moulding cycle.

(11) No support has been provided for the hot-runner manifold block (M1) directly opposite the sprue bush (R3). It is normal to incorporate a support pad in this position.

(12) No method of securing the hot-runner unit in position has been shown.

(13) No support has been provided for the manifold block (M1) directly opposite the secondary nozzles. It is normal practice to incorporate support pads in this position.

(14) A radial gap should have been provided in both the back plate and the register ring around the sprue bush to minimise the transmission of heat to these plates.

(15) The cavity inserts (C5) are not supported by a backing plate.

(16) The extraction of the moulded bead at the open end of the component cannot be accomplished with the stripper plate design as shown. Mould designs for internal undercut type components which can be 'jumped off' are based upon the principle that the moulding can expand during ejection. As this design shows that the bead is formed partially in the stripper plate, expansion during ejection is not possible.

(17) The O-ring (O1) grooves situated close to the parting surface are too deep with respect to the cavity. Failure is likely.

(18) The coolant flow-way passes from the core plate (C12) into the core insert (C11) without any method of sealing being incorporated to prevent leakage.

(19) No method has been incorporated to prevent the core insert (C11) from being incorrectly assembled. A locating pin is necessary to ensure that the flow-ways are in line.

(20) No method of controlling the temperature of the stripper plate (S19) has been shown.

(21) The horizontal coolant flow-way shown in the core insert (C11) is blind at both ends. This hole cannot be drilled, nor can the blade be fitted.

15.9 STEPPED PARTING SURFACE MOULD DESIGN

Problem: The drawings shown in Figures 15.8 and 15.9 are a plan and cross-sectional view of a four-impression injection mould for a channel bracket (Figure 15.10). Check this design with respect to both design and drawing errors.

CHECKING MOULD DRAWINGS

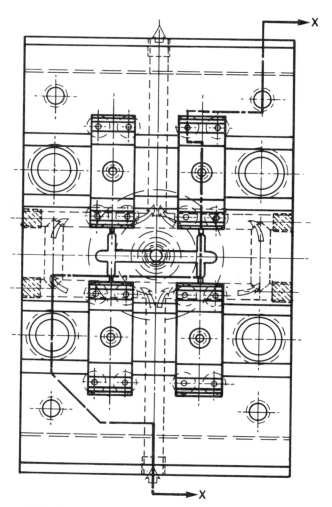

Figure 15.8—Mould checking problem 8 (plan view). Stepped parting surface type mould. (Check this drawing with respect to both design and drawing errors)

15.9.1 Design Errors
(1) The minor diameter of the cold slug well is less than the diameter of the sprue pin (S16). The sprue pin, therefore, cannot function.
(2) The ejector space (between the rear face of the cavity plate (C6) and the front face of the retaining plate (R4)) is excessive for ejecting this component.
(3) The ejector pins (E4) are rather slender. Stepped ejector pins should have been used.
(4) The ejector pin (E4) is too close to the impression wall.

STEPPED PARTING SURFACE MOULD DESIGN

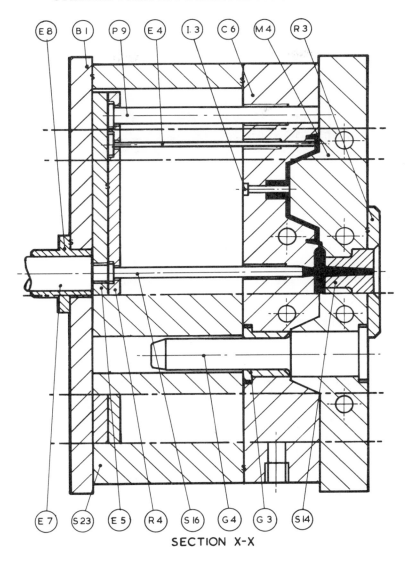

Figure 15.9—Mould checking problem 8. (Cross-sectional view)

(5) The head of the ejector pin is too large in diameter.
(6) The ejector plate (E5) depth is too thin. Deflection of this plate is likely during ejection.
(7) The ejector plate is too large in plan view. The push-back pins (P9) could have been positioned closer to the impression.
(8) Sleeve type ejection should have been incorporated to ensure positive ejection of the boss.

CHECKING MOULD DRAWINGS

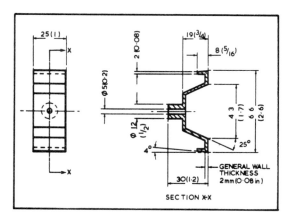

Figure 15.10—Channel bracket for which a two-impression mould is required (refer to Figures 15.8 and 15.9)

(9) The local insert, core pin (I3), which forms the hole in the boss, is unsupported.

(10) The support blocks (S23) are positioned too far apart. (Refer to the plan view; the cross-sectional view is misleading.) A major part of the cavity plate (C6) is unsupported and this plate could be deflected by the applied injection force.

15.9.2 Drawing and Projection Errors

(1) The section cutting plane does not pass through the centre of the mould, therefore such items as sprue pin, sprue bush, ejector rod, ejector rod bush, register plate, etc., should not appear in section.

(2) While a rectangular side gate machined into the moving half (C6) is indicated in the plan view, the sectional view shows an overlap gate machined into the fixed mould plate (M4).

(3) The side-sectional view indicates a support block (S23) situated directly behind the guide bush (G3). In plan view, however, the only support for the mould plate (C6) is at each end. See hidden detail lines in plan view. The sectional view is, therefore, incorrect.

(4) The section cutting plane line at the bottom of the plan view passes through the ejector plate assembly (E5/R4) and the support block (S23). This feature is shown incorrectly in the side-sectional view.

(5) The diameter of the runner in the plan and sectional views do not match.

(6) The direction of the hatching for the ejector plate (E5) and retaining plate (R4) is different in the two planes indicated in the sectional view.

(7) Too many hidden detail circles are shown about the main horizontal and vertical centre-lines in plan view.

(8) The lower horizontal flow-way hole machined in the moving mould plate (C6) is not on the same centre-line in the two views.

STEPPED PARTING SURFACE MOULD DESIGN

(9) The ejector pins (E4), which are close to the horizontal centre-line in plan view, pass through the central coolant flow-way system.

(10) The fluid circulation system incorporated in the cavity plate (C6) is inefficient. If one side of the circuit becomes blocked, the coolant will bypass this side.

(11) The two short vertical flow-way holes (which connect the two main horizontal flow-way holes in plan view) cannot be drilled as shown.

(12) The two main horizontal flow-way holes are not shown to be on the same centre-line as the vertical flow-way hole in the sectional view.

(13) The guide pillars, as positioned, enter the ejector space (see projection error (3)).

(14) Parting surface relief is not indicated on either plan or cross-sectional view. Parting surface relief is particularly important with respect to stepped parting surfaces so that the time spent in 'bedding' the two mould surfaces together is minimised.

(15) The clamping ledge on both mould halves is too narrow.

(16) A guide pillar (G4) has not been offset to avoid the two mould halves from being assembled incorrectly.

(17) The cavity could have been incorporated in the design as an insert rather than an integer mould plate, as shown. Both methods have advantages and disadvantages and the final decision will depend upon the toolmaking facilities available.

16

Worked Examples of Simple Injection Moulds

16.1 GENERAL

This chapter is included primarily to give the novice some simple injection mould designs as a reference on which to base his own designs. In all cases a general cross-sectional drawing is given together with a plan view of the moving half. To help the beginner understand the intricacies of the stepped parting surface, an isometric view of the moving half is also given.

In the following designs the three basic types of ejector system are included, namely the pin, sleeve and stripper plate designs.

16.2 EXAMPLE 1 (PIN EJECTION TYPE MOULD)

16.2.1 Problem. Design a two-impression injection mould to produce a pin box, as shown in Figure 16.1. The box is to be manufactured in high-impact polystyrene. The drawing should include a plan view of the moving half and a side-sectional view of both mould halves.

Before starting the design, however, decide upon the main features that you intend to incorporate. The following design notes indicate one approach.

Drawing references: Plan view of the moving half (Figure 7.14); side-sectional view of both mould halves (Figure 7.15); isometric view of the moving half for reference purposes (Figure 7.16). A detailed procedure for designing this mould is given in Chapter 7.

16.2.2 Design Notes

(1) *General considerations.* The component is a rectangular box which is to be moulded in high-impact polystyrene. This is a simple design problem in which pin ejection is adopted as the ejection method. The cavities and cores are positioned in the fixed and moving halves, respectively.

(2) *Parting surface.* The parting line of the component is on one plane and is situated at the open end of the pin box. Likewise, the mould's parting surface is on one plane.

(3) *Impression layout.* The impressions for the mouldings are positioned on the vertical centre-line of the mould and are equidistant from the sprue.

EXAMPLE 1 (PIN EJECTION TYPE MOULD)

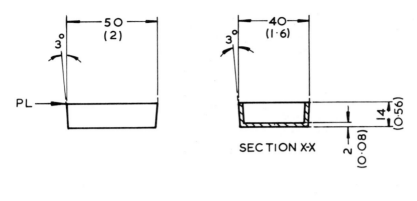

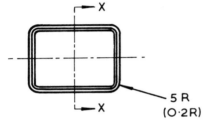

Figure 16.1—Example 1: pin box for which a two-impression mould is required

(4) *Mould plate layout.* As there are two impressions, an insert/bolster mould plate design is adopted for both mould halves. The inserts are fitted into a large rectangular recess machined into the bolster plate (B5). The inserts (in both mould plates) are separated by the bridge piece (B6).

(5) *Ejection.* Ejector pins are the most practicable form of ejection for this type of component. The number of ejector pins (E4) required depends upon the size of component, and for this particular box three ejector pins are adequate. From the novice's viewpoint the cross-sectional drawing is simplified if one of the ejector pins is positioned on the vertical centre-line, as shown.

(6) *Ejector assembly and ejector grid.* A conventional ejector plate assembly is adopted for this design. Because of the impression layout it is convenient to adopt a long, narrow, rectangular ejector plate (E5). This permits the associated support blocks (S23) to be positioned relatively close together and thereby give maximum support to the core plate. Additional support is obtained by incorporating end support blocks.

Four push-back pins (P9) are incorporated and the distance between each pair (horizontally) is kept to a minimum. The ejector assembly is guided and supported by the ejector rod (E7), which passes through the ejector rod bush (E8); this latter component being fitted in the back plate.

WORKED EXAMPLES OF SIMPLE INJECTION MOULDS

(7) *Feed system.* The two impressions are fed via the sprue, runner and gate as shown in cross-section. A fully round runner is adopted in conjunction with a rectangular side gate. This design is an example of the balanced feeding technique as the length of the flow-path to each impression is the same.

(8) *Fluid circulation system.* For simplicity, a basic cooling system consisting of two drillings only is chosen for this design. The drillings pass through the bolster (B5) on each side of the inserts in both mould plates. The two drillings could be connected externally at the top of the mould if required to provide a continuous flow-path for the coolant. The ends of the drillings are threaded to accommodate standard adaptors.

(9) *Mould alignment.* Guide pillars (G4) are incorporated in the moving half and guide bushes (G3) in the fixed half. This arrangement allows the guide pillars to act as a means of protecting the cores when the mould is removed from the machine. Care must be exercised in the positioning of the pillars, however, to ensure that the free fall of the mouldings is not obstructed during the ejection phase.

(10) *Sprue pulling.* An undercut type of cold slug well, with associated sprue pin (S16), is provided directly opposite the sprue.

(11) *Parting surface relief.* The surface contact between the two mould faces is restricted to the small area adjacent to each impression, plus four rectangular areas situated one at each corner of the mould plate. This feature is clearly seen in the isometric view (Figure 7.16).

16.3 EXAMPLE 2 (SLEEVE EJECTION)

16.3.1 Problem. Design a four-impression injection mould to produce the knob shown in Figure 16.2. The knob is to be manufactured in CAB. The drawing should include a plan view of the moving half and a front sectional view of both mould halves.

Drawing references: Plan view of the moving half (Figure 16.3); side-sectional view of both mould halves (Figure 16.4); isometric view of the moving half for reference purposes (Figure 16.5).

16.3.2 Design Notes

(1) *General considerations.* The component is cylindrical in shape and the beginner should notice particularly where the parting line occurs. This is indicated in Figure 16.2 by the letters PL. The location of the parting surface means that a major part of the impression will be incorporated in one mould half only, and that the moulding wall is entrapped between the cavity and core in that half. This feature invariably necessitates incorporating a very efficient ejector system in the design. For this particular component sleeve ejection is adopted.

(2) *Parting surface.* The parting line on the component occurs on the line round the position of maximum dimension when viewed in plan. Therefore the parting line occurs on the ridge designated PL in Figure

EXAMPLE 2 (SLEEVE EJECTION)

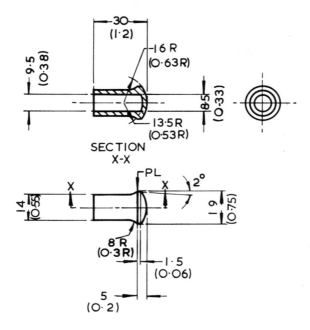

Figure 16.2—Example 2: cylindrical knob for which a four-impression mould is required

16.2. The parting line is on one plane, and likewise the mould's parting surface is on one plane.
(3) *Impression layout.* The impressions are positioned on the horizontal and vertical centre-lines at an equal distance from the sprue.
(4) *Mould plate design.* As there are four impressions, an insert/bolster design is adopted for both mould plates. As the component is cylindrical in shape, cylindrical flange type inserts (C5) can be used. These are fitted into a frame type bolster (B5). The inserts are pinned (X9) to ensure correct alignment is maintained between the runner machined into the bolster and the runner machined into the insert.
(5) *Ejection.* For the reasons stated above, sleeve type ejection is adopted for this design. The sleeve (S6) is mounted in a conventional ejector plate assembly (E5/R4a). The core pin (C10) which passes through the centre of the sleeve is retained by the core retaining plate (R4).
(6) *Ejector grid and ejector plate assembly.* Because of the impression layout it is convenient to adopt a narrow ejector plate assembly (E5/R4a) for this design. This permits the associated support block (S23) to be positioned relatively close together and thereby give maximum support to the mould plate. (Refer to the dotted lines shown in Figure 16.3.) It is advantageous to delay deciding upon the horizontal position of the push-back pins (P9) in this design until a

WORKED EXAMPLES OF SIMPLE INJECTION MOULDS

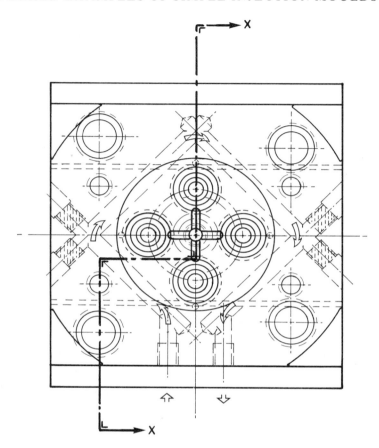

Figure 16.3—Plan view of moving half of four-impression injection mould for cylindrical knob shown in Figure 16.2

suitable configuration for the coolant flow-way system has been finalised.

The ejector plate assembly is guided and supported by the ejector rod (E7) which passes through the ejector rod bush (E8). This latter component is securely fitted in the back plate (B1).

(7) *Feed system.* The four impressions are fed via the sprue, runner and gate as shown in cross-section (Figure 16.4). A fully round runner is adopted in conjunction with a rectangular type side gate. Note that the gates are machined into the fixed mould plate, and for this reason they are not shown in plan view (Figure 16.3).

The rectangular type gate is not ideal for feeding cylindrical type components, and is adopted here for reasons of simplicity. It should be noted, however, that this side feed will give rise to a weld line where the two fronts of the melt meet opposite the gate.

EXAMPLE 2 (SLEEVE EJECTION)

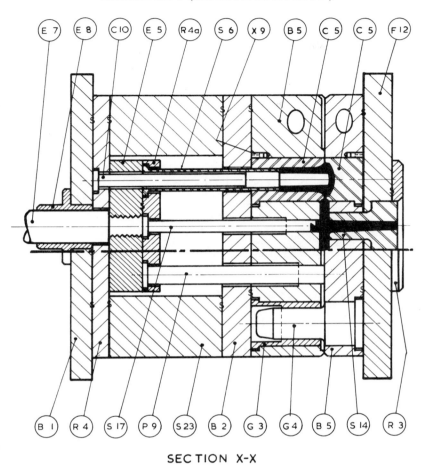

SECTION X-X

Figure 16.4—Cross-sectional view of both mould halves of injection mould for cylindrical knob shown in Figure 16.2

(8) *Fluid circulation system.* For reasons of simplicity, a basic cooling circuit consisting of six drillings is adopted for this design. An identical circuit is used for both mould halves. As shown by the dotted lines in the plan view (Figure 16.3), the four main holes are drilled at 45° to the major centre-lines. This diamond-shaped configuration ensures that the coolant fluid flows fairly close to each insert. The inlet and outlet holes interconnect with the above circuit and are drilled from the bottom of the respective mould plates (B5).

Note that the angled holes are counterbored to facilitate the tapping operation and for the fitting of plugs. (Plugs, by convention, are shown as dotted cross-hatched lines in plan view.)

WORKED EXAMPLES OF SIMPLE INJECTION MOULDS

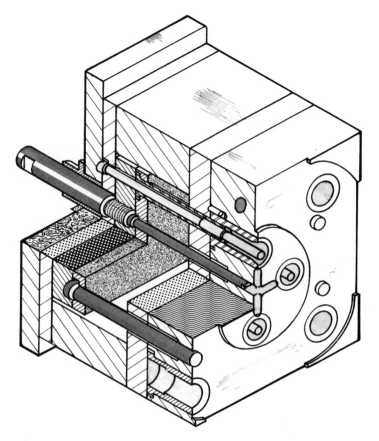

Figure 16.5—Isometric view of the moving half of injection mould for cylindrical knob shown in Figure 16.2

(9) *Mould alignment.* Guide pillars (G4) are incorporated in the fixed half and guide bushes (G3) in the moving half. This arrangement ensures that the guide pillars do not obstruct the free fall of mouldings during ejection. (Refer to the corresponding design note in Section 16.2.)

(10) *Sprue pulling.* A Z-type sprue puller (S17) is provided directly opposite the sprue. The undercut created by the puller ensures that the sprue is pulled directly the mould is opened.

(11) *Parting surface relief.* The surface contact between the two halves of the mould at the parting surface is restricted to the circular area which encompasses the four impressions (see isometric view, Figure 16.5), plus four fillet-shaped areas situated one at each corner of the moving mould plate. Note that this type of relief can readily be machined on a lathe.

EXAMPLE 3 (STRIPPER PLATE EJECTION)

16.4 EXAMPLE 3 (STRIPPER PLATE EJECTION)

16.4.1 Problem. Design a four-impression injection mould to produce the threaded cap shown in Figure 16.6. The form of the thread is such that thread jumping is a practical proposition. The threaded cap is to be

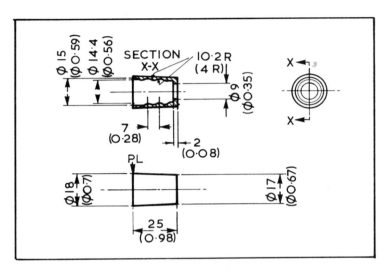

Figure 16.6—Example 3: threaded cap for which a four-impression mould is required

produced in L.D. polyethylene. The drawing should include a plan view of the moving half, and a side-sectional view of both mould halves.

Drawing references: Plan view of the moving half (Figure 16.7); side-sectional view of both mould halves (Figure 16.8); isometric view of the moving half for reference purposes (Figure 16.9).

16.4.2 Design Notes

(1) *General considerations.* The component to be moulded is basically a circular box, and as the problem specifies that thread jumping is permissible, a stripper plate type ejector system is the logical choice. A direct actuation stripper plate design is adopted, based upon the assumption that a number of ejector positions are available on the injection machine.

(2) *Parting surface.* The parting line of the component is on one plane (reference PL in Figure 16.6). Likewise, the mould's parting surface is on one plane.

(3) *Impression layout.* The impressions are positioned on the horizontal and vertical centre lines at an equal distance from the sprue; that is, on a pitch-circle diameter (PCD).

(4) *Mould plate design.* As there are four impressions, an insert/bolster design is adopted for both mould plates. As the component is

WORKED EXAMPLES OF SIMPLE INJECTION MOULDS

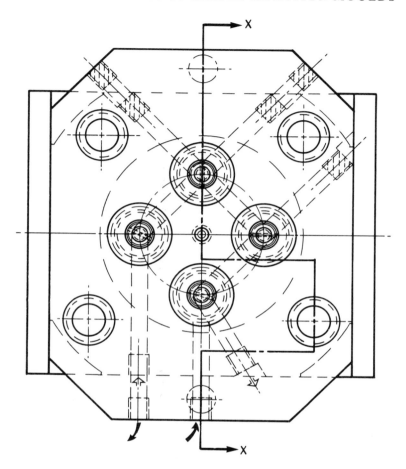

Figure 16.7—Plan view of the moving half of four-impression injection mould for threaded cap shown in Figure 16.6

cylindrical in shape, cylindrical flange type inserts (C5) can be used. These are fitted into a frame type bolster (B5/F12). The cavity inserts are pinned in order to maintain runner alignment.

(5) *Ejection.* Stripper plate ejection is adopted for this design to ensure that an even ejector force is applied to the base of the component. To successfully jump-out internally threaded components, it is essential that the moulding is free to expand slightly during the ejection phase to ride over the threaded form machined on the core insert (C11).

With reference to Figure 16.8, the stripper plate (S19) is sandwiched between the cavity plate (B5/F12) and the core plate (B5/B1). The stripper plate is longer than the other mould plates in the vertical direction to allow for direct actuation by the machine's actuator rods (indicated by the chain-dotted line).

EXAMPLE 3 (STRIPPER PLATE EJECTION)

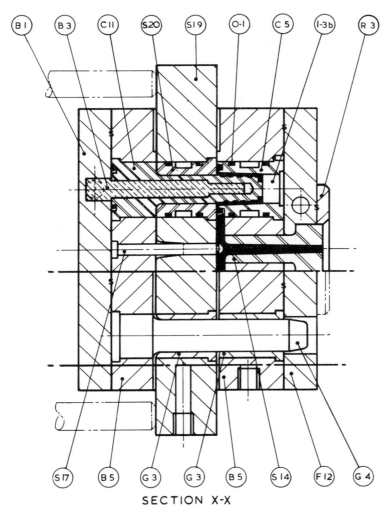

SECTION X-X

Figure 16.8—Cross-sectional view of both mould halves of injection mould for threaded cap shown in Figure 16.6

The apertures in the stripper plate through which the cores pass, are bushed. This feature facilitates replacement should wear increase the clearance between the two sliding parts to a point where plastics material can creep down.

To facilitate the machining operation, the outside diameter of the stripper plate bush (S20) is the same as the fitting diameter of the core insert (C11) and the cavity insert (C5). The feature permits the core bolster plate (B5), the cavity bolster plate (B5) and the stripper plate (S19) to be bored and ground at one machine setting.

WORKED EXAMPLES OF SIMPLE INJECTION MOULDS

The bushed aperture is slightly larger in diameter than the core impression to avoid the possibility of the stripper plate bush scoring the tops of the threads during ejection.

(6) *Ejector assembly and ejector grid.* As the stripper plate is directly operated by the machine's actuator rods, neither an ejector assembly nor an ejector grid is required with this design.

(7) *Feed system.* A ring type of gate is adopted to avoid weld lines. (Compare this feed system design with the example shown in Section 16.3.) As this is an internally threaded component, weld lines and their associated inherent weaknesses are best avoided. The runner, annular in plan, is trapezoidal in cross-section and is machined into the cavity insert (C5). The gate is formed by connecting the annular runner to the impression by a shallow recess (i.e. ring gate).

The individual runner annuli of the four impressions are interconnected by a runner system in the form of a cross. The cross-sectional shape of this runner is also trapezoidal. Refer to Figure 16.10 which illustrates the feed system as moulded.

(8) *Fluid circulation system.* As the component incorporates a variable wall section, an efficient cooling system is required. To achieve this, coolant circulation systems are provided in the core plate, stripper plate, cavity plate and cavity backing plate.

Core plate circuit. A baffled hole system is adopted to permit coolant to pass through each of the core inserts. The individual baffled holes are interconnected by a series of drillings in the back plate (B1). Leakage from between the core insert and the back plate is prevented by incorporating O-seals into the base of the core insert (C11) as shown. A line drawing of the coolant flow-path is shown in Figure 16.11.

Stripper plate circuit. The external annulus method is adopted for controlling the temperature of the individual stripper plate bushes. The individual circulation annuli are interconnected by a drilled hole system as shown in Figure 16.9. The precise flow-path is more readily seen from the flow diagram, Figure 16.12. O-seals fitted on each side of the annulus prevent fluid leakage.

Cavity plate circuit. The external annulus method is adopted for cooling the cavity plate as well. The flow-path for the circulating circuit is identical to the stripper plate circuit as illustrated by the flow diagram (Figure 16.12).

Front plate circuit. This is basically a sprue bush cooling circuit consisting of three drillings forming a U-type configuration. A flow diagram is shown in Figure 16.13.

(9) *Mould alignment.* Guide pillars (G4) are incorporated in the moving half, and guide bushes (G3) in the fixed half. An additional requirement for the guide pillar in this design is to guide and support the stripper plate. Bushes are fitted in the stripper plate (S15) as shown (Figure 16.8). The length of the guide pillar must be suitable to allow for the complete stripper plate movement.

EXAMPLE 3 (STRIPPER PLATE EJECTION)

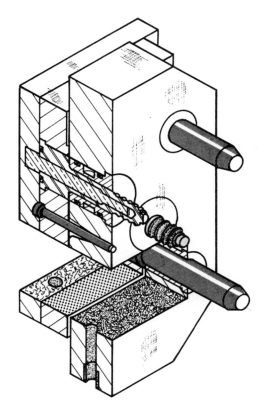

Figure 16.9—Isometric view of the moving half of injection mould for threaded cap shown in Figure 16.6

As the guide pillars are mounted in the moving half, care must be exercised in the positioning to ensure that the fall of the moulding is not obstructed during ejection.

(10) *Sprue pulling.* A mushroom-headed sprue puller is provided directly opposite the sprue. The undercut created by the puller ensures that the sprue is pulled as the mould is opened. Note that neither a Z-type sprue puller nor the undercut cold slug well design is suitable for use in a stripper plate design.

(11) *Parting surface relief.* The surface contact between the two mould halves at the parting surface is restricted to a circular area which encompasses the four impressions plus four irregular-shaped areas situated one at each corner of the cavity plate. The core plate is similarly relieved to reduce the contact area with the stripper plate. The shape of the relief is indicated by short dashes on the plan view (Figure 16.7).

WORKED EXAMPLES OF SIMPLE INJECTION MOULDS

Figure 16.10—The feed system from threaded cap mould

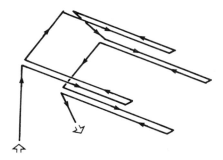

Figure 16.11—Coolant flow path circuit for core plate

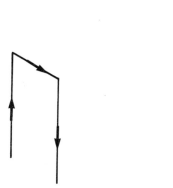

Figure 16.13—Coolant flow path circuit for front plate

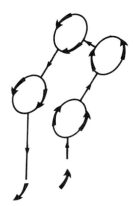

Figure 16.12—Coolant flow path circuit for cavity plate and stripper plate

Terms Used in Injection Mould Design

There is a tendency for mould designers to use different terms when referring to the same mould part. To avoid confusion the definitions given here apply to the terms as used in this monograph. A figure reference follows each definition.

The mould drawings in Chapters 15 and 16 include leader lines and balloons to identify each component part. The reference number shown in each balloon precedes each term listed below. Minor mould parts are grouped together under the designatory letter 'X' at the end of this section.

A

A1   ACTUATOR   A component part of a hydraulic or pneumatic circuit, attached to a mould to provide facilities for power actuation of splits, side cores and side cavities. (Figures 8.20, 9.20, 11.27.)

A2   ADAPTORS, WATER   A standard pipe fitting, fitted to the inlet and outlet holes of the water cooling system. (Figure 6.37.)

A3   ANTE-CHAMBER   A shaped recess, machined adjacent to the impression, to accommodate an extended nozzle or secondary nozzle. The design permits the nozzle to be positioned relatively close to the gate entry. (Figure 13.34.)

A4   ANTE-CHAMBER BUSH   A hardened steel bush which contains the ante-chamber. (Figure 13.5.)

B

B1   BACK PLATE   The rear plate of a mould in which facilities are provided for the mould half's attachment to the injection machine. (Figure 3.1.)

B2   BACKING PLATE   A steel plate situated directly behind the frame of a frame type bolster. (Figure 2.13.)

B3   BAFFLE   A metal strip, plate, or plug fitted in the coolant flow-way circuit to restrict the flow to a prescribed path. (Figure 6.31.)

TERMS USED IN INJECTION MOULD DESIGN

B4   BLADE EJECTOR   A rectangular ejector element used for the ejection of very slender parts. (Figure 3.39.)

B5   BOLSTER   A steel plate or block, machined to accommodate inserts. The insert-bolster assembly constitutes a mould plate. (Figure 2.14.)

B6   BRIDGE PIECE   A steel block incorporated centrally between sets of inserts to contain the runner system. (Figure 2.14.)

C

C1   CAM   A hardened steel member fitted to one mould plate for the purpose of operating splits and side cores. (Figures 8.8, 8.11, 8.15.)

C2   CARRIAGE   The part of the side core or side cavity assembly which provides for the guiding and operating functions of the design. (Figure 9.11.)

C3   CARTRIDGE HEATERS   Cylindrical, electrical resistance type heating elements. (Figure 13.53.)

C4   CAVITY   The female portion of a mould which gives to the moulding its external form. (Figure 2.1.)

C5   CAVITY INSERT   A metal block in which the cavity form is sunk. The insert is fitted into a bolster to form the cavity plate. (Figure 2.11.)

C6   CAVITY PLATE   A plate or block of steel which contains the cavity. The cavity may be sunk directly into the plate (the integer method) or be of a two-part construction consisting of a bolster and cavity inserts. (Figure 2.1.)

C7   CHASE BOLSTER   A steel plate or block machined to accommodate splits. (Figures 2.18, 8.26, 8.27.)

C8   CIRCUIT   The complete system of internal flow-ways within a mould. (Figure 6.9.)

C9   COLUMNS   Cylindrical metal rods which form part of the outrigger system. (Figure 9.16.)

C10   CORE   The male portion of a mould which forms the internal shape of the moulding. (Figure 2.1.)

C11   CORE INSERT   A metal block which incorporates the core. The insert is fitted into a bolster to form a core plate. (Figure 2.11.)

TERMS USED IN INJECTION MOULD DESIGN

C12   CORE PLATE   A plate or block of steel which incorporates the core. The core may be machined directly from a solid plate (the integer method) or be of two-part construction (the insert-bolster method). (Figures 2.1, 2.14.)

C13   CORE SHAFT   The rear portion of a rotating core. (Figure 11.13.)

C14   CORE SHAFT PLATE   A steel plate situated directly behind the gear plate in an unscrewing mould. (Figure 11.40.)

C15   CROSS-HEAD   A steel block forming part of the carriage of an externally operated side core or side cavity assembly. (Figure 9.16.)

C16   CORE PIN   A circular hardened steel pin which forms the internal shape of a moulding. (Figure 9.16.)

D

D1   DAYLIGHT   The space between the mould plates when the mould is open. (Figure 12.1.)

D2   D-SHAPED EJECTOR PIN   A flat-sided ejector element. (Figure 3.33.)

D3   DIAPHRAGM GATE   A particular type of gate for feeding circular mouldings. (Figure 4.21.)

E

E1   EJECTION TECHNIQUES   A general term for the various ejection methods. These include pin, sleeve, bar, blade, air, valve, and stripper plate ejection methods. (Figures 3.21, 3.50.)

E2   EJECTOR ELEMENTS   The part of the ejector system which applies the ejection force to the moulding. An ejector pin is a typical example of an ejector element. (Figure 3.21.)

E3   EJECTOR GRID   The part of the mould assembly which supports the mould plate and provides a space in which the ejector plate assembly can be fitted and operated. (Figure 3.1.)

E4   EJECTOR PIN   A hardened steel circular rod fitted to the ejector assembly, used to apply the ejection force to the moulding. (Figure 3.22.)

E5   EJECTOR PLATE (BAR)   A steel plate or bar incorporated in the design for the purpose of transmitting the ejection force to the element. (Figure 3.11.)

TERMS USED IN INJECTION MOULD DESIGN

E6 EJECTOR PLATE ASSEMBLY The part of the mould to which the ejector element is fitted. The assembly normally consists of an ejector plate, retaining plate, and ejector rod. (Figure 3.11.)

E7 EJECTOR ROD A circular steel rod which in small moulds provides the actuation and guiding functions of the ejector plate. (Figure 3.11.)

E8 EJECTOR ROD BUSH A hardened steel bush fitted in the back plate, through which the ejector rod operates. (Figure 3.11.)

E9 EXTENDED NOZZLE An injection machine nozzle which projects deep into the mould plate to permit reduction in sprue length. (Figure 12.10.)

F

F1 FAN GATE A particular type of side gate. (Figure 4.19.)

F2 FEED PLATE The first plate in an underfeed mould design. In the majority of such designs the plate contains the sprue and runner system. (Figure 12.3.)

F3 FEED STRIPPER PLATE A steel plate situated between the front plate and the floating cavity plate of an underfeed type mould. (Figure 12.11.)

F4 FEED SYSTEM An internal passageway in the mould which provides a flow path for the plastic material from the machine's nozzle to the impression. The feed system normally comprises sprue, runner and gate. (Figure 4.1.)

F5 FINGER CAM A hardened steel rod, fitted at an angle to the mould plate for the purpose of operating splits and side cores. (Figure 8.9.)

F6 FILM GATE A particular type of side gate. (Figure 4.23.)

F7 FLASH A thin wafer of plastic material that escapes from between the mating parting surfaces of the mould or through small crevices within the impression. (Figure 9.24.)

F8 FLUID CIRCULATION SYSTEM Interconnected drillings, slots, chambers, etc., through which a fluid is passed with the object of maintaining the mould plates at a constant temperature. (Figure 6.9.)

F9 FLOATING CAVITY PLATE A steel plate which incorporates the cavity form and which is the centre plate of the basic underfeed design. (Figure 12.1.)

TERMS USED IN INJECTION MOULD DESIGN

F10 FLOW-WAYS (WATERWAYS) Holes or channels within the mould through which a fluid (usually water) is passed to maintain the mould's temperature. (Figure 6.15.)

F11 FORM PIN A circular hardened steel pin which incorporates moulding form and is used for the moulding of internal undercuts. It may have either a straight or an angled action. (Figures 10.3, 10.4.)

F12 FRONT PLATE A steel plate forming part of the frame type bolster assembly, and which by virtue of its position becomes the front plate in the conventional two-part (plate) mould design. Facilities are provided in the plate for the attachment of the fixed mould half to the injection machine. (Figure 7.8.)

G

G1 GATE A channel or orifice connecting the runner to the impression. (Figure 4.1.)

G2 GEAR PLATE A steel plate machined to accommodate the sun-and-planet gear system of an unscrewing mould. (Figure 11.40.)

G3 GUIDE BUSHES Hardened steel bushes which provide a wear-resisting surface for the guide pillars in a mould plate. (Figure 2.22.)

G4 GUIDE PILLARS Hardened steel circular members incorporated in the design to align the two mould halves. They may also support the weight of other mould members such as stripper plates and floating cavity plates on multi-plate moulds. (Figure 2.22.)

G5 GUIDE PINS Hardened cylindrical steel rods which are fitted at an angle in the moving mould plate to control the direction of movement of split cores. (Figure 10.11.)

G6 GUIDE STRIP A flat steel bar attached to the mould plate for guiding splits, side cores or side cavities. (Figure 8.4.)

H

H1 HEATED-SPRUE BUSH A sprue bush which incorporates a heating element in order to keep the material in the sprue aperture at melt temperature. (Figure 13.6.)

H2 HOT-RUNNER MOULD A mould which contains a heated runner manifold block within its structure. (Figure 13.9.)

H3 HOT-RUNNER UNIT A heated runner manifold block maintained at a closely controlled elevated temperature. (Figure 13.9.)

TERMS USED IN INJECTION MOULD DESIGN

I

I1   IMPRESSION   The part of the mould which imparts shape to the moulding and is formed by two mould members, namely the cavity and core. (Figure 2.2.)

I2   INNER CORE   The inner element of a multiple-core assembly.

I3   INSERT
(a) *Cavity and core insert*   A metal block in which the cavity or core form is incorporated. (Figure 2.11.)
(b) *Local insert*   This is a relatively small block of steel incorporated into a cavity or core plate primarily to facilitate the machining function. (Figures 2.8, 2.9, 2.10.)
(c) *Moulding insert*   This is normally a metal part which becomes an integral part of the moulding. This can be achieved either by incorporating the moulding insert at the moulding state, or by pressing the moulding insert into the moulding as a separate operation. (Figure 11.1.)

I4   INSERT–BOLSTER ASSEMBLY   A method of mould plate construction. (Figure 2.14.)

I5   INTEGER CAVITY   A cavity produced by machining the form directly from a plate or block of steel. (Figure 2.5.)

I6   INTEGER CORE   A core produced by machining the form directly from a plate or block of steel. (Figure 2.5.)

I7   INTERNAL UNDERCUT   See 'undercut'.

J

J1   JOINT LINE   The line which is visible on an external undercut type of moulding formed at the junction of a pair of splits. (Figure 8.1.)

J2   JUMPING   A method of extracting an internal undercut type moulding. The moulding is caused to expand slightly during the ejection phase to ride-over the undercut. (Figure 10.15.)

L

L1   LAND   That part of the parting surface immediately adjacent to the impression and runner. (Figure 5.12.)

L2   LAP GATE   A particular type of rectangular gate. (Figure 4.18.)

TERMS USED IN INJECTION MOULD DESIGN

L3   LAYOUT   The particular way the individual impressions are arranged when viewed in plan. In general the impressions are arranged either in line or on a pitch circle diameter. (Figure 4.9.)

L4   LENGTH-BOLTS   Steel bolt-and-nut assemblies used to actuate the floating plate in a multi-plate design. (Figure 3.54.)

L5   LOCKING HEEL   The projecting portion of a chase bolster incorporated in the design to lock the splits, side cores and side cavities in the moulding position. (Figures 8.8, 9.14.)

L6   LONG-REACH NOZZLE   See 'extended nozzle'.

L7   LOOSE THREADED CORE   A loose, hardened steel member fitted into the mould prior to moulding, to form internal moulded threads. (Figure 11.6.)

M

M1   MANIFOLD BLOCK   A heated steel block, maintained at a closely controlled elevated temperature, forming part of a hot-runner unit. (Figure 13.21.)

M2   MOULD HALF
(a) *Fixed* A plate, or an assembly of plates, normally incorporating the cavity, attached to the stationary platen of the machine. (Figure 2.1.)
(b) *Moving* A plate, or an assembly of plates, normally incorporating the core attached to the moving platen of the injection machine. (Figure 2.1.)

M3   MOULD, INJECTION   An assembly of parts containing an impression within it, into which plastic material is injected and cooled. (Figures 16.7, 16.8.)

M4   MOULD PLATE   A plate or block of steel which contains either a cavity or a core form. In certain designs the mould plate constitutes a complete mould half. (Figure 2.1.)

M5   MULTI-DAYLIGHT MOULD   A mould in which more than one daylight occurs when it is opened. This term refers to underfeed moulds in particular. (Figure 12.1.)

M6   MULTI-PART (-PLATE) MOULD   A mould assembly which consists of more than two parts (plates). For example, a three-part mould. (Figure 12.1.)

TERMS USED IN INJECTION MOULD DESIGN

N

N1   NOZZLE   The part of the injection machine through which the molten plastic travels from the injection cylinder to the sprue bush of the mould. (Figure 2.32.)

O

O1   O-RING   A synthetic rubber ring fitted into a suitable groove in the mould to prevent leakage of water from the cooling system. (Figure 6.31.)

O2   OUTRIGGER   An assembly of parts mounted on an external wall of a mould to provide facilities for the external operation of splits, side cores and side cavities. (Figure 9.20.)

O3   OVERLAP GATE   A particular type of gate used to prevent jetting of the melt into the impression. (Figure 4.18.)

P

P1   PAD   A local mould insert incorporated in the design to facilitate a change of engraving. (Figure 2.10.)

P2   PARTING LINE   A mark or line produced on the moulding, formed at the junction of the parting surfaces. (Figure 5.7.)

P3   PARTING SURFACE   That part of both mould plates, adjacent to the impression, which butt together to form a seal and prevent loss of plastic material from the impression. (Figure 5.7.)

P4   PIN EJECTION   A particular ejection technique. (Figure 3.22.)

P5   PIN GATE   A particular type of circular gate. (Figure 4.24.)

P6   PINION   A small cog-wheel associated with a rack and pinion transmission system. (Figure 11.23.)

P7   PLATFORM   A steel plate, mounted on columns, to support an actuator for the external actuation of side cores, side cavities, etc. (Figure 9.20.)

P8   PLUGS   Metal screws used to block the ends of drilled waterways. (Figure 6.17.)

P9   PUSH-BACK PIN (RETURN PIN)   A hardened circular steel pin incorporated in the design to return the ejector assembly to its rear position as the mould is closed. (Figure 3.16.)

TERMS USED IN INJECTION MOULD DESIGN

R

R1 RACK A steel bar with gear teeth cut into one side to operate in conjunction with a pinion in unscrewing type moulds. (Figure 11.23.)

R2 RECTANGULAR EDGE GATE A particular type of edge gate. (Figure 4.15.)

R3 REGISTER RING (LOCATING RING) A flat circular steel plate fitted to the front face of the mould to locate the mould in its correct position with respect to the injection machine's nozzle. (Figures 2.4, 2.36.)

R4 RETAINING PLATE
(a) *Ejector* A steel plate securely attached to the ejector plate for the purpose of retaining the ejector elements.
(b) *Side core* A steel plate securely attached to the carriage for the purpose of retaining the side core element.

R5 RETURN PIN See 'push-back pin'.

R6 RING GATE A particular type of gate for feeding circular type mouldings. (Figure 4.22.)

R7 ROTATING CAVITY DESIGN An unscrewing mould in which the cavity inserts are caused to rotate and thereby unscrew the mouldings from the cores. (Figure 11.17.)

R8 ROTATING CORE DESIGN An unscrewing mould in which threaded core inserts are caused to rotate and thereby automatically unscrew the mouldings. (Figures 11.9, 11.12.)

R9 RUNNER A channel, machined in one or both mould plates, to connect the sprue with the entrance (gate) to the impression. (Figure 4.1.)

R10 RUNNER PLATE A steel plate incorporating the runner, inserted into the feed plate in an underfeed design.

S

S1 SCREW JACKS A threaded steel member incorporated in the mould to control the rate of opening of the moving half in a rotating cavity or rotating core design. (Figure 11.18.)

S2 SECONDARY EJECTOR PLATE A steel plate incorporated in a two-stage ejection design for the purpose of transmitting the ejector force to either a core or ejector element for the secondary ejection stroke.

TERMS USED IN INJECTION MOULD DESIGN

S3   SECONDARY NOZZLE   A steel or beryllium–copper nozzle which forms part of the hot-runner unit assembly. (Figure 13.27.)

S4   SIDE CAVITY
(a) A general term for a part of a cavity which can be withdrawn at right angles to the mould's axis. This feature permits the moulding of a certain type of component which has a projection below its parting surface.
(b) *Side cavity element* A steel member which contains the impression form.
(c) *Side cavity assembly* A side cavity element fitted to a carriage. (Figure 9.36.)

S5   SIDE CORE
(a) A general term for a local core, normally mounted and operated at right angles to the mould's axis, to permit a hole or recess to be formed in the side face of a moulding.
(b) *Side core element* A steel member which contains the impression form.
(c) *Side core assembly* The side core element fitted to a carriage. (Figure 9.11.)

S6   SLEEVE EJECTOR   A hardened steel hollow ejector element. (Figure 3.36.)

S7   SPACING PAD   A short cylindrical rod which is used to position and support the hot-runner manifold. (Figure 13.56.)

S8   SPLIT CORE   A steel core manufactured in two or more parts to facilitate the moulding of internal undercut components. (Figure 10.9.)

S9   SPLIT MOULD   A mould in which the cavity is formed by two or more parts held together by a chase bolster during the injection phase. (Figure 8.3.)

S10   SPLITS   Two or more steel blocks containing the impression which can be opened, normally at right angles to the mould's axis, to facilitate the moulding of external undercut type components. (Figure 8.8.)

S11   SPRING DETENT   A safety device used in split and side core moulds. (Figure 8.29.)

S12   SPRING-LOADED SYSTEM   A safety method used in split and side core moulds. (Figure 8.30.)

S13   SPRUE   The plastic material formed in the tapered passage which connects the nozzle to the mould's parting surface. (Figure 2.3.)

TERMS USED IN INJECTION MOULD DESIGN

S14 SPRUE BUSH A hardened steel bush which incorporates the tapered sprue passageway. (Figure 2.3.)

S15 SPRUE GATE A direct feed into the impression by the sprue. (Figure 4.14.)

S16 SPRUE PIN A hardened steel circular rod fitted directly behind the cold slug well in order to extract it. (Figure 3.60.)

S17 SPRUE PULLER An undercut recess or pin situated opposite the sprue entry to ensure that the sprue is removed from the sprue bush during the opening stroke. (Figure 3.64.)

S18 STRIPPER BAR An ejection technique. (Figure 3.49.)

S19 STRIPPER PLATE A steel plate situated between the core and cavity plate in a multi-part design for the purpose of ejection. (Figure 3.50.)

S20 STRIPPER PLATE BUSH A hardened steel bush mounted in the stripper plate, coaxial with the core. (Figure 3.57.)

S21 STRIPPER RING A hollow steel disk incorporated in a mould design to serve as a local stripper plate. (Figure 3.58.)

S22 SUBSURFACE GATE A particular type of gate. (Figure 4.27.)

S23 SUPPORT BLOCKS Steel blocks, normally of a general rectangular shape, incorporated in the design to support the mould plate. They form part of the ejector grid and the hot runner unit grid. (Figures 3.1, 13.59.)

S24 SUPPORT BLOCKS, LOCAL Steel blocks, normally of a circular cross-section, the purpose of which is to give additional support to that already provided by the main support blocks. (Figure 3.4.)

T

T1 TAB GATE A particular type of side gate. (Figure 4.20.)

T2 TAPERED SLEEVE A tapered hollow cylinder made from insulating material, used to seal the open end of the ante-chamber. (Figure 13.34.)

T3 TIE-RODS Steel rods used to couple together various mould members for actuation purposes. (Figure 3.49.)

T4 THERMOCOUPLE A temperature-sensing device incorporated in hot-runner unit designs. (Figure 13.9.)

TERMS USED IN INJECTION MOULD DESIGN

T5   THREE-PART (-PLATE) MOULD   A mould plate assembly consisting of three main mould parts, for example an underfeed mould. (Figure 12.1.)

T6   TRANSMISSION PLATE   A steel plate through which the transmission system is operated in an unscrewing mould. (Figure 11.41.)

T7   TRANSMISSION SYSTEM   The intermedium between the power source and the operating gear system of an unscrewing mould. (Figure 11.22.)

T8   TWO-PART (-PLATE) MOULD   A basic mould consisting of two mould halves. (Figure 2.1.)

U

U1   UNDERCUT
(a) *External* Any recess or projection on the outside surface of the component which prevents its removal from the cavity. (Figure 8.1.)
(b) *Internal* Any restriction which prevents a moulding being extracted from the core in line of draw. (Figure 10.1.)

U2   UNDERFEED MOULD   A mould in which the feed system is arranged to feed into the underside (or, occasionally, the inside) of the component. (Figure 12.3.)

U3   UNSCREWING MOULD   A mould in which threaded components are automatically unscrewed from the mould. (Figure 11.30.)

V

V1   VALVE EJECTOR   A valve-headed ejector element. (Figure 3.43.)

V2   VALVE EJECTOR BUSH   A hardened steel bush mounted in the core plate, and coaxial with the valve ejector for guidance purposes.

V3   VENT   A shallow recess or hole incorporated in the design to permit entrapped air and other gases to escape freely. (Figure 5.12.)

W

W1   WATER WAYS   Holes or channels within the mould through which water is passed to maintain the mould's temperature. (Figure 6.15.)

W2   WEAR PLATE   A hardened steel plate, fitted to the locking heel of a split or side core type mould, incorporated in the design to provide facilities for adjustment should wear occur. (Figure 8.19.)

TERMS USED IN INJECTION MOULD DESIGN

X (minor items)

X1 BUSH	X7 PACKING PLATE	X13 SPACING PAD
X2 CIRCLIP	X8 PAD	X14 SPRING
X3 GASKET	X9 PIN	X15 STEEL DISC
X4 GRUBSCREW	X10 PIPE	X16 STOP BOLT
X5 KEY	X11 PLUG	X17 WASHER
X6 NUT	X12 POT MAGNET	

Index

actuator, 95, 108, 225, 258, 262, 314, 317
   rod, 69, 95, 102, 106
adaptors, 181
air ejection, 96, 108
angled hole cooling system, 161, 164
angled lift splits, 236–43
   angled guide dowel actuation, 237
   cam track actuation, 239
   spring actuation, 242
ante-chamber, 138, 351
   nozzle, 236, 531
Archimedean screw, 313
axially fixed core, 299

back plate, 64–8, 73, 413, 416
backing plate, 46, 49
baffle, 158, 177
baffled-straight hole cooling system, 177
bar ejection, 98-100
barb nozzle, 350–1
bedding down, 29
bench fitting, 27–30
beryllium copper, 23–9
blade ejection, 91–3
bolster, 3–4, 40, 45–50
   chase, 49
   cooling, 44, 163–5
   frame-type, 44
   fundamental requirements, 45
   machining, 47
   material, 45
   solid, 46
   strip-type, 48
   survey, 50
   types, 45
bolster plate, 50
boring, 4
bridge piece, 41

bubbler cooling system, 177

cam, 213–25
   dog-leg, 217, 255, 265, 326
   finger, 215–17, 253, 265
   plate, 220–3, 265
cam track actuation, 220–3
casting, 17–21, 23–4
   investment, 18
   pressure, 23
cavity, 3, 16, 20, 22, 24, 26, 28, 31, 34
   alignment, 28, 44
   form copying, 14–17
   incorporation, 34
   machining, 13–16
   threaded, 6, 322–7
cavity insert, 34, 38–45, 307
   cooling, 165-74
cavity plate, 31, 34–8, 45, 54, 100, 101, 147, 148, 195
   floating, 324, 330–5, 338, 339, 340, 402
   integer type, 34–8, 157–63
chain and sprocket transmission, 309, 329
chase bolster, 46, 49, 210, 224, 228, 236
   circular, 231
   enclosed channel, 49, 230
   open-channel, 49, 224, 228, 241
checking mould drawings, 444–63
   hot runner mould, 457–9
   pin ejection mould, 445–6
   side-core type mould, 452–4
   sleeve ejection type mould, 447–8
   splits type mould, 450–2
   stepped parting surface, 459–63
   stripper plate type mould, 448–50
   underfeed type mould, 454–7
circuits, 153–83
   See also water cooling

INDEX

circulation system, 153
clamping force, 150
cold hobbing, 21
cold slug well, 109–11, 112–13
compressed air ejection, 98, 108
cooling, *see* mould cooling
cooling plate, 160
copper pipe cooling system, 167
core, 17, 18, 28, 31, 34, 38, 283, 299, 303
   alignment, 28, 44
   incorporation, 34
   manufacture, 18, 19, 23, 28
   threaded, 299–303
   see also split cores
core insert, 34, 38, 40–5, 305
   cooling, 171–80
core machining 16–17
core pin, 88–90
core plate, 3, 34–8, 100
   integer, 34
core shaft, 301
cost, 43
curved holes, 272–3

D-shaped ejector pin, 85–9
damaged parts, 44
daylights, 101, 330–45
diaphragm gate, 134
distortion, 44, 83
dog-leg cam, 217, 254, 255, 267, 326
draft angles, 126, 148

edge forms, 143–7
ejection, 63–111, 234
   air, 96, 108
   blade, 79, 91, 93
   compressed air, 96, 108
   from fixed half, 107
   pin, 78–84, 108, 239
      worked example, 464
   sleeve, 79, 88, 108
      worked example, 466
   stripper bar, 79, 98, 109
   stripper plate, 100-9, 234
      chain actuation, 105
      direct actuation, 105
      from fixed half, 108
      length bolt actuation, 104
      tie-rod actuation, 102
      worked example, 471
   stripper ring, 107
   techniques, 77

valve, 79, 93, 94
ejector bar, 72
ejector elements, 77–107
ejector grid, 63–8, 95, 102, 108
   circular support blocks, 68
   frame-type, 63, 65
   in-line, 63
ejector pin, 78–84, 108, 239
   D-shaped, 85
   stepped, 85
ejector plate, 65, 70, 108
ejector plate assembly, 68–77
   guiding and supporting, 73–6
   return systems, 76–7
      push-back pins, 77
      spring, 77
      stop pins, 77
ejector rod, 74
ejector rod bush, 74
ejector strokes, 68
ejector valve, 79, 93, 94
   cooling, 181
electrically powered unscrewing system, 315
electro-deposition, 19
end mill, 13, 14, 47
explosion moulding, *see* precompressed moulding
extended nozzle, 342–5, 354
extension piece, 166
external undercut, 207–9

facing, 4
fan gate, 132
feed plate, 332–3, 338
feed system, 112–38
   triple daylight mould, 343
   underfeed mould, 333–45
film gate, 136
finger cam, 215–17, 253, 265
finishing, hand, 14
impression, 27
fitting, 27–30
   ejector system, 29
   guide pillar, 28
   register ring, 30
   sprue bush, 30
fixed half, 33
fixed threaded cavity, 322
flash, 120, 148, 150, 266
form pin, 277–83
   angled action, 279
   forward movement, 280

491

INDEX

spherical headed, 281
  straight action, 279
frame-type ejector grid, 63, 65
front plate, 339

gate, 32, 122–39, 300
  balanced, 125
  cross-sectional area, 129, 138
  diaphragm, 134
  fan, 132
  film, 136
  overlap, 125, 130
  pin, 136
  positioning, 123
  rectangular edge, 127
  ring, 135
  round, 138
  size, 129, 138
  sprue, 126
  subsurface, 138
  tab, 133
  types, 138
gate dimensions, 119, 128, 129, 131–4, 138
gate mark, 126, 128
gate open time, 128
gear train, 312
grinding machine, 7–9
  cylindrical, 7
  surface, 9
guide bush, 28, 33, 51, 52
guide dowel, 237–8
guide pillar, 6, 28, 33, 51, 52, 333, 345
  fitting, 28, 51
  function, 53
  manufacture, 5
  positioning, 56
  tapered location, 54
guide pin, 285
guide strip, 210–11, 287

heat treatment, 30
heal block, 216, 218, 228–30
hobbing, cold, 21
hole
  above the parting line, 266
  below the parting line, 266, 279
  curved, 266, 272
  in-line-of-draw, 247
hot-runner moulds, 355–417
hot-runner plate, 405
hot-runner plate mould, 404–7
hot-runner unit grid, 400

hot-runner unit mould, 355–405
  advantages, 357
  configuration of melt flow-way, 369
  expansion of unit, 372–4
  hot-runner unit, 355
    inlet aperture, 370
    internally heated manifold bushing, 371
    manifold block, 363
  melt flow-way, 367
    secondary nozzle, *see* secondary nozzle
    worked example, 401
hydraulic actuation, 223, 314, 318

impression, 31–4
  finishing, 33
  number, 44
injection, worked examples, 464–76
injection pressure, 150
in-line ejector grid, 63
in-line layout, 306, 315, 319, 329
insert, 34, 38–45, 305, 307
  circular, 40
  dimensions, 40
  method of fitting, 40, 41
  rectangular, 41
  shape and type, 39
  split, 39
  see also cavity insert; core insert; local insert; splits type mould
insert-bolster, 38–43
  cooling, 163–65
insulated runner system, 401–4
  with heater probes, 403
integer cavity, manufacture, 34–8
integer cavity plate, 34, 43
integer core, manufacture, 34
integer core plate, 34, 43
integer, method, 34
integer mould plate cooling, 155–63
internal undercuts, 274–89
  examples, 277
  form-pin, 277–81
    angled action, 279
    forward movement, 280
    spherical headed, 281
    straight action, 279
  side cores, 287
  split cores, 283
    actuation, 285
    angled action, 285
    maximum forward movement, 286

INDEX

stripper plate design, 288
stripping, 288
valve ejector method, 289
investment casting, 17
Italian nozzle, 350

joint line, 323, 325, 326

key plate, 279

land, 150–1
latching system, 235
lathe, 4–7
layout, 119–21, 306, 307, 315, 317
lifting eyebolt, 230–1
local insert, 35–8
locking chase, 46, 49, 210, 224, 228, 236
locking heel, 216, 218, 228–30, 256, 257
loose threaded core, *see* threaded components

machine power unscrewing system, 313, 318
machine tools, 4–13
  see also under specific types
machining, spark, 25
manually powered unscrewing system, 309–12
milling cutter, 11–13
milling machine, 11–16
  copy, 14
  horizontal, 13
  profile tracing, 15
  three-dimensional tracing, 16
  tracer controlled, 14
  vertical, 11, 13
mould
  construction, 31–62, 327
  multi-daylight, 330–45
  parting surface, *see* parting surface
  refractory, 17–18
  split, see splits-type mould
  stripper plate, 100–9
  terminology, 477–89
  see also under specific types
mould alignment, 28
mould attachment to platen, 60–2
mould construction, 31–62
mould cooling, 153–83
  adaptors, 181
  angled hole system, 161
  baffled-straight hole system, 177
  bolster, 163

cavity inserts, 165, 168
circulation system, 15
cooling plate system, 160
copper pipe system, 168
core inserts, 171
deep integer-type cavities, 159
drillings, 155
insert-bolster assembly, 163–79
inserts, 165, 168, 171
integer-type cavity plate, 155
integer-type core plate, 161
integer-type mould plate, 155–63
mould plate, 155–63
plugs, 154, 181
rectangular inserts, 166
sprue bush, 180
stepped circuit, 162
valve-type ejectors, 179
water system, 153
mould plate, 31, 34–50, 80, 88, 100, 102, 107, 112, 141–52, 155–63, 179, 210, 330, 332, 333, 373, 402–41
  cooling, 155–63
  integer-type cooling, 155–63
mould size, 44
mould surfaces, balancing, 148
mould temperature, 153–83
moulding defects, 151
moulding face pin, 81
moving half, 33, 220
multi-daylight mould, 101, 330–45
multi-impression mould, 95
multi-nozzle manifold, 354–5

nozzle, 346, 349–55
  ante-chamber, 346, 351
  barb, 350
  extended, 346, 349
  internally heated, 346, 352–4
  multi-, 354–5
  standard, 57–8

open channel chase bolster, 228, 240
opening force, 150
outrigger, 260–2
overlap gate, 130

parting line, 141–51
  complex edge form, 144
  hole below, 266, 269
  hole on, 266
  stepped, 147

493

INDEX

parting surface, 141–51
  angled, 144
  balancing forces, 148
  comlex, 144
  flat, 141
  locally stepped, 147
  nature of, 141
  non-flat, 142
  position, 141
  profiled, 143, 147
  relief, 150
  side cores below, 266, 269
  side cores, on, 266
  stepped, 147
pin ejection, 78, 109, 197, 445, 464–6
  from fixed half, 107
  worked example, 464
pin gate, 134, 304, 336
pitch circle layout, 306, 307
planing machine, 9
platen, mould attachment to, 60–2
plugs for cooling system, 183
polishing, 27, 30
power and transmission system, 307
power source, 299
precompressed moulding, 392
pressure casting, 17–19
procedure for designing an injection mould, 185–204
projections, 249
push-back return system, 75, 77

quick release latches, 401

rack and pinion, 307, 311, 313, 317
ratchet spanner, 311
recess, 244
rectangular edge gate, 125
rectangular hole, 247
register ring, 32, 58, 112
relief of parting surfaces, 150
replacement of damaged parts, 44
retaining plate, 72
Reuleaux's formula, 335
ring gate, 135
rotating cavity, 304
round edge gate, 138
runner, 32, 112–21
  balanced, 119
  cross-sectional shape, 112
  diameter, 116
  layout, 119–21
    multi-impression moulds, 121

    single impression moulds, 119
    three impression moulds, 121
    two impression moulds, 120
  requirements, 113
  size, 116
runner stripper plate, 339–43
runnerless mould, 346–407
  evolution, 347–8
  general considerations, 346
  hot-runner, 355–407
    insulated runner mould, 401–4
    plate, 404
    plate mould, 404–7
    types, 355
    unit, 357
    unit grid, 400
    unit mould, 357–415
    valve gated, 387–92
  nozzle types, 346–49
    ante-chamber, 351
    barb nozzle, 350–1
    direct (standard) feed, 349
    extended, 349
    multi-nozzle manifold, 348, 354

screw jack, 305, 323
secondary nozzle, 374–87
  ante-chamber, 381
  direct feed, 375
    annulus-type, 378
      core rod design, 378
      torpedo design, 379
    sliding-type, 377
  internally heated, 382
    direct feed, 383, 384
      core-rod design, 384
      torpedo design, 383
      projecting core-rod design, 385
      projecting torpedo design, 386
  standard, 375–6
  summary, 386–7
  valve system, 388–92
secondary sprue, 336
  reverse tapered, 336
secondary sprue gate, 315
sequential impact moulding, 392
servomotor, 315
shaping machine, 8
Shaw investment casting process, 17–19
shouldered plug, 362
shrinkage, 6, 63, 78
side cavity, 244–76
  carriage, 253

INDEX

design features, 252
design requirements, 252
examples of use, 245, 246, 250
external side cavity, 258
    actuation methods, 254, 262
    assembly detail, 258
    carriage, 258
    outrigger, 260
internal side cavity, 252
    assembly details, 253
    carriage fitting, 253
    guiding arrangements, 254
    locking the carriage assembly, 257
    method of actuation, 264
mounted on cavity side, 274
mounted on core side, 275
multiple, 276
principle, 244
projections, 249
types, 266
side core, 244–76, 287
    angled withdrawal, 271
    below parting surface, 269
    carriage, 273
    comparison with split mould, 249
    component types, 246
    curved, 260
    design requirements, 252
    examples of the use, 245
    external side core, 258
        actuation, 262
        assembly details, 258
        carriage, 258
        outrigger, 260
    externally actuated, 258, 264, 269
    internal side core, 252
        actuation, 254
        assembly details, 253
        carriage fitting, 253
        guiding arrangements, 254
        locking the carriage assembly, 257
    internally mounted, 271
    mouldings, 246
    on parting surface, 268
    principle, 244
    semi-location of, 266
    types, 266–73
side holes, 244
Sindanyo insulation, 398, 399
single daylight mould, 330
sleeve, 88
    stepped, 90

sleeve ejection, 79, 88, 98, 108
    worked example, 466–70
sliding splits, 210–36
slots, 244–5
    machining, 13
spark erosion machine, 25
spark machining, 25
split-core, 283–6
    actuation, 285
    angled action, 285
    maximum forward movement, 286
    straight action, 285
split insert, 209, 210
splits, 34, 207–43
    actuation, 213–28
    angled guide dowel actuation, 237
    angled lift, 236–43
    balancing system, 230
    cam actuation, 220–3, 239
    cam track actuation, 220
    dog-leg cam actuation, 217–20, 326
    finger cam actuation, 215–19, 265
    guiding and retention, 210–13
    hydraulic actuation, 225–8
    latching system, 235
    locking method, 228–31
    multi-impression, 240
    operation, 215–28
    safety arrangements, 231
    sliding, 211
    spring actuation, 225–6
    stripper plate, 234–6
    threaded, 325
splits-type mould, 207–43
    actuation methods, 213–18
    comparison with side core, 248, 249
spring actuation, 224, 242, 255
    angled lift splits, 242
    side cores, 255
    sliding splits, 229
spring detent, 231
spring return systems, 233
sprue, 31, 57, 58, 112, 113, 343–5
    reverse tapered, 336
    secondary, 336
sprue bush, 31, 57, 58
    cooling, 180
    heated, 341–5, 353
sprue gate, 31, 125, 126, 336, 341, 355
    secondary, 336, 341
sprue puller, 110, 111, 335, 338, 341
    grooved, 110
    mushroom headed, 110

INDEX

reverse taper, 110
Z-type, 110
standard mould unit, 396–443
  advantages, 410
  double ejection, 442–3
  DME standard mould unit, 419–20, 421–2
  DMS standard mould unit, 416–19
  Hasco standard mould unit, 423–4
  hot-runner mould, 438, 440, 441
  insulated runner mould, 420
  limitations, 411
  manufacturers, 411–12, 416–24
  multi-daylight mould, 430
  sleeve ejection mould, 425, 426
  stripper plate mould, 425, 427, 429
  splits-type mould, 428–30
  two-part standard mould unit, 412–24
  Uddform-Sustan standard mould unit, 420, 423
  underfeed-type mould, 433–7, 439
  unscrewing-type mould, 428, 431, 432
standard nozzle, 346
stepped circuit cooling system, 162
stepped ejector pins, 79, 85
stepped parting surface, 147
stepped sleeve, 90
stop pins, 77, 78, 233
stripper bar ejection, 98
stripper plate, 100–9
stripper plate ejection, 100–8
  chain actuation, 105
  direct actuation, 105
  from fixed half, 107
  length bolt actuation, 104
  tie-rod actuation, 102
  worked example, 471
stripper plate mould, 100–9
stripper ring, 107
stripper ring ejection, 107
stripping
  external threads, 323, 324
  internal threads, 295
  internal undercuts, 277–89
sub-surface gate, 138
sun and planet system, 306, 312
support blocks, 73

tab gate, 133
tapered location, 54
terms used in injection mould design, 477–89
thread forms, 290–3

threaded components, 290–329
  design, 290
  examples, 208, 290, 293
  externally threaded, 208, 321–7
    automatic unscrewing, 311
    fixed threaded cavity, 322
    general considerations, 290–2
    splits, 325–7
    stripping, 323–4
  internally threaded, 292–321
    automatic unscrewing, 312, 316, 318–21
    electrically powered, 315–19
    fixed core design, 293
    general considerations, 290–2
    hydraulic powered, 314, 318
    impression layouts, 306
    loose threaded cores, 295–8
    machine powered systems, 313, 318
    manually powered systems, 309–12, 316
    power and transmission systems, 307–21
    stripping, 295, 324
  mould construction, 327–9
  unscrewing moulds, 298–321, 323–7
three-plate mould, 331–42
tie-rod actuation, 102
toggle actuator, 263
transmission system, 299
triple-daylight mould, 343–5
try-out, 30
turning, 4, 5

undercut moulding, 207
undercut runner system, 388, 389
undercuts, 207, 274
  external, 207
  internal, 274–89
underfeed mould, 331–43
  basic design, 332–5
  feed system, 335
unscrewing moulds, 298–321, 323–7
  see also threaded components

valve ejection, 93, 94, 179
valve gated hot runner units, 389–92
venting, 151, 152

water cooling, 153–83
  see also mould cooling

INDEX

wear plates, 228
withdrawing rotating core, 301
witness mark, 123, 128
worked examples of injection moulds, 464–76
  sleeve ejection, 466–70

stripper plate ejection, 471–6
pin ejection, 202–4, 464–6
triple daylight mould, 344
underfeed-type mould, 334

Z-type sprue puller, 111